Neither Physics nor Chemistry

Transformations: Studies in the History of Science and Technology

Jed Z. Buchwald, general editor

Red Prometheus: Engineering and Dictatorship in East Germany, 1945–1990, Dolores L. Augustine

A Nuclear Winter's Tale: Science and Politics in the 1980s, Lawrence Badash

Jesuit Science and the Republic of Letters, Mordechai Feingold, editor

Ships and Science: The Birth of Naval Architecture in the Scientific Revolution, 1600–1800, Larrie D. Ferreiro

Neither Physics nor Chemistry: A History of Quantum Chemistry, Kostas Gavroglu and Ana Simões

H.G. Bronn, Ernst Haeckel, and the Origins of German Darwinism: A Study in Translation and Transformation, Sander Gliboff

Isaac Newton on Mathematical Certainty and Method, Niccolò Guicciardini

Weather by the Numbers: The Genesis of Modern Meteorology, Kristine Harper

Wireless: From Marconi's Black-Box to the Audion, Sungook Hong

The Path Not Taken: French Industrialization in the Age of Revolution, 1750–1830, Jeff Horn

Harmonious Triads: Physicists, Musicians, and Instrument Makers in Nineteenth-Century Germany, Myles W. Jackson

Spectrum of Belief: Joseph von Fraunhofer and the Craft of Precision Optics, Myles W. Jackson

Lenin's Laureate: Zhores Alferov's Life in Communist Science, Paul R. Josephson

Affinity, That Elusive Dream: A Genealogy of the Chemical Revolution, Mi Gyung Kim

Materials in Eighteenth-Century Science: A Historical Ontology, Ursula Klein and Wolfgang Lefèvre

American Hegemony and the Postwar Reconstruction of Science in Europe, John Krige

Conserving the Enlightenment: French Military Engineering from Vauban to the Revolution, Janis Langins

Picturing Machines 1400–1700, Wolfgang Lefèvre, editor

Heredity Produced: At the Crossroads of Biology, Politics, and Culture, 1500–1870, Staffan Müller-Wille and Hans-Jörg Rheinberger, editors

Secrets of Nature: Astrology and Alchemy in Early Modern Europe, William R. Newman and Anthony Grafton, editors

Historia: Empiricism and Erudition in Early Modern Europe, Gianna Pomata and Nancy G. Siraisi, editors

Nationalizing Science: Adolphe Wurtz and the Battle for French Chemistry, Alan J. Rocke

Islamic Science and the Making of the European Renaissance, George Saliba

Crafting the Quantum: Arnold Sommerfeld and the Practice of Theory, 1890–1926, Suman Seth

The Tropics of Empire: Why Columbus Sailed South to the Indies, Nicolás Wey Gómez

Neither Physics nor Chemistry

A History of Quantum Chemistry

Kostas Gavroglu and Ana Simões

The MIT Press
Cambridge, Massachusetts
London, England

© 2012 Massachusetts Institute of Technology

All rights reserved. No part of this book may be reproduced in any form by any electronic or mechanical means (including photocopying, recording, or information storage and retrieval) without permission in writing from the publisher.

For information about special quantity discounts, please email special_sales@mitpress.mit.edu

This book was set in Stone Sans and Stone Serif by Toppan Best-set Premedia Limited. Printed and bound in the United States of America.

Library of Congress Cataloging-in-Publication Data

Gavroglu, Kostas.
Neither physics nor chemistry : a history of quantum chemistry / Kostas Gavroglu and Ana Simões.
 p. cm. — (Transformations : studies in the history of science and technology)
Includes bibliographical references and index.
ISBN 978-0-262-01618-6 (hardcover : alk. paper)
1. Quantum chemistry—History. I. Simões, Ana. II. Title.
QD462.G38 2012
541'.28—dc22

2011006506

Photographs of Linus Pauling at the blackboard and the 1948 Colloque published in this book are from the Ava Helen and Linus Pauling Papers, Special Collections, Oregon State University.

10 9 8 7 6 5 4 3 2 1

Contents

Preface   vii

Introduction   1

1   Quantum Chemistry *qua* Physics: The Promises and Deadlocks of Using First Principles   9

2   Quantum Chemistry *qua* Chemistry: Rules and More Rules   39

3   Quantum Chemistry *qua* Applied Mathematics: Approximation Methods and Crunching Numbers   131

4   Quantum Chemistry *qua* Programming: Computers and the Cultures of Quantum Chemistry   187

5   The Emergence of a Subdiscipline: Historiographical Considerations   245

Notes   263
Bibliography   287
Index   335

Preface

The Windows

In these dark rooms where I live out
empty days, I circle back and forth
trying to find the windows.
It will be a great relief when a window opens.
But the windows are not there to be found—
or at least I cannot find them. And perhaps
it is better that I don't find them.
Perhaps the light will prove another tyranny.
Who knows what new things it will expose?

Constantine P. Cavafy (1863–1933). Cavafy lived most of his life in Alexandria, Egypt, and wrote his poetry in Greek. (From: Edmund Keeley. *C.P. Cavafy*. Copyright © 1975 by Edmund Keeley and Philip Sherrard. Reprinted by permission of Princeton University Press.)

All Is Symbols and Analogies

Ah, all is symbols and analogies!
The wind on the move, the night that will freeze,
Are something other than night and a wind.
Shadows of life and of shiftings of mind.

Everything we see is something besides
The vast tide, all that unease of tides,
Is the echo of the other tide—clearly
Existing where the world there is is real

Everything we have's oblivion.
The frigid night and the wind moving on—
These are shadows of hands, whose gestures are the
Illusion which is this illusion's mother

Fernando Pessoa (1888–1935) (November 9, 1932, excerpt from notes for a dramatic poem on Faust). Pessoa lived mostly in Lisbon, Portugal, but spent part of his youth in Durban, South Africa. He wrote in Portuguese and English and used several heteronyms. (From: E.S. Schaffer, ed. *Comparative Criticism,* Volume 9, *Cultural Perceptions and Literary Values* [University of East Anglia, CUP, 1987]. Copyright © 1987 Cambridge University Press. Reprinted by permission of Cambridge University Press.)

Like many other books, this book has had a long period of gestation. We first met years ago on the other side of the Atlantic, in 1991 in Madison, Michigan, when one of us was writing the scientific biography of Fritz London and the other completing her Ph.D. thesis about the emergence of quantum chemistry in the United States. Since then, on and off, we have been discussing various aspects of quantum chemistry—of a subdiscipline that is not quite physics, not quite chemistry, and not quite applied mathematics and that was referred to as mathematical chemistry, subatomic theoretical chemistry, quantum theory of valence, molecular quantum mechanics, chemical physics, and theoretical chemistry until the community agreed on the designation of quantum chemistry, used in all probability for the first time by Arthur Erich Haas (1884–1941), professor of physics at the University of Vienna, in his book *Die Grundlagen der Quantenchemie* (1929).

Progressively, we became more and more intrigued by the emergence of a culture for doing quantum chemistry through the synthesis of the various traditions of chemistry, physics, and mathematics that were creatively meshed in different locales. We decided to look systematically at the making of this culture—of its concepts, its practices, its language, its institutions—and the people who brought about its becoming. We discuss the contributions of the physicists, chemists, and mathematicians in the emergence and establishment of quantum chemistry since the 1920s in chapters 1, 2, and 3. Chapter 4 deals with the dramatic changes brought forth to quantum chemistry by the ever more intense use of electronic computers after the Second World War, and we continue our story until the early 1970s. To decide when one stops researching, to decide what not to include is always a decision involving a dose of arbitrariness. Necessarily and naturally, a lot has been left out.

The first work that had convincingly shown that quantum mechanics could successfully deal with one of the most enigmatic problems in chemistry was published in 1927. It was a paper by Walter Heitler and Fritz London, who discussed the bonding of two hydrogen atoms into a molecule within the newly formulated quantum mechanical framework. Thus, we start our narrative *after* the advent of quantum mechanics and try to read the unevenly successful attempts to explain the nature of bonds that were made by different communities of specialists within different institutional settings and supported by different methodological and ontological choices.

The narrative about the development of quantum chemistry should not be considered only as the history of the way a particular (sub)discipline was formed and established. It is, at the same time, "part and parcel" of the development of quantum mechanics. The formation of the particular (sub)discipline does, indeed, have a *relative autonomy*, with respect to the development of quantum mechanics, but this kind of autonomy can only be properly appreciated when it is embedded within the overall framework of the development of quantum mechanics. The history of quantum mechanics is, certainly, not an array of milestones punctuated by the "successes" of the applications of quantum mechanics. Such applications should not only be considered either as extensions of the limits of validity of quantum mechanics or as "instances" contributing to its further legitimation, as in any such "application" we can think of—be it nuclear physics, quantum chemistry, superconductivity, superfluidity, to mention a few—new concepts were introduced, new approximation methods were developed, and new ontologies were proposed. The development of quantum mechanics "proper" and "its applications" are historically a unified whole where, of course, each preserves its own relative autonomy.

In a couple of years after the amazingly promising papers of Heitler, London, and Friedrich Hund, Paul Adrien Maurice Dirac made a haunting observation: that quantum mechanics provided all that was necessary to explain problems in chemistry, but at a cost. The calculations involved were so cumbersome as to negate the optimism of the pronouncement. It appears that until the extensive use of digital computers in the 1970s, the history of quantum chemistry is a history of the attempts to devise strategies of how to overcome the almost self-negating enterprise of using quantum mechanics for explaining chemical phenomena.

We tried to write this history by weaving it around six clusters of relevant issues.

During these nearly 50 years, many practitioners proceeded to introduce semiempirical approaches, others concentrated on rather strict mathematical treatments, still others emphasized the introduction of new concepts, and nearly everyone felt the need for the further legitimization of such a theoretical framework—in whose foundation lay the most successful physical theory. This composes our first cluster, one where the epistemic aspects of quantum chemistry were being slowly articulated. The second cluster is related to all the social issues involved in the development of quantum chemistry: university politics, impact of textbooks, audiences at scientific meetings, and the consolidation of alliances with practitioners of other disciplines. The contingent character in the development of quantum chemistry is the third cluster, as at various junctures during its history, many who were working in this emerging field had a multitude of alternatives at their disposal—making their choices by criteria that were not only technical but also philosophical and cultural. The progressively extensive use of computers brought about dramatic changes in quantum chemistry. "Ab initio calculations," a phrase synonymous with impossibility, became a perfectly realizable prospect. In a few years a single instrument, the electronic computer, metamorphosed the subdiscipline itself, and what brought about these changes composes our fourth cluster. The fifth cluster is about philosophy of chemistry, especially because quantum chemistry has played a rather dominant role in much of what has been written in this relatively new branch of philosophy of science. Our intention is not to discuss philosophically the host of issues raised by many scholars in the field but to raise a number of issues that could be clarified through philosophical discussions. Among these issues, perhaps the most pronounced is the role of mathematical theories in chemistry, including their descriptive or predictive character. Different styles of reasoning, different ways of dealing with constraints, and different articulations of local characteristics have been all too common in the history of quantum chemistry. These compose the sixth cluster.

Throughout the book, the references to these clusters are not always explicit, but they are certainly present in our narrative all the time. In this manner, we hope to have been able to put forth a historiographical perspective of the way one can approach the history of an in-between subdiscipline such as quantum chemistry.

We keep on reminding our students that they should never forget that any history, including history of science, is fundamentally about people. There are many such figures in the history of quantum chemistry, and we hope to have been able to bring out how the specificity of each and his or her education and role in various institutions shaped the culture of quantum chemistry. The complex processes of negotiations concerning all sorts of technical and conceptual issues that molded the flexible and at times elusive identity of quantum chemistry may be traced in the multifarious activities of these people.

One of the truly difficult parts of writing about the history of the physical sciences is the extent of the technical details to be included. It is one of those "standard" problems, which, nevertheless, needs to be clarified and specified every time. The problem becomes even more difficult when the interpretation of the technical parts of the works involved in such a history does not have any "grand" implications and, hence, cannot be intelligibly put into plain language. Time dilation, length contraction, the curvature of space, the discreteness of atomic orbits, the uncertainty principle, and the reduction of the wave packet are exceedingly complex notions that, nevertheless, can be reasonably well described and discussed without, in a first approximation, having to resort to the mathematical details behind them. It is obviously the case that we do not imply that whoever decides to write about these subjects without the heavy use of mathematics is guaranteed to do a good job. Quite the opposite is the case, and the misunderstandings and myths around these subjects are mostly due to such popular writings. Popularization does require the effective use of language—but it also requires much more. Nevertheless, there have been excellent popular accounts of these developments, and what is more important, there have been superb scholarly works where use of the technical background was optimal for comprehension of the implications of the theory. How, though, does one go about to explain the work of scientists whose extremely significant contributions are inextricably tied up with the understanding of the technical details? If one knows nothing about the subject and does not have any training in the general area of the subject matter, then it is impossible to learn the subject by just reading the history of the area, no matter how conscientiously the authors present the technical details. In contrast, for those readers who either know the subject or can follow the technical details because of their training, what is included may appear to be a rather watered down version that does not do much justice to the wealth of a particular formulation. There is, obviously, no standard rule or prescription of how to get out of this Sisyphean deadlock. The decisions we took as to how to present the technical details depended on what we believed to be pertinent every time such a problem arose while keeping in mind that whoever will be interested in reading the book should be able to read it without having to follow closely the technical details.

By the time of the 1970 Conference on Computational Support for Theoretical Chemistry, which discussed how computational support for theoretical chemistry could be efficiently achieved, it was clear to all quantum chemists that a long way had been traversed since the publication of the Heitler and London paper in 1927. The "theory of resonance" proposed by Linus Pauling and the molecular orbital approach developed by Hund and Robert Sanderson Mulliken had been systematically elaborated, a host of new concepts had come into being, and many and powerful approximation methods were being extensively used in a complementary manner. Many quantum chemists started dealing with large and complicated molecules. Chemistry, it appeared, might not have acquired its "own" theory by the physicists' standards, but certainly, quantum mechanics did provide the indispensable framework for dealing with chemical problems. Dirac, after all, might have turned out to be right.

The computer had forced many practitioners to rethink the status of theory vis-à-vis inputs from empirical data and more or less approximate calculations, and visual imagery acquired a new physical support and heralded new applications. Experiments took on new meanings: Many ab initio calculations "substituted" for experiments, and mathematical laboratories became part of the new sites of quantum chemistry. Institutionally, the discipline became truly international, and its new cohesive strength arose from a successful networking crossing continents, generations, practitioners' research areas, and different and at times antagonistic modes of reasoning. In a very short time, the possibilities provided by the new instrument brought about a realization that the future of the subdiscipline would be radically different than its past: Gone were the days of discussions and disputes about conceptual issues and approximation methods, and the promised future was full of numbers expressing certainties rather than signifying semiempirical approaches.

Our historical and historiographical considerations have been shaped through a "dynamic conversation" with a number of historical works. John Servos's *Physical Chemistry from Ostwald to Pauling* (1990), Mary Jo Nye's *From Chemical Philosophy to Theoretical Chemistry* (1993), and aspects in Thomas Hager's biographical studies (1995, 1998) on Linus Pauling represent some of the first works where historical issues of quantum chemistry began to be discussed. A number of Ph.D. dissertations have dealt with facets of the history of quantum chemistry: Robert Paradowski (1972) offered a comprehensive analysis of Pauling's structural chemistry; Buhm Soon Park (1999a) concentrated on the study of the role of computations and of computers in reshaping quantum chemistry; Andreas Karachalios (2003, 2010) offered a detailed study of Erich Hückel; Martha Harris (2007) argued that the chemical bond, as explained quantum mechanically, became a signifier of disciplinary change by the 1930s, distinguishing the new quantum chemistry from the older physical chemistry; and Jeremiah James (2008) has discussed Pauling's research program at the California Institute of Technology during the 1920s and 1930s.

Scholars, including many colleagues and various chemists, who wrote papers, chapters in books, dictionary entries, recollections, biographical memoirs, autobiographies, obituary notices, or gave interviews have provided us with a wealth of information often following different methodologies. Furthermore, there are a number of works where some historiographical issues have been tackled. The discussion of the emergence and development of quantum chemistry in different national contexts has been given considerable attention. Studies offering comparative assessments of some protagonists' views and practices include analyses of Pauling and George W. Wheland's views on the theory of resonance; of the different contexts of the simultaneous discovery of hybridization by Pauling and John Clarke Slater; of the contrasting teaching strategies of Charles Alfred Coulson and Michael J. Dewar; as well as of Pauling and Coulson as seen through their famous textbooks *The Nature of the Chemical Bond* and *Valence*, respectively. The period after the Second World War has not yet been systematically studied, except for preliminary assessments of the impact of computers in the methodological, institutional, and organizational reshaping of quantum chemistry. Furthermore, quantum chemistry has provided ample material for much of the discussion in the philosophy of chemistry, and various problems pertinent to philosophy of chemistry, most prominently that of reductionism, have been addressed from a historical perspective.

Over the years, a number of scholars have worked on topics related to the history of quantum chemistry. Their work and the conversations with some have been an inspiration and an immense help for us. We especially acknowledge the work of Steven G. Brush, who introduced one of us to the history of quantum chemistry, on Hückel and benzene; of Andreas Karachalios on Hückel and Hellmann; of Helge Kragh on Bohr, Hund, and Hückel; of Mary Jo Nye on the history of theoretical chemistry; of Buhm Soon Park on the different genealogies of computations; of Sam Schweber on Slater; and of J. van Brakel, Robin Findlay Hendry, Jeff Ramsey, Eric Scerri, Joachim Schummer, and Andrea Woody on the philosophical considerations of issues in quantum chemistry. While writing the book we received many comments and much advice and support from many colleagues and friends. We thank Jürgen Renn for his hospitality at the Max Planck Institute for the History of Science (MPIWG) and for the use of the services of its excellent library. Robert Fox and José Ramon Bertomeu Sanchez have contributed in different ways to hasten us in the period that gave way to the last stage of this long journey. Theodore Arabatzis read the manuscript and offered valuable comments. Jed Z. Buchwald was particularly supportive of our project from the very beginning and accepted our proposal to include the book in the series he directs. Patrick Charbonneau made a number of incisive comments. Referees made perceptive comments and very useful suggestions. We thank them all.

Along this journey, various chemists and scientists have contacted us, offering their memories and comments. We thank them all, and especially J. Friedel, who commented on the sections about French quantum chemists. The oral interviews assembled on the Web page created by Udo Anders have been very helpful, as well as Anders Fröman's and Jan Lindenberg's recollections. The last year of research depended on the constant support of Urs Schoeflin, the librarian of the MPIWG, and his staff, as well as on Lindy Divarci, who took care of our requests; on the librarian Halima Naimova from the Astronomical Observatory of Lisbon; on Michael Miller, technical archivist at the American Philosophical Society; and on Daniel Barbiero, manager of archives and records at the National Academy of Sciences. We thank them all.

Our professional lives in Greece and Portugal are interlaced with all kinds of activities for the further entrenchment of our discipline, and, thus, often we had to stop the project to get involved with time-consuming yet necessary undertakings in the precarious institutional environment for such subjects as history of science and technology. But in all these instances, we have been privileged to be surrounded by colleagues who are truly excellent scholars with whom we share the same views as to the ways our discipline will continue to be strengthened within our local conditions and with whom we have good friendships. We specifically thank Ana Carneiro, Luís Miguel Carolino, Maria Paula Diogo, Henrique Leitão, Marta C. Lourenço, Tiago Saraiva, Theodore Arabatzis, Jean Christianidis, Manolis Patiniotis, Faidra Papanelopoulou, and Telis Tympas. We have also been involved in many projects that did not intersect with quantum chemistry. Perhaps the most satisfying and enjoyable was the creation and a fruitful first decade of the activities of the international group Science and Technology in the European Periphery (STEP).

We thank the families of Fritz London and Charles Alfred Coulson, who have kindly provided us with photographs, and Mariana Silva for preparing the diagrams for publication. We also thank Professor W. H. E. Schwarz for his help. At long last, writing a joint book, kilometers apart, in two extremities of Europe emerged from the world of dreams into the real world. We hope our readers will find this book useful. We enjoyed each and every step of the convoluted process leading to it, from e-mail discussions to phone conversations to a very long discussion ironing out all the difficult problems related to the book at "another" in-between site—a cafe situated between Hagia Sophia and the Blue Mosque in Istanbul.

The shaping of scientific disciplines is mediated by people, their choices, allegiances, and conflicts, as well as by their changing networks of interactions. But certainly, identity search and identity crises are neither primarily nor exclusively associated with them. During a dinner in Lisbon with our partners Eleni Stambogli and Paulo Crawford, we talked about the movie *When Cavafy Met Pessoa* (directed by Stelios Charalambopoulos), which is about the amazingly similar lives of these two contemporaneous poets, exquisite explorers of the human nature, so prized in Greece and Portugal and who had never met. The choices that led to the poems at the beginning of the book are, perhaps, the only thing that each author has done independently. Otherwise, what is in the book has been untirelessly discussed and reflects the views of both.

Some of what has already appeared in a few of our published works has been expanded and reworked in this book. In chapters 1 and 2, we drew from our papers "The Americans, the Germans and the Beginnings of Quantum Chemistry: The Confluence of Diverging Traditions" (*Historical Studies in the Physical Sciences* 1994;25:47–110); "One Face or Many? The Role of Textbooks in Building The New Discipline of Quantum Chemistry" (in Anders Lundgren, Bernadette Bensaude-Vincent, eds. *Communicating Chemistry. Textbooks and their Audiences, 1789–1939*, Science History Publications, 2000, pp. 415–449); and "In Between Words: G.N. Lewis, the Shared Pair Bond and Its Multifarious Contexts" (*Journal of Computational Quantum Chemistry* 2007;28:62–72).

In chapter 3, we drew from our papers "Quantum Chemistry *qua* Applied Mathematics. The Contributions of Charles Alfred Coulson 1910–1974" (*Historical Studies in the Physical Sciences* 1999;29:363–406); and "Quantum Chemistry in Great Britain: Developing a Mathematical Framework for Quantum Chemistry" (*Studies in the History and Philosophy of Modern Physics* 2000;31:511–548).

Introduction

Although it is relatively easy to relate what something is not, it is always challenging to be clear about what something is. The first part of the title of our book clearly delineates what quantum chemistry is not. The rest of the title is a promise to tell what this discipline is and how it developed.

One year before the year we chose to end our narrative—with the Conference on Computational Support for Theoretical Chemistry in 1970—at a symposium on the "Fifty Years of Valence," Charles Alfred Coulson, one of the protagonists of our story and Rouse Ball Professor of Applied Mathematics at the University of Oxford at the time, talked of chemistry as a discipline that is concerned with explanation and cultivates a sense of understanding. "Its concepts operate at an appropriate depth and are designed for the kind of explanation required and given" (Coulson 1970, 287). He noted that when the level of inquiry deepens, then a number of older concepts are no longer relevant. And then, Coulson emphatically declared that one of the primary tasks of the chemists during the initial stage in the development of quantum chemistry was to *escape from the thought forms of the physicists* (Coulson 1970, 259, emphasis ours). Indeed. Among the many and, at times, insurmountable barriers during the development of quantum chemistry, perhaps the one hurdle that was the most incapacitating was the prospect of problems of (self)identity the new subdiscipline would have: It appeared that whatever was done to lead to the establishment of quantum chemistry as a subdiscipline in chemistry would, in effect, be indistinguishable from whatever was needed to establish *it* as a subdiscipline of physics! Hence, escaping the thought forms of the physicists was a strategic choice in developing the culture of the new subdiscipline and in articulating its practices—not consciously pursued by all, but, surely, in the minds of those whose work eventually established the subdiscipline. And Coulson, more than anyone else, turned out to be particularly sensitive to the almost imperceptible borderline between physics and chemistry when one decided to "deepen the level of inquiry."

Nearly at the same time, the Swedish quantum chemist Per-Olov Löwdin, professor of quantum chemistry at the University of Uppsala and the founder of the *International*

Journal of Quantum Chemistry in 1967, wrote in the editorial of the first issue that quantum chemistry "uses physical and chemical experience, deep going mathematical analysis and high speed electronic computers to achieve its results." He acknowledged that quantum mechanics was offering a framework for the unification of all the natural sciences—including biology. And, as for quantum chemistry, he emphasized that it was a young field "which falls between the historically developed areas of mathematics, physics, chemistry, and biology" (Löwdin 1967, 1).

Both Coulson and Löwdin, though they were clear about the kinds of problems quantum chemistry tackled, were, somewhat uncertain as to the signifying characteristics of its culture and practices. Coulson tells us that chemistry explains and gives insight and a sense of understanding—but this is the case in a host of other disciplines. We are told that its concepts operate at an appropriate depth and they cater for the kind of explanation we seek—again, something all too common in many other disciplines. It is noted that these concepts are no longer relevant when our inquiry deepens—again, as it happens in many other disciplines. Two outstanding quantum chemists such as Coulson and Löwdin, despite their thoughtful comments about the status of quantum chemistry, were, in effect, expressing their *uneasiness* when it came to delineate the methodological, philosophical, and disciplinary boundaries of quantum chemistry, echoing what was discussed in meetings, what was stated in papers, what was implied in textbooks, throughout the four decades since the 1927 paper of Walter Heitler and Fritz London who showed in no uncertain terms that the covalent bond—a kind of mystery within the classical framework—could be mathematically tackled and physically understood by using the recently formulated quantum mechanics. In a way, our narrative is the unfolding of this uneasiness while at the same time it displays the variety of strands whose synthesis gave rise to quantum chemistry: the different methodological traditions that came to the fore, the decisions of the leaders of each tradition to consolidate a framework of practices, the rhetorical strategies and the processes of legitimization, the role of textbooks, journals, and conferences in building the relevant scientific community, the ways major institutions accommodated the rise of the new subdiscipline, and the theoretical and philosophical issues raised through the multitude of practices within the subdiscipline. And, thus, quantum chemistry acquired the status of a (sub)discipline situated "somewhere between the historically developed areas of mathematics, physics, chemistry, and biology" and whose fundamental characteristics were brought about by physicists, chemists, biologists and mathematicians who tried to "escape from the thought forms of the physicists" (Coulson 1970, 259).

In this book, the development of an "in-between" discipline such as quantum chemistry is narrated through six interrelated clusters of issues to be analyzed below, that manifest the particularities of its evolving (re)articulations with chemistry, physics, mathematics, and biology, as well as institutional positioning.

The first cluster involves issues related to the historical becoming of the epistemic aspects of quantum chemistry: the multiple contexts that prepared the ground for its appearance; the ever present dilemmas of the initial practitioners as to the "most" appropriate course to choose between the rigorous mathematical treatment, its dead ends, and the semiempirical approaches with their many promises; the novel concepts introduced and the intricate processes of their legitimization. The source of these dilemmas lies in what appeared from the very beginning to be a doomed prospect: the Schrödinger equation, used in any manner for the explanation of a chemical bond, could not provide analytical solutions except for the case of hydrogen and helium! Quantum chemistry appears to have been formed through the confluence of a number of distinct trends, with each one of them claiming to have been the decisive factor in the formation of this discipline: Neither the relatively straightforward quantum mechanical calculations of Fritz London and Walter Heitler in 1927, nor the rules proposed by Robert Sanderson Mulliken to formulate an *aufbau* principle for molecules, nor Linus Pauling's reappropriation of structural chemistry within a quantum mechanical context, nor Coulson's and Douglas Rayner Hartree's systematic but at times cumbersome numerical approximations—by themselves and in a manner isolated from each other—could be said to have given quantum chemistry its epistemic content. Though it may appear that there is a consensus that quantum chemistry had always been a "branch" of chemistry, this was not so during its history, and different (sub)cultures (physics, applied mathematics) attempted to appropriate it. The historical development of quantum chemistry has been the articulation of its relative autonomy both with respect to physics as well as with respect to chemistry, and we will argue for the historicity of this relative autonomy.

The second cluster of issues is related to disciplinary emergence: the naming of chairs, university politics, textbooks, meetings, networking, as well as alliances quantum chemists sought to build with the practitioners of other disciplines were quite decisive in the formation of the character of quantum chemistry. To stress this and the former cluster of issues, the book intercalates the analysis of the contributions of the various participants, whether belonging to the same or different local/national contexts. It also intercalates the analysis of their work with the discussion of their specific activities as community builders. This entangled narrative aims at giving the reader a feeling for the complexities of the various interactions at the individual, community, and institutional levels. The emergence of quantum chemistry in the institutional settings of Germany, the United States, and Britain, and later on in France and Sweden, and a number of conferences and meetings of a programmatic character helped to mold its character: a marginal activity at the beginning, it had the good luck to have gifted propagandists and able negotiators among its practitioners. The strong pleas of Heitler, London, and Friedrich Hund for chemical problems to yield to quantum mechanics, Mulliken's tirelessness in familiarizing physicists and chemists with the attractiveness of the molecular orbital approach, Pauling's aggressiveness to project resonance theory as the only way to do quantum chemistry, Coulson's incessant attempts to popularize his views in order to explain the character of valence, the research of Raymond Daudel and of Bernard and Alberte Pullman into molecules with biological interest, and Löwdin's founding of a new journal, all these contributed toward the gradual coagulation of the language of the emerging subdiscipline and of its social presence as well.

The third cluster of issues is related to a hitherto totally neglected aspect of quantum chemistry; that is, its contingent character. Quantum chemistry could have developed differently, and it will be shown that the particular form it took was historically situated, at times being the result of not only technical but also of cultural and philosophical considerations. The historiographic possibilities provided by the category of contingency for the development of the natural sciences have been intensely discussed among historians and philosophers of science. Our elaboration of this issue is not to make partisan points but to argue that, perhaps, "in-between" (sub)disciplines provide a privileged context in which to investigate the interpretative possibilities provided by the notion of contingency. Contingency is not an invitation to do hypothetical history. It is not an invitation to ruminate about meaningless "what if" situations, but rather to realize that at every juncture of its development, quantum chemistry had a number of paths along which it could have developed. What is important to understand is not what different forms quantum chemistry could or might have taken, but, rather, the different possibilities open for developments and the set of difficulties that at each particular historical juncture formed those barriers that dissuaded practitioners from pursuing these possibilities. Throughout this 50-year period, the criteria for assessing the "appropriateness" of the schema being developed gravitated among a rigorous commitment to quantum mechanics, a pledge toward the development of a theoretical framework where quasi-empirical outlooks played a rather decisive role in theory building, and a vow to develop approximation techniques for dealing with the equations. Such criteria were not, strictly speaking, solely of technical character, and the choices adopted by the various practitioners at different times had been conditioned by the methodological, philosophical, and ontological commitments and even by institutional considerations. The development of quantum chemistry appears, also, to have been the result of an attitude by many physicists, chemists, mathematicians, biologists, and computer experts who did not feel constrained by any orthodoxy and were thus not discouraged from proposing idiosyncratic ways to circumvent the cul-de-sacs brought about by the impossibility of exact solutions. Thinking in terms of contingency may bring to the surface the disparate ways the culture and practices of quantum chemistry were formed.

The fourth cluster of issues is related to a rather unique development in the history of this subdiscipline: the rearticulation of the practices of the community after the early 1960s, which was brought about by an instrument—the electronic computer. The fundamental liability of quantum chemistry, the impossibility to perform analytical calculations, was, all of a sudden, turned into an invaluable asset that also contributed to the further legitimization of electronic computers. In the early 1960s, it appeared that a whole subject depended on this particular instrument in order to produce trustworthy results. In a very short while, a particular instrument undermined most of the fundamental criteria with respect to which the practitioners were making their choices since the late 1920s. All of a sudden, ever more scientists started to realize that "quantum chemistry is no longer simply a curiosity but is contributing to the mainstream of chemistry" (National Academy of Sciences 1971, 1). The prospect of ab initio calculations, which did not use experimental data built in the equations in any way, seemed to offer the promise of new and reliable results, and apt to reach a sophistication and accuracy dependent on the needs of each quantum chemist. The members of a whole disciplinary community, who, through a historically complicated process had attained a consensus about the coexistence of different approaches for doing quantum chemistry, became in a relatively short time subservient to the limitless possibilities of computations provided by a particular instrument. Fostered by the use of computers, applied to ab initio but also to semiempirical calculations, members of the community of quantum chemists recognized that a new culture of doing quantum chemistry was asserting itself and vying for hegemony among the more traditional ones. The increasing complexity of molecular problems was dealt with by means of mathematical modeling and a burst of activities in relation to the writing and dissemination of computer programs. There were even cases where it became unnecessary to perform expensive experiments because calculations would provide the required information!

The fifth cluster of issues is related to philosophy of science. It is undoubtedly the case that in recent years there has been an upsurge of scholarship in the philosophy of chemistry. The issues that have been raised throughout the history of quantum chemistry played a prominent role in these philosophical elaborations and discussions: reductionism, scientific realism, the role of theory, including its descriptive or predictive character, the role of pictorial representations and mathematics, the role of semiempirical versus ab initio approaches, and the status of theoretical entities and of empirical observations (Woolley 1978; Primas 1983, 1988; Vermeeren 1986; Gavroglu 1997, 2000; Ramsey 1997; Scerri 1997; Scerri and McIntyre 1997; Janich and Psarros 1998; van Brakel 2000; Woody 2000; Hendry 2001, 2003, 2004; Early 2003; Baird 2006). Throughout the development of quantum chemistry, it appears that almost all its practitioners were aware that apart from the technical problems they had to deal with, they were also encountering a host of "other" problems as well. These problems were, in fact, philosophical problems. But almost none of these practitioners was thinking of formulating the answers in philosophical terms, as no one, really, thought of these problems as *philosophical* problems. Yet they all considered the answers to these thorny issues as a necessary procedure toward the establishment of quantum chemistry. In discussing these issues, many quantum chemists were, in effect, negotiating the ways to "escape the thought forms of the physicists." Notably, most of the first generation of quantum chemists became strong allies to the philosophers of science, who, long after these people were gone, attempted to establish a new subdiscipline.

The sixth cluster is of a quasi-methodological and quasi-cultural character. The history of quantum chemistry displays instances that we suggest to discuss in terms of "styles of reasoning." To specify the notion of style, Ian Hacking asserted that the style of reasoning associated with a particular proposition p determines the way in which p points to truth or falsehood. "Hence we cannot criticize that style of reasoning, as a way of getting to p or to not-p, because p simply is that proposition whose truth value is determined in this way" (Hacking 1985, 146). A style, in other words, brings into being candidates for truth.

The types of styles are introduced as categories of possibilities, the range of possibilities depending upon that style. Summarizing his views on styles of scientific reasoning, Hacking (1985, 162) noted that "many categories of possibility, of what may be true or false, are contingent upon historical events, namely the development of certain styles of reasoning." A style can be further understood in terms of a network of constraints and the kind of reasoning imposed by these constraints, which could delineate the conceptual boundaries that determine the types of problems that are posed as well as the type of their solutions.

A style, and the subsequent discourse formed within it, possesses a peculiarly self-referential character about the criteria it sets and against which it assesses its own coherence. It is a conceptual coherence characteristic of a set of propositions that become the allowable possibilities of a particular type of discourse. These propositions can, in fact, be accommodated within another type of discourse, and there are obviously ways for understanding their meaning as well as deciding their truth value within this second type of discourse. But, as a whole, they will not seem to be coherent within this second type of discourse. It is rather the case that, again as a whole, these propositions do not appear to establish an affinity with the latter discourse. This discourse is "indifferent" toward them, exactly because these propositions, as a whole, do not offer any clues for tracing out the categories of possibilities of the second discourse—even though they were decisive in doing just that in the original discourse. What Heitler and London did by introducing group theory for the study of valence, Mulliken's extension of Bohr's *aufbau* principle to molecules and the articulation of molecular orbitals, and what Pauling did with his resonance theory, all these could be considered as different discourses, each characteristic of a different style. The crucial point to have in mind is that our aim is not to substitute "theory" or "models" by "style." Our aim is to consider the developments within a variety of theoretical frameworks so that we can have as many multifaceted insights into the developments as possible. It can be shown how decisive the "style" of a researcher was for discovering new phenomena, developing effective methods, or proposing novel explanatory schemata. The various developments in quantum chemistry can also help us to provide some answers to questions like: How can styles be differentiated from one another? Is the difference in styles merely an expression of personal idiosyncrasies? Is one justified to even talk about different styles of scientific inquiry when discussing the physical sciences, as the "objective" nature of what is being investigated seems to require a methodological uniformity? Is it at all meaningful to compare two different types of discourse? And, if it is, how are those differences to be signified? In coming to understand the various developments in terms of types of discourse, one realizes a truly liberating lesson: There are no good or bad styles, nor are there any correct and wrong types of discourse. It is rather the categories of possibilities each one offers and the attempts to explicate the possibilities of each discourse that are so significant in examining the development of theories. And it is exactly for that reason that understanding failures becomes as intriguing as appreciating successes. In the case of quantum chemistry, participants seem to have understood these constraints to the fullest becoming wizard explorers of the possibilities they offered. The ongoing discussions about the significance of the semiempirical approaches were, in effect, discussions related to the legitimacy of the semiempirical approach and, hence, the legitimacy of a particular style of doing quantum chemistry.

These six clusters of issues—the epistemic content of quantum chemistry, the social issues involved in disciplinary emergence, the contingent character of its various developments, the dramatic changes brought about by the digital computer, the philosophical issues related to the work of almost all the protagonists, and the importance of styles of reasoning in assessing different approaches to quantum chemistry—form the narrative strands of our history. Such an approach may be a useful way to deal with the development of in-between subdisciplines—electrochemistry, biochemistry, biophysics. It is, however, certainly the case that these clusters of issues appear to be indispensable for understanding how quantum chemistry developed during its first 50 years.

1 Quantum Chemistry *qua* Physics: The Promises and Deadlocks of Using First Principles

In the opening paragraph of his 1929 paper "Quantum Mechanics of Many-Electron Systems," Paul Adrien Maurice Dirac announced that:

> The general theory of quantum mechanics is now almost complete, the imperfections that still remain being in connection with the exact fitting in of the theory with relativity ideas. These give rise to difficulties only when high-speed particles are involved, and are therefore of no importance in the consideration of atomic and molecular structure and ordinary chemical reactions, in which it is, indeed, usually sufficiently accurate if one neglects relativity variation of mass with velocity and assumes only Coulomb forces between the various electrons and atomic nuclei. *The underlying physical laws necessary for the mathematical theory of a large part of physics and the whole of chemistry are thus completely known, and the difficulty is only that the exact application of these laws leads to equations much too complicated to be soluble.* It therefore becomes desirable that approximate practical methods of applying quantum mechanics should be developed, which can lead to an explanation of the main features of complex atomic systems without too much computation. (Dirac 1929, 714, emphasis ours)

For most members of the community of physicists, it appeared that the solution of chemical problems amounted to no more than quantum-mechanical calculations. Physicists came under the spell of Dirac's reductionist program, and quantum chemistry came to be usually regarded as a success story of quantum mechanics. Although it took some time for physicists to realize that Dirac's statement was a theoretically correct but practically meaningless dictum, the first attempts to solve chemical problems in the "proper way"—that is, in the physicists' way—appeared to be rather promising. These attempts started before the publication of Dirac's paper, and they may have provided some kind of justification for such a generalized statement.

The Old Quantum Chemistry: Bonds for Physicists and Chemists

The prehistory of quantum chemistry has its beginnings in the 1910s with various attempts, both by physicists and chemists, to explain the nature of bonds within two essentially disparate theoretical traditions—physical chemistry and molecular spectroscopy—and two conflicting views of atomic constitution. For Gilbert Newton Lewis, the emblematic albeit idiosyncratic representative of the first group, the starting point was the static atom of the chemists. For Niels Bohr whose views were closer to those of the second tradition, the starting point was his dynamical atom, soon appropriated by the physicists and used to explain the complexities of molecular spectra.

In the last part of his trilogy "On the Constitution of Atoms and Molecules," Bohr considered systems containing several nuclei and suggested that most of the electrons must be arranged around each nucleus in such a way "as if the other nucleus were absent." Only a small number of the outer electrons would be arranged differently, and they would be rotating in a ring around the line connecting the nuclei. This ring, which "keeps the system together, represents the chemical 'bond'" (Bohr 1913, 862).[1] According to these general guidelines, in the hydrogen molecule the two electrons were rotating in a ring in a plane perpendicular to the line joining the nuclei. Although Bohr tentatively suggested a model for the water molecule,[2] it was in the case of the hydrogen molecule that he ventured to prove quantitatively its mechanical stability, offering a value for the molecular heat of formation twice as large as the experimental one (Langmuir 1912). Thus, the chemical consequences of Bohr's molecular model conflicted with experimental data for the simplest molecule, and the calculations were much too complicated to be carried through in the case of more complex molecules.

The exploration of another molecular model—the Lewis model with the shared electron pair, a topic we address in chapter 2—was, however, to give a satisfactory, albeit qualitative, answer to the problem of chemical bonding. The translatability of Lewis's picture into Bohr's dynamical language was found by "transforming" Lewis's static shared electrons into orbital electrons revolving in *binuclear* trajectories (Kemble et al. 1926). In the simplest case of diatomic molecules, and reasoning by analogy with the hydrogen molecule, the binding orbits of shared electrons were thought to fall into two distinct classes. In the class most directly associated with the Lewis model, shared orbital electrons were thought to move in binuclear orbits around both nuclei, providing the necessary interatomic binding "glue" on the assumption that electrons spent most of their time in the region between nuclei. In the second class, following Bohr's suggestion, shared electrons moved either in a plane perpendicular to the line joining the two nuclei or in crossed orbits. Similar models were explored in the case of the hydrogen molecule ion with the difference that only one electron was involved (Pauli 1922).

Again, agreement with experimental values for the few cases where quantitative calculations could be carried on could not be achieved.

Quite independently from considerations related to atomic spectroscopy, quantization was applied to molecules 2 years before it was applied to atoms (Jammer 1966;

Kuhn 1978; Hiebert 1983; Barkan 1999). But whereas Bohr's revolutionary assumption related radiation frequencies to energy changes accompanying electronic jumps between allowed orbits, in the case of the molecule, the more conservative Niels Bjerrum (a physical chemist and compatriot and friend of Bohr) accepted the classical electrodynamical identity between the frequency of emitted radiation and the mechanical frequency of motion. His hybrid model assumed simply the quantization of rotational energy, in conjunction with classical electrodynamics and the equipartition theorem. Starting with a simple model of the molecule as a vibrating rotator, Bjerrum provided a model to explain the infrared molecular spectra of some simple diatomic molecules and confirmed the long-sought interdependence between kinetic theory and spectroscopy within the framework of a very "restricted" quantum theory.

The close agreement between theory and experiment provided a strong argument in favor of quantization of rotational energies/frequencies. Such was the opinion of Bohr in a letter to Carl W. Oseen: "I do not know what your point of view of the quantum theory really is; but to me it seems that its experimental reality can hardly be doubted, this is perhaps most evident from Bjerrum's beautiful theory, and Eva von Bahr's papers almost seem to offer direct proof of the quantum laws or at least of the impossibility of treating the rotation of molecules with anything resembling ordinary mechanics."[3]

The interpretation of infrared molecular spectra proved to be so successful that atomic and molecular spectroscopy developed as quite separate branches until 1919–1920. Then, Torsten Heurlinger (a graduate student of Johannes Rydberg who held one of the chairs of experimental physics at the University of Lund) and Adolf Kratzer (Arnold Sommerfeld's former Ph.D. student and assistant), completing the work started by physicist Karl Schwarzschild, showed that Bohr's frequency condition could be extended beyond the motion of electrons and applied to the interpretation of the rotational and vibrational motions of molecules in such a way that Heurlinger and Kratzer managed to unite atomic and molecular spectroscopy under the same theoretical umbrella. The American physicist and expert on molecular spectra Edwin Crawford Kemble noted that the interpretation of band spectra by the Einstein–Bohr hypothesis that spectroscopic frequencies are the measures of energy differences and are not identical to the frequencies of the motion of the emitting system undermined the semiclassical theory of Bjerrum, despite its many successes. "The abandonment of the initially successful Bjerrum theory has been brought about primarily by the necessity of unifying our interpretation of line and band spectra" (Kemble et al. 1926, 107). From then on, spectroscopists calculated the frequencies of the emission/absorption in molecular spectra by using the quantization of energy *plus* the Einstein–Bohr frequency relation, now applied to all frequency regions, whether in the infrared, red, visible, or ultraviolet part of the electromagnetic spectrum.

Walter Heitler and Fritz London: Outlining a Program for Quantum Chemistry

The Heitler and London Paper of 1927

The stability of the hydrogen molecule within the newly developed quantum mechanics was first successfully explained by Walter Heitler and Fritz London in their paper of 1927 (Gavroglu and Simões 1994; Gavroglu 1995; Karachalios 2000).[4] In April of that year, Heitler and London, both recipients of a Rockefeller Fellowship, decided to go to the University of Zürich where Erwin Schrödinger was—they both felt more at ease with his more intuitive approach than with Werner Heisenberg's matrix mechanics. Schrödinger agreed to their stay, but there was not much collaboration with him.

Fritz London (1900–1954) was born in Breslau to a Jewish family. His father was professor of mathematics at the University of Breslau. In 1921, the year he graduated from the University of Munich, he wrote a thesis under the supervision of Alexander Pfänder (one of the best known phenomenologists) dealing with deductive systems. It was among the very first attempts to investigate ideas about philosophy of science expressed by the founder of the phenomenological movement in philosophy, Edmund Husserl. It was a remarkable piece of work by a 21-year-old who developed an antipositivist and antireductionist view. In fact, London's first published paper in a professional journal was in philosophy. He published his thesis in 1923 in the *Jahrbuch für Philosophie und phanomenologische Forschung*, and Pfänder, along with Moritz Geiger and Max Scheler, was one of the co-editors of the *Jahrbuch*, whose editor in chief was Husserl himself. London first went to work with Max Born at the University of Göttingen, but Born could not dissuade him from working in philosophy and sent him to Arnold Sommerfeld at the University of Munich. He did his first calculations in spectroscopy, and, in 1925, he published his first paper in physics with H. Honl on the intensity of band spectra.

Concerning his approach to philosophy, London did not follow the practice of a lot of physicists who were either among the founders of quantum mechanics or among its first practitioners (Everitt and Fairbank 1973; Gavroglu 1995). Most of these physicists wrote some kind of a philosophical piece *after* having made those contributions by which they established their reputations in the community. Some of these pieces are texts for a rather sophisticated audience, but most are popularized accounts—explanations of the implications of quantum mechanics and relativity, historico-philosophical accounts of the development of what is called "modern physics," attempts to present in a systematic manner a series of philosophical issues within the context of the new developments. London followed a different path. His work in philosophy, never mentioned by others when there is reference to the philosophical writings of this generation, was of the professional kind and was impressively ambitious: He wanted to discuss the status of a deductive theory and the conditions for the existence of such a theory. In a thoughtful essay examining Husserl's philosophy of science, Thomas Mormann (1991) considers London's thesis together with Husserl's ideas concerning philosophy of science as having anticipated the semantic approach to the philosophy of science.

London's first academic appointment, starting in October 1925, was as Paul Peter Ewald's assistant at the Technische Hochschule in Stuttgart. Ewald was the director of the Institute for Theoretical Physics, and it was in this environment that London started working on quantum theory. In fact, instead of continuing to work in spectroscopy as the "Sommerfeld culture" stipulated, London, as soon as he reached Stuttgart, plunged into matrix mechanics. He first used Carl Gustav Jacob Jacobi's classical transformation theory of periodic systems and "adopted" it for matrix mechanics proving that energy conservation was independent of the combination principle of atomic theory. This he proved after showing that the two definitions of the matrix derivative in the famous "three-man paper" of Born, Heisenberg, and Pascual Jordan followed from his proposal of a more general definition of the matrix derivative (Jammer 1966; Hendry 1984; Kragh 1990).

His next two papers were quite significant in what came to be known as the transformation theory of quantum mechanics, a theory that was independently and much more fully developed and completed by Dirac and Jordan in 1926–1927. Eventually, transformation theory allowed quantum mechanics to be formulated in the language of Hilbert spaces. In this new framework, quantum mechanics could be treated in a mathematically more satisfactory way, and its results could acquire a consistent physical interpretation, dependent less on visualizability and on a description in space-time and giving more emphasis on underlining the novel foundational characteristics of quantum mechanics.

Walter Heitler (1904–1981) was born in Karlsruhe to a Jewish family, and his father was a professor of engineering. His interest in physical chemistry grew while he attended lectures on the subject at the Technische Hochschule, and through these lectures he came into contact with quantum theory. He had also acquired a strong background in mathematics. Wishing to work in theoretical physics, he first went to the Humboldt University of Berlin but found the atmosphere not too hospitable especially because a student was left to himself to choose a problem and write a thesis. Only after its completion would the "great men" examine it. After a year in Berlin he went to the University of Munich and completed his doctoral thesis with Karl Herzberg on concentrated solutions. The writing of his thesis coincided with the development of the new quantum mechanics, but because of the kind of problems he was working on, he never had the opportunity to study the new developments in any systematic manner. After completing his thesis, Sommerfeld helped him to secure funding from the International Education Board, and he went to the Institute for Theoretical Physics at Copenhagen to work with Bjerrum on a problem about ions in solutions. He was not particularly happy in Copenhagen. Determined to work in quantum mechanics, he convinced Bjerrum, the International Education Board, and Schrödinger to spend the second half of the period for which he received funding in Zürich (Heitler 1967; Gavroglu 1995).

About a month after arriving in Zürich, Heitler and London decided to calculate the van der Waals forces arising from weak attractive interactions between two hydrogen atoms considering the problem to be "just a small 'by the way' problem." Nothing indicates that London and Heitler were either given the problem of the hydrogen molecule by Schrödinger or that they had detailed talks with him about the paper. Linus Pauling, who was also in Zürich during the same time as Heitler and London, noted that neither he nor Heitler and London discussed their work with Schrödinger,[5] despite the fact that Schrödinger knew what they were all working on as witnessed by Robert Sanderson Mulliken's visit to Zürich in 1927. Schrödinger (figure 1.1) told Mulliken that there were two persons working in his institute who had some results "which he thought would interest me very much; he then introduced me to Heitler and London whose paper on the chemical bond in hydrogen was published not long after" (Mulliken 1965, S7). Ewald thought that the question of the homopolar bond was in London's mind before going to Zürich, and Pauling remembered discussions with Heitler about bonding when he was in Munich in 1926.

Figure 1.1
Erwin Schrödinger and Fritz London in Berlin in 1928.
Source: Courtesy of Edith London.

Heitler and London's initial aim was to calculate the interaction of the charges of two atoms "without even thinking of the exchange." They were not particularly encouraged by their result because the Coulomb integral, which represents the energy that an electron would have in the diatomic molecule if it occupied one atomic orbital, could not account for the van der Waals forces: "So we were really stuck and we were stuck for quite a while; we did not know what it meant and did not know what to do with it,"[6] Heitler remembered. Heisenberg's work on the quantum mechanical resonance phenomenon, which had already been published, was not of particular help to Heitler and London, as the exchange was part of the resonance of two electrons, one in the ground state and the other excited, but both in the same atom (Carson 1996).

Years later, Heitler would still remember the hot afternoon, "the picture before me of the two wave functions of two hydrogen atoms joined together with a plus and minus and with the exchange in it." He called London and they started to work on the idea, and by daybreak they had resolved the problem of the formation of the hydrogen molecule. They had also realized that there was a second type of interaction, a repulsive one between the two hydrogen atoms, something they were unaware of but that was nothing particularly new, as a number of chemists were aware of the old electrochemical hypothesis as to the nature of the chemical bond. And though they were able to complete the calculation, they had "to struggle *with the proper formulation of the Pauli principle*, which was not at that time available, and also the connection with spin ... There was a great deal of discussion about the Pauli principle and how it could be interpreted."[7]

Heitler and London started their calculations by considering the two hydrogen atoms coming slowly close to each other. They assumed electron 1 to belong to atom *a* and electron 2 to atom *b* or electron 2 to belong to atom *a* and electron 1 to atom *b*. Because the electrons were identical, the total wave function of the system was the linear combination of the wave function of the two cases,

$$\Psi = c_1 \Psi_a(1)\Psi_b(2) + c_2 \Psi_a(2)\Psi_b(1).$$

The problem now was to calculate the coefficients c_1 and c_2. This they did by minimizing the energy,

$$E = \frac{\int \Psi H \Psi d\tau}{\int \Psi^2 d\tau}.$$

They found two values for the energy,

$$E_1 = 2E_0 + \frac{C+A}{1+S_{12}}; \quad E_2 = 2E_0 + \frac{C-A}{1-S_{12}}.$$

S_{12} is the overlap integral and measures the extent to which the two atomic wave functions overlap one another ($\int \psi_a \psi_b d\tau$). The integral C is the Coulomb integral ($\int \psi_a H \psi_a d\tau$), and A is the exchange integral ($\int \psi_a H \psi_b d\tau$). Both C and A had negative values, but A was larger than C. E_1 implied $c_1/c_2 = 1$, and E_2 implied $c_1/c_2 = -1$. Hence the wave function of the system could now be written as

$$\Psi_I = \Psi_a(1)\Psi_b(2) + \Psi_a(2)\Psi_b(1)$$

$$\Psi_{II} = \Psi_a(1)\Psi_b(2) - \Psi_a(2)\Psi_b(1).$$

Up to now, the spin of the electrons was not taken into consideration. The symmetry properties required by the Pauli exclusion principle were satisfied only by Ψ_I. This was the case when the electrons had antiparallel spins. But Ψ_I corresponded with E_1. E_1 was less than $2E_0$, the sum of the energies of the two separate hydrogen atoms, and, hence, it signified attraction. Ψ_{II}, which when spin was taken into consideration was a symmetric combination, corresponded with E_2. But E_2 was greater than $2E_0$, and it implied repulsion. The bonding between the two neutral hydrogen atoms became possible only when the relative orientations of the spins of the electrons were antiparallel. They noted that this was the justification for the electron pairing that Walter Kossel had talked about, but they did not refer to Gilbert Newton Lewis (Kohler 1971, 1973). To form an electron pair it did not suffice to have only energetically available electrons; they also had to have the right orientations. The homopolar bonding could, thus, be understood as a *pure quantum effect*, as its explanation depended wholly on the electron spin, which had no classical analogue. Heitler and London (1927, 472) found the energy to be 54.2 kcal/mole (2.4 eV/molecule) and the internuclear distance 0.8 Å.[8]

William M. Fairbank, who was London's colleague at Duke University in the early 1950s and the co-author with C. W. Francis Everitt of the entry on Fritz London in the *Dictionary of Scientific Biography*, recalled London telling him that Schrödinger was pleasantly surprised because he did not expect that his equation could be used to solve chemical problems as well. Born and James Franck were very enthusiastic about the paper. Sommerfeld had a rather cool reaction, but he also became very enthusiastic once Heitler met him and explained certain points.

The exchange force remained a mystery. Heitler and London were not expecting to find any such force, as London had told Alfred Brian Pippard, because they had started working on the problem as a problem in van der Waals forces.[9] They soon realized that the proposed exchange mechanism obliged them to be confronted with a fundamentally new phenomenon. They had to answer questions posed by experimental physicists and chemists about what was exchanged: Were the two electrons being *actually* exchanged? Was there any sense in asking what the frequency of exchange is? It was eventually realized by both that the exchange was a fundamentally new phenomenon with no classical analogue. "I think the only honest answer today is that the exchange is something typical of quantum mechanics, and should not be interpreted—or one should not try to interpret it—in terms of classical physics."[10] Both London and Heitler in all their early writings repeatedly stressed this "non visualizability" of the exchange energy. It is one aspect of their work that, in the early stages, was consistently misrepresented.

Though it appeared that the treatment of the homopolar bond of the hydrogen molecule was an "extension" of the methods successfully used for the hydrogen molecule ion by Olaf Burrau (1927), there was a difference between the two cases that led to quite radical implications. It was the role of the Pauli principle. John Heilbron in his penetrating study of the origins of the exclusion principle talked about "one of the oddest of the instruments of microphysics" and that Wolfgang Pauli's first enunciation in December 1924 had the form not of a dynamical principle but of the Ten Commandments (Margenau 1944; van der Waerden 1960; Heilbron 1983). During the ceremony at the Institute for Advanced Study at Princeton University to honor Pauli's receipt of the Nobel Prize in Physics for 1945, Hermann Weyl talked of the Pauli principle as something that revealed a "general mysterious property of the electron" (Pauli 1946; Weyl 1946).

During the stay of Heitler and London in Zürich, Pauli's paper on spin appeared.[11] Though they greatly appreciated it, they thought that it was not particularly satisfactory, because it was "a sort of hybrid between a wave equation and some matrix mechanics superposed on it. It was, so to speak, glued together, but not naturally combined together."[12] In the case of the hydrogen molecule ion, its solution was a successful application of Schrödinger's equation where the only forces determining the potential are electromagnetic. A similar approach to the problem of the hydrogen molecule leads to a mathematically well defined but physically meaningless solution—there can be no accounting of the attractive forces. There was, then, a need for an additional constraint, so that the solution would become physically meaningful. An interesting aspect of the theoretical significance of the original work of Heitler and London was that this additional constraint was not in the form of further assumptions about the forces involved. Invoking the exclusion principle as a further constraint led to a quite amazing metamorphosis of the physical content of the mathematical solutions. Under the new constraint, the terms formerly giving strongly repulsive forces gave strongly attractive forces. These terms became now physically meaningful, and their interpretation in terms of the Pauli principle led to a realization of the new possibilities provided by the electromagnetic interaction.

Later on, London proceeded to a formulation of the Pauli principle for cases with more than two electrons that was to become more convenient for his later work in group theory: The wave function can, at most, contain arguments symmetric in pairs; those electron pairs on which the wave function depends symmetrically have antiparallel spin. He considered spin to be the constitutive characteristic of quantum chemistry. And because two electrons with antiparallel spins are not identical, the Pauli principle did not apply to them, and one could, then, *legitimately*, choose the symmetric solution (Heitler and London 1927; London 1928).

With the Pauli principle, it became possible to comprehend "valence" saturation: It seemed reasonable to suppose that whenever two electrons of different atoms combine to form a symmetric Schrödinger vibration, a bond will result. As it will be repeatedly argued in the work of both Heitler and London, spin would become one of the most significant indicators of valence behavior and would forever be in the words of John Hasbrouck Van Vleck (a physicist from Harvard) "at the heart of chemistry" (Van Vleck 1970, 240).

Reactions to the 1927 Paper
Right after its publication, it became quite obvious that the Heitler–London paper was opening a new era in the study of chemical problems. The fact that the application of quantum mechanics led to the conclusion that two hydrogen atoms form a molecule and that such was not the case with two helium atoms was particularly significant. Such a "distinction is characteristically chemical and its clarification marks the *genesis of the science of sub-atomic theoretical chemistry*" remarked Pauling (1928, 174), who later became one of the dominating figures in quantum chemistry. A similar view with a slightly different emphasis was put forward by Van Vleck (1928, 506): "Is it too optimistic to hazard the opinion that this is perhaps the *beginnings of a science of 'mathematical chemistry'* in which chemical heats of reaction are calculated by quantum mechanics just as are the spectroscopic frequencies of the physicist?"

In their book on quantum mechanics for chemists, Pauling and E. Bright Wilson hailed the paper as the "greatest single contribution to the clarification of the chemists' conception of valence" (Pauling and Wilson 1935, 340) that had been made since Lewis's ingenious suggestion in 1916 of the electron pair (see chapter 2). Heisenberg in an address to the Chemical Section of the British Association for the Advancement of Science in 1931 considered the theory of valence of Heitler and London to "have the great advantage of leading exactly to the concept of valence which is used by the chemist" (Heisenberg 1932, 247). A. David Buckingham quoted William McCrea, who recalled his own attempts to solve the problem of the hydrogen molecule bond, when one day in 1927, McCrea told Ralph Howard Fowler that a paper by Heitler and London apparently solved the problem in terms of a new concept: a quantum mechanical exchange force. Fowler thought it was an interesting idea and asked McCrea to present the paper in the next colloquium—"which is how quantum chemistry came to Britain" (McCrea 1985; Buckingham 1987).

A meeting where questions related to chemical bonding and valence were exhaustively discussed was quite suggestive of the changes occurring among the chemists. This was the "Symposium on Atomic Structure and Valence" organized by the Division of Physical and Inorganic Chemistry of the American Chemical Society and held in 1928 at St. Louis. Chemists attending the meeting of the American Chemical Society appeared to be sufficiently fluent in the ways of the new physics. George L. Clark's opening remarks are quite remarkable in that respect.

He talked of certain modes of behavior in a way ingrained among chemists and physicists. The former failed to test their well-founded conceptions with the facts of physical experimentation, and the latter did not delve critically into the facts of chemical combination. He criticized the firm entrenchment, as he called it, of chemists and physicists in their own domains, so that no comprehensive channels of communication between the two had been established nor had a language that would be accepted by both been developed. "The position of the Bohr conception has seemed so convincing that perhaps the majority of thinking chemists were coming to accept the dynamic atom, which is fully capable of visualization" (Clark 1928, 362).

Without denying one of the cardinal characteristics of the chemists' culture—that of visualizability—Clark was courageous enough to talk not of the majority of chemists but of the majority of *thinking* chemists. It was a small yet telling sign of the problems that were encountered at the beginning to convince the chemists about the importance and the legitimacy of using quantum mechanics.

Clark was not alone in attempting to specify the problematic relationship between the physicists and the chemists. Worth Rodebush, one of the first to receive a doctorate in 1917 from the newly established Department of Chemistry at the University of California at Berkeley under the chairmanship of Lewis, went a step further than Clark. The divergent paths of physicists and chemists had started being drawn together after the advent of quantum theory and especially after Bohr's original papers. But in this process "the physicist seems to have yielded more ground than the chemist. The physicist appears to have learned more from the chemist than the chemist from the physicist. The physicist now tells the chemist that his ways of looking at things are really quite right because the new theories of the atom justify that interpretation, but, of course, the chemist has known all the time that his theories had at least the justification of correspondence with a great number and variety of experimental facts" (Rodebush 1928, 511).[13] He gracefully remarked that it was to the credit of the physicist that he can now calculate the energy of formation of the hydrogen molecule by using the Schrödinger equation. But the difficulty in a theory of valence was not to account for the forces that bind the atoms into molecules. The outstanding task for such a theory was to predict the existence and absence of various compounds and the unitary nature of valence that can be expressed by a series of small whole numbers leading to the law of multiple proportions. The "brilliant theories" of Lewis accounted for the features of valence "in a remarkably satisfactory manner, at least from the chemist's point of view" (Rodebush 1928, 513). London's group theoretical treatment of valence—to which we refer in the next section—was considered as an important piece of work even though it did not answer all the queries of the chemist such as, for example, the differences in degree of stability between chemical compounds. He was afraid that the rule of eight—the number of electrons in a closed shell—was being threatened, but there again it may be a kind of "chemical correspondence principle" because of the qualitative character of the chemical methods.

Van Vleck's review of quantum mechanics presented at the symposium concentrated on explaining the principles and the internal logic of the new theory. He was quite sympathetic to matrix mechanics. He gave full credit to the work of Heitler and London, something found in most of Van Vleck's papers through 1935, before he was convinced to use the more "practical" methods of Pauling and Mulliken (Van Vleck 1928). Van Vleck fully accepted Dirac's attitude that the laws for the "whole of chemistry are thus completely known" and thought that the dynamics that was so successful in explaining atomic energy levels for the physicist should also be successful in calculating molecular energy levels for the chemist. The actual calculations may be formidable indeed, but the mathematical problem confronting the chemist was "to investigate whether there are stable solutions of the Schrödinger wave equation corresponding to the interaction between two (or more) atoms, using only the wave functions which have the type of symmetry compatible with Pauli's exclusion principle." Such a program for examining the implications of quantum mechanics for chemistry "has been made within the past few months in important papers by London and by Heitler. Although this work is very new, it is already yielding one of the best and most promising theories of valence" (Van Vleck 1928, 500). And he drew attention to the crucial feature of such an approach, lest the chemists "get the wrong idea." The non-occurrence of certain compounds was not because the calculations yielded energetically unstable combinations, but because the corresponding solutions to the Schrödinger equation did not satisfy the symmetry requirements of the Pauli principle. The achievements of quantum mechanics in physics were summarized in ten points, and the section about chemistry was appropriately titled "What Quantum Mechanics Promises to do for the Chemist." Great emphasis was placed on the importance of spin for chemistry, and it was shown that the Pauli exclusion principle could provide a remarkably coherent explanation of the periodic table. Its extreme importance was stressed elsewhere as well: "The Pauli exclusion principle is the cornerstone of the entire science of chemistry" (Van Vleck and Sherman 1935, 173). Nevertheless, if quantum mechanics was to be of any use in chemistry, one should go further than the periodic table and understand which atoms can combine and which cannot.

Among the reviews published at the time, Pauling's article published in *Chemical Reviews* did much to propagandize quantum mechanics, explicitly aiming at the "education" of chemists in the ways of the new mechanics (Pauling 1928).[14] Pauling presented the details of the calculation by Burrau (1927) of the electron charge density distribution of the hydrogen molecule ion, because the original article was published in a journal "which is often not available." Burrau was the first to integrate successfully the wave equation for the simplest molecule—the hydrogen molecule ion. He found a numerical expression for the electronic wave function in the field of the two nuclei; that is, he obtained the first numerical expression of a molecular (binuclear) orbital, together with values for the equilibrium internuclear distance, total energy, and vibrational energy of the lowest state.

The Heitler–London treatment of the structure of the hydrogen molecule was considered as "most satisfactory," and it was repeatedly stated that in a few years, spin and resonance—which Pauling had, in the meantime, formulated, and which would eventually become his trademark—will provide a satisfactory explanation of chemical valence (Pauling 1928a, 1931, 1931a; Pauling and Sherman 1933, 1933a; Pauling and Wheland 1933) (see chapter 2).

Perhaps the most cogent manifestation of the characteristic approach of the American chemists was Harry Fry's contribution in the symposium on Atomic Structure and Valence. He attempted to articulate what he called the pragmatic outlook. He started by posing a single question that should be dealt with by the (organic) chemists. What would be the kind of modifications to the structural formulas so as to conform to the current concepts of electronic valence? This, he insisted, should by no means lead to a confusion of the fundamental purpose of a structural formula, which was to present the number, the kind, and the arrangement of atoms in a molecule as well as to correlate the manifold chemical reactions displayed by the molecule.

It should here be noted that no theory in any science has been so marvelously fruitful as the structure theory of organic chemistry . . . When we are considering methods of modifying this structure theory of organic chemistry, by imposing upon its structural formulas an electronic valence symbolism, are we not, as practical chemists, obligated to see to it that such system be one that is calculated to elucidate our formulas rather than render them obscure through the application of metaphysically involved implications on atomic structure which are extraneous to the real chemical significance of the structural formulas, per se . . . The opinion is now growing that the structural formula of the organic chemist is not the canvas on which the cubist artist should impose his drawings which he alone can interpret . . . Many chemists believe that the employment of a simple plus and minus polar valence notation is all that is necessary, at the present stage of our knowledge, to effect the further elucidation of structural formulas. *On the grounds that practical results are the sole test of truth, such simple system of electronic valence notation may be termed 'pragmatic.'* (Fry 1928, 558–559, emphasis ours)

"Chemical pragmatism" resisted the attempts to embody in the structural formulas what Fry considered to be metaphysical hypotheses: questions related to the constitution of the atom and the disposition of its valence electrons. It was the actual chemical behavior of molecules that was the primary concern of the pragmatic chemist, rather than the imposition of an electronic system of notation on these formulas that was further complicated by the metaphysical speculations involving the unsolved problems about the constitution of the atom. Fry had to admit the obvious fact that as the chemists will know more about the constitution of the atom, they would be able to explain more fully chemical properties. He warned, though, that premises lying outside the territory of sensation experience are bound to lead to contradictory conclusions, quoting Immanuel Kant and, surely, becoming the only chemist to use Kant's ideas to convince other chemists about an issue in chemistry!

Group Theory and Problems of Chemical Valence
The first indications that the work they started in their joint paper could be continued by using mathematical group theory involving molecular symmetry elements and operations are found in a letter from Heitler to London in late 1927.[15] By September, Heitler had gone to Göttingen as Born's assistant and London to Berlin as assistant to Schrödinger, who had succeeded Max Planck. Heitler was very excited about physics at Göttingen and especially about Born's course in quantum mechanics where everything was presented in the matrix formulation and then one derived "God knows how, Schrödinger's equation."[16] He believed that the only way the many-body problem could be dealt with was with group theory and outlined his program to London in two long letters.

His first aim was to clarify the meaning of the line chemists drew between two atoms. His basic assumption was that every bond line meant exchange of two electrons of opposite spin between two atoms. He examined the case with the nitrogen molecule and, in analogy with the hydrogen case, among all the possibilities, the term containing the outermost three electrons of each atom with spins in the same direction (i.e., ↑↑↑ and ↓↓↓) was picked out as signifying attraction.

He became convinced that only by using group theory was it possible to proceed to a general proof. But if one assumed "that the two atomic systems ↑↑↑↑↑ . . . and ↓↓↓↓↓ . . . are always attracted in a homopolar manner. We can, then, eat Chemistry with a spoon."[17]

This overarching program to explain all of chemistry got Heitler into trouble more than once. Eugene Wigner used to tease Heitler, because Wigner was skeptical that the whole of chemistry had been explained. Wigner would ask Heitler: "'[W]hat chemical compounds would you predict between nitrogen and hydrogen?' And of course, since he did not know any chemistry he couldn't tell me."[18] Heitler confessed as much in his interview: "The general program was to continue on the lines of the joint paper with London, and the problem was to understand chemistry. This is perhaps a bit too much to ask, but it was to understand what the chemists mean when they say an atom has a valence of two or three or four . . . . Both London and I believed that all this must be now within the reach of quantum mechanics."[19]

Heitler, then, went on to work out in detail the methane molecule CH_4. C is in ↑↑↑↑. (C has to be excited from its ground state in order to be ↑↑↑↑. But this is consistent with experience.) There are exactly four different "lives" in the L-shell for four electrons that are antisymmetrically combined. The four H atoms would be accordingly ↓↓↓↓. Methane could be, therefore, reduced to the simple formula: The four atoms are attracted in a homopolar manner to the C atom, without, however, any repulsion among them. The tetrahedral arrangement resulted from this. The prospects from all these preliminary thoughts were quite promising "if it were possible to approximate better the whole damn thing."[20]

London was in agreement with Heitler that group theory may provide many clues for the generalization of the results derived by perturbation methods. The aim was quite obvious: to prove that quantum mechanics stipulates that among all the possibilities resulting from the various combinations of spins between atoms, only one term provides the necessary attraction for molecule formation. Nevertheless, London was not carried away by the spell of the new techniques—as Heitler was in the company of Wigner and Hermann Weyl at Göttingen. London "did not join in my studies of group theory. He thought it was too complicated and wanted to get on in his own more intuitive way."[21]

In Göttingen, Heitler started to study group theory intensively. Wigner's papers had already appeared, and there was a realization that group theory could be used for classifying the energy values in a multibody problem as well as for calculating perturbation energies. The theory of the irreducible representations of the permutation group provided the possibility of dealing *mathematically* with the problems of chemical valence in view of the difficulties involved in dealing with the many-body problems. The unavailability of reliable methods for tackling many-body problems haunted London all his life, yet years later, this difficulty became peculiarly liberating for London, helping him to articulate the concepts related to macroscopic quantum phenomena such as superconductivity and superfluidity.

After moving to Göttingen, Heitler started publishing a series of papers dealing with the question of valence by using group theoretical methods. As described in a significant paper with Georg Rumer (Heitler and Rumer 1931), they were able to study the valence structures of polyatomic molecules and find the closest possible analogue in quantum mechanics to the chemical formula that represented the molecule by fixed bonds uniting two adjoining atoms. They found that the emerging quantum mechanical picture was more general and that the bonds were not strictly localized. Nevertheless, the dominant structure was, in general, the one corresponding with the chemical formula. But there were other structures that were also significant, and these structures were quite useful in understanding chemical reactions. He recollected that London "was the first [a long time before the Heitler–Rumer paper] who showed that the activation energies in the treatment of the three hydrogen atoms could be understood in quantum mechanics, and this method gave us then a general understanding for it."[22] Later, Pauling called this a resonance between several structures. "A point which was violently objected to by the chemists was that both London and I stated that the carbon atom with its 4 valences must be in an excited state . . . all this was later accepted by the chemists, but at that time I don't think the chemists did find this of much use for them."[23]

Convinced that it was impossible to continue his work in chemical valence by analytic methods, London also turned to group theory. By the middle of 1928, he drew a program to tackle "the most urgent and attractive problem of atomic theory: the mysterious order of clear lawfulness, which is the basis for the immense factual knowledge of chemistry and which has been expressed symbolically in the language of chemical formulas" (London 1928, 60). London's group theoretical approach to chemical valence was formed around three axes. First, anything that may give a rather strong correlation between qualitative assessments of a theoretical calculation and the "known chemical facts" provided a strong backing for the methodological correctness of the approach chosen by expressing the observed regularities as rules. Second, because analytic calculations were hopelessly complicated and in most cases impossible, the use of group theoretical methods was especially convenient when one was dealing with the valence numbers of polyelectronic atoms, as the outcome was expressed either as zero or in natural numbers. Third, the overall result was that the interpretation of the chemical facts was compatible with the conceptual framework of quantum mechanics. Using group theoretical calculations, one could hope "to discover in the quantum mechanical description conceptual facts which in chemistry have proven themselves in complicated cases as a guide through the diversity of possible combinations, and see them in their connection with the structure of atoms" (London 1928a, 459). Hence, he attempted to give the valence numbers of the homopolar combinations an appropriate interpretation that "rests on the conceptual representations" of wave mechanics. Within such a program, London intended to deal with the problem of the mutual force interactions between the atoms; to examine whether it was possible to decipher the meaning of the rules that the chemists had found in semiempirical ways and to place those on a "sound" theoretical basis; and to determine the limits of these rules and if possible to initiate a quantitative treatment of them.

But he was not at all certain that the principles considered so far in atomic theory could, in fact, be used for the realization of such a program. This was because the characteristic interaction of the chemical forces deviated completely from other familiar forces: These forces seemed to "awake" after a previous "activation," and they suddenly vanished after the "exhaustion" of the available "valences." By making use of elementary symmetry considerations, it was known that the mode of operation of the homopolar valence forces could be mapped onto the symmetry properties of the Schrödinger eigenfunction of the atoms of the periodic system and could be interpreted as quantum mechanical resonance effects. This interpretation was formally equivalent to its chemical model, that is, it produced the same valence numbers and it satisfied the same formal combination rules, as they were expressed in the symbolic representation of the structural formulas of chemistry, that followed within the group theoretical possibilities as an immediate consequence of the Pauli principle in connection with the two valuedness of the electron spin. In particular, the fact that the valences were "saturated" proved in this context to be an expression of the restriction that the Pauli ban denotes for the occupation of equivalent states. Through group theory, London realized that the "uniqueness of the chemical symbolism is actually a consequence of the most fundamental theorems of the theory of the representations of the symmetric group" (London 1928b, 48).

London's "spin theory of valence" dealt mainly with those cases where each electron in a pair comes from a different atom. He examined the conditions whereby electrons from different atoms can pair with each other so that the resultant spin of the pair was zero. An electron already paired with another electron in the same atom was not considered in this schema of pair formation for bonding. Two electrons in the same atom were said to be paired if they had opposite spins and all their other quantum numbers were the same. But such an electron that was already paired could become available for bond formation with an electron from another atom if it could be unpaired without the expenditure of too much energy. London claimed that an electron can be unpaired provided that the total quantum number n of that electron does not change. Such an unpairing was considered by London as an intermediate step in the formation of a compound (London 1928, 1928a, 1928b, 1929).

Erich Hückel: Nonvisualizability and the Quantum Theory of the Double Bond

Heitler and London were led to tackle the problem of the chemical bond through their attempt to study the van der Waals forces. Their approach showed in no uncertain terms that the newly developed quantum mechanics would also be the appropriate framework for chemical problems. They attempted to bypass the calculational difficulties by using group theory and, most importantly, by *not* being faithful to one of the chemists' cardinal "principles"—that of visualizability. Another parallel approach to chemical bonding was being developed in Germany. From the start it attempted to cater to the community of organic chemists despite its strong grounding in quantum mechanics.

For a long time, the work of Erich Hückel (1896–1980) and his role in establishing quantum chemistry has not been given the attention it deserves. This is no longer the case, and we owe it especially to the systematic and perceptive work of Andreas Karachalios (Parr 1977; Hartmann and Longuet-Higgins 1982; Brock 1992; Berson 1996, 1996a, 1999; Park 1999a; Kragh 2001; Karachalios 2003, 2010). Hückel's contributions were mainly in the area of organic chemistry and more specifically on aromatic molecules and he had—through his talks and review papers—attempted to "talk" especially to the organic chemists, trying to convince them of the possibilities arising from quantum mechanics.

Hückel's studies at the University of Göttingen were interrupted by the First World War, and he spent some time as an aid to Ludwig Prandtl, who was then involved in his ground-breaking studies in aerodynamics. Hückel completed his doctoral dissertation at Göttingen under Peter Debye in 1921, and he studied the properties of anisotropic fluids trying to detect the kind of structures occurring in liquid crystals using Debye's method of X-ray interference developed for the study of the atomic structure of crystals. No particularly pronounced space lattice structure was detected, but Hückel acquired significant experience in the use of physical methods for the study of chemical problems. His work on the bonding of the unsaturated and aromatic compounds followed the first papers he published together with Debye where they discussed issues in the theory of strong electrolytes.

For a year after receiving his doctorate, he was David Hilbert's assistant at Göttingen. At the time, Hilbert was lecturing on the special and general theories of relativity. The next offer was from Born, who had accepted a post at Göttingen, while the promised assistantship from Debye, who had in the meantime moved to Zürich, was being delayed in the cogwheels of the Swiss bureaucracy, which had very strict laws for the employment of foreigners. With Born he published a paper on the quantum theory of polyatomic molecules (Born and Hückel 1923), which involved rather complicated mathematical computations.

His participation in the "Bohr Festival" (Mehra and Rechenberg 1982, ch. 3) during summer 1922, where various issues in quantum theory had been intensely debated, underlined his conviction that the new theory had a lot to offer for chemistry. Starting in fall 1922, he officially became an assistant to Debye at the Eidgenössische Technische Hochschule Zürich (ETH). His main research topic was the theory of strong electrolytes, as experimental results, especially those concerning electrical conductance, yielded unexplainable deviations from the predictions of the theory. In their joint paper published in 1923, Debye and Hückel, using statistical thermodynamics, developed a new function for the ion distribution, and their theory gave satisfactory results for the freezing-point depressions and the limiting law of the electric conductance in dilute solutions.

Soon after receiving his appointment at Zürich, Hückel started working on his habilitation, which he completed at the end of 1924. In this work, he was able to extend the treatment of dilute solutions to solutions with high concentrations of electrolytes (Hückel 1925). This work in colloidal chemistry led to the writing of a book on adsorption and capillary condensation of gases and vapors on solid surfaces and porous bodies (Hückel 1928). The momentous developments of quantum mechanics took place while Hückel continued to be absorbed by his book, and it was due to Debye's pressure that he shifted his attention to the systematic involvement with quantum mechanics. In 1928 in Leipzig, Debye had organized a meeting on "Quantum Theory and Chemistry." The lectures were delivered by different people, including some of the protagonists in the developments of quantum mechanics, and Hückel attended the meeting. He spent 3 months in 1929 in Copenhagen, and it was finally Bohr who directed him to the study of double bonding.

Starting in fall 1929, he received a fellowship from the Emergency Association of German Science (*Notgemeinschaft der Deutschen Wissenschaft*). Through this fellowship, it became possible to work at the Department of Theoretical Physics of the University of Leipzig, where Heisenberg and Friedrich Hund were already professors of theoretical physics and mathematical physics, respectively. In fall 1930, Hückel was appointed a dozent for teaching "chemical physics" at the Polytechnic in Stuttgart in Ewald's group where he stayed until 1937. In March 1934, he decided to join the National Socialist People's Welfare organization (*Nationalsozialistische Volkswohlfahrt*, or NSV) (Karachalios 2003).

Hückel's tenure at Stuttgart marked the beginning of a slightly different research program. He proceeded to study the binding state of alternating single and double bonds, something that, in effect, meant the study of the electron configuration of the carbon atoms in benzene and other aromatic compounds (Brush 1999). Mulliken later referred to the first of these papers (Hückel 1931) as "monumental" (Mulliken 1965, 8). This research was, in fact, his second habilitation (1931), as the one he had written on strong electrolytes was considered by the authorities as unsatisfactory because he was employed to teach chemical physics—considered part of physics and not of chemistry. The faculty regulations stipulated the submission of a separate thesis before he could make such a disciplinary transition from physical chemistry to chemical physics.

When Ewald emigrated, Hückel assumed teaching his lectures on theoretical physics between April 1 and September 30, 1937. In May 1937, Hückel became a member of the National Socialist German Workers' Party (*Nationalsozialistische Deutsche Arbeiterpartei*, or NSDAP), and toward the end of year he was offered the position of extraordinary professor of theoretical physics at the University of Marburg—an offer that, despite the favorable assessments of his work by people like Sommerfeld and Heisenberg, may not have been independent of his joining the party. In fact, that was the reason Hückel gave, later on, for having joined the party.

The Quantum Theory of Double Bonding
The concept of a double bond is as old as the proposal of the tetrahedral carbon atom—the bonding of two tetrahedrons connected along one edge. The experimentally observed rigidity of the double bonds could not be explained, and there was no quantum mechanical treatment for this kind of bonding. The problem had been discussed during the Leipzig meeting. Hückel had started to tackle this problem while in Copenhagen and continued to study it in Leipzig (Hückel 1930).

Hückel attempted to use classical interactions—such as dipole interaction—between the substituents to explain the observed rigidity against rotation, and he first tried to understand the stability of double bonding according to classical physics, but to no avail. He noticed, however, the unexpected way in which double bonds absorbed ultraviolet light and proposed that this anomaly may be due to the electronic structure of the double bond and, hence, the necessity for a quantum mechanical treatment of the phenomenon. He directed his efforts toward understanding the stabilization brought about due to the charge distribution of the electrons. Hückel moved along the following lines: He considered the oxygen molecule as a kind of "algorithmic device" and proceeded from O=O to formaldehyde (O=CH$_2$) to ethylene (CH$_2$=CH$_2$) by substituting the oxygen nucleus with a carbon nucleus with the removal of two hydrogen nuclei, and the resulting CH$_2$ was then linked to oxygen. Through the same process, he "formed" ethylene.

Hückel started his treatment by choosing the electronic configuration of the ground state of the oxygen molecule proposed by John Edward Lennard-Jones (1929), even though he knew about the alternative suggestion by Hund and Mulliken:

$(1s)^2 \ (1s)^2 \ (2s)^2 \ (2s)^2 \ (2p_+)^2 \ (2p_-)^2 \ (2p\sigma)^2 \ \{2p\pi_+, 2p\pi_-\}$.

This arrangement implied that the oxygen molecule in its ground state involved a double bond with four valence electrons: two in the $2p\sigma$ state and two in the $\{2p\pi_+, 2p\pi_-\}$ state. The former gave rise to a homopolar valence bond (because of the antisymmetry of the electrons) and comprised one of the valence lines. Each one of the other two electrons occupied one of the two degenerate orbitals, hence the ground state of the oxygen molecule involved a triplet state, which is responsible for the experimentally observed paramagnetism of oxygen. Considering the ground state of formaldehyde (O=CH$_2$) and by making a number of simplifying assumptions (such as, for example, neglecting the σ–π coupling), he found that both the two σ as well as the two π electrons formed homopolar valence bonds. But the perturbation brought about by the two hydrogen substituents formed a splitting of the doubly degenerate π one-electron state into two different states. Then, there resulted two polyelectronic states. One is a singlet state and is diamagnetic (due to the orbital motion of electrons). The other is a triplet state and paramagnetic (due to electron spin and associated with unpaired electrons). It was not possible to calculate which state had a lower potential energy, and, hence, he could not reach a criterion for choosing the more stable configuration. He suggested that experimental results—not yet available—on the magnetic susceptibility of formaldehyde would be able to distinguish between the two. Nevertheless, through some analogical thinking he came to the conclusion that the O=C double bond was not the same as that of the oxygen molecule, and this was due to the hydrogen substituents. He continued in the same manner to "reach" ethylene. His overall conclusion for a quantum mechanical treatment of the double bond was that it consisted of two different types of bonds, one σ bond and one π bond. The former had no stabilizing effect on the molecule's planar arrangement, whereas the latter was responsible for the fact that the substituents could not rotate freely around the double bond (Karachalios 2010).

Hückel could not be indifferent to the interpretative particularities of his theoretical treatment formulating, as it were, a serious epistemological challenge to the approaches of almost all the protagonists of quantum chemistry. He noted that his treatment of the double bond appeared to be coherent with Jacobus van't Hoff's overall approach for the rigidity of the double bond against rotation. But he was emphatic that his "theory does not completely conform with this picture because no *real meaning* is attached to the four valence directions depicted by van't Hoff; only the plane on which they lie in van't Hoff's picture has such a meaning" (Karachalios 2003, 77, emphasis ours).

The phrase "real meaning" refers to physical significance and, perhaps, physical reality—even though the notion of what is real, or of what has physical materiality, was not something quantum chemists dealt with in any systematic and philosophically strict manner. Yet, Hückel's theory shifted the emphasis on what was physically significant in van't Hoff's model from the four valences to the plane on which they lie, without, however, altering the model itself. As it often happened among quantum chemists, quantum mechanics reassessed one of the interpretative cornerstones of chemistry, that of visualizability. The latter, so closely attached to the classical worldview, could no longer take advantage of its heuristic role when quantum mechanics started to be widely used in the treatment of chemical problems. There have been many cases when quantum chemists would opt for the nonvisualizable representations in configuration space in order to stress that, perhaps, the strong affinity chemists had with visualizable entities and the tradition to present results in terms of visualizable entities may have become a rather serious hurdle in their attempt to adapt to quantum mechanics. But *at the same time*, pictorializing "entities" was not something to be dispensed with altogether because, especially in view of the impossibility to have analytical solutions, "pictorialization" had continued to be a particularly useful and, at times, effective way of dealing with chemical problems. This was not always so straightforward, and Hückel's interventions had underlined the difficulties involved: To talk of the changes brought about by the use of quantum mechanics in the stereochemical model—perhaps the "most visual" of all models—without disfiguring it altogether was a challenge in itself.

Comparing, in fact, Hückel's treatment of the double bond with that of Pauling and of John Clarke Slater, there appeared some nontrivial differences. As we show in the next chapter, Pauling's main theoretical entity for dealing with the two C=C bond was his directed sp^3 hybrid orbitals. The overlapping of two such tetrahedrally directed orbitals gave the sought for stability of the double bond against rotation. The maximum overlapping of the two eigenfunctions occurs when the two tetrahedral carbon atoms share an edge. Thus, Pauling's approach gave much credence to the original van't Hoff model, explaining at the same time the stability against rotation (Pauling 1931). Not much later, Mulliken developed a more satisfactory quantum theory of double bonding using group theory (Mulliken 1932c, 1932d, 1933).

Pauling and Slater justified the visual models of traditional chemists through skillful application of the mathematical language of quantum mechanics. Hückel's model, on the contrary, underpinned the organic chemist's classical visualizations while placing them at the same time on a new foundation that would ultimately articulate a new epistemological framework.

Quantum Theory of Aromaticity
Hückel was able to provide a quantum mechanical explanation of why a number of properties of aromatic compounds were associated with a group of six electrons like in benzene—the "aromatic sextet" first noticed in 1925 by J. W. Armit and Robert Robinson, who had used Lewis's valence scheme (Armit and Robinson 1925). Hückel showed that these electrons, which formed a ring, formed in fact a complete closed shell, and he also explained a number of experimental observations on benzene (Hückel 1931, 213). Hückel's shift in emphasis was evident from his paper. He claimed that aromaticity was due to the number of electrons forming a complete electron shell, rather than the number of atoms forming a ring, thus providing an explanation of the stability of the aromatic cyclic systems.

Hückel improved his treatment of aromaticity in a second paper, through a method (referred to by Hückel himself as method II) that still bears his name: the HMO method (Hückel's molecular orbital method) (Hückel 1932; Pullman and Pullman 1952; Dewar 1969; Coulson, O'Leary, and Mallion 1978). The end result was a quantum mechanical treatment of aromaticity, through the $4n+2$ (where $n = 0, 1, 2, \ldots$) rule as the criterion for aromaticity. The formula referred to the number of π electrons in a given organic cyclic compound, which would be classified as aromatic if the number was 2, 6, or 10. Furthermore, this explained the stability of the aromatic molecules with respect to reactions. One important result of his approach was that for aromatic and unsaturated compounds, the number of possible valence structures is not the same as the number of different states of determined energy, nor is a state of determined energy necessarily identified with a specific valence structure.[24]

Hückel's papers included rather involved mathematical calculations and did not make easy reading—especially for chemists, despite the fact that his work had been deemed as particularly significant for organic chemistry. The indifference of many chemists in Germany was not even shaken with Hund's 1933 survey article in the *Handbuch der Physik*. Hückel was indeed in a rather peculiar situation. His case was exceptional for while he worked in physical chemistry, he was able to overcome the deficiencies of his training as an organic chemist by taking advantage of his brother Walter Hückel's expertise in the field, which probably helped him in asking the pertinent questions in organic chemistry to be answered in the framework of quantum mechanics.[25] But his successes were not appreciated either by physicists who did not care much about problems of organic chemistry or by organic chemists who were unfamiliar with the new techniques and their mathematical framework, despite a few who thought that such ignorance was not for the benefit of their craft.[26] By 1937, Hückel abandoned the field unable to challenge a scientific establishment in which German physicists and chemists were unwilling to accept research on the quantum mechanical properties of the chemical bond.

Hans Hellmann: Fundamental Theorems and Semiempirical Approaches

Hans Hellmann (1903–1938) was born in Wilhelmshaven in Germany, and his father was a noncommissioned officer in the navy. He chose to leave Hitler's Germany because of his Jewish wife, settled in the Soviet Union, and was executed in the great purges. In 1922, he started attending classes at the Institute of Technology in Stuttgart planning to major in electrical engineering but soon changed to physics, and in 1925 at the University of Kiel he attended lectures by Kossel, who was among the first to discuss the electronic theory of the covalent bond. While at Kiel, he worked on experiments measuring the frequency-dependent dielectric constants of conducting hydrous salt solutions. He moved to Berlin where he worked with Otto Hahn and Lise Meitner on experiments synthesizing "radioactive preparations for physical research." In 1929 at the University of Stuttgart and working with Erich Regener, he received his doctorate with the thesis "On the Occurrence of Ions from the Decomposition of Ozone and the Ionization of the Stratosphere." This work showed that, contrary to the speculation that pairs of ions appeared when ozone was decomposed, what actually happened was the production of an extremely small number of such pairs (Schwarz et al. 1999, 1999a).

In the same year that Hellmann defended his dissertation, Erwin Fues was appointed to the chair of theoretical physics and applied mathematics at the Institute of Technology in Hannover and offered Hellmann an assistantship, which Hellmann accepted. In Hannover, Hellmann had the opportunity to discuss issues in various branches of chemistry, as there were many chemists in the Faculty of General Science. This turned out to be particularly useful for his future researches in quantum chemistry.

In 1931, he was appointed lecturer in physics at the Veterinary College at Hannover. In 1933, 1 month apart and independent of each other, Hellmann and Slater proved what came to be known as the molecular virial theorem—the possibility to calculate exactly the kinetic and potential energies for a stationary system if its total energy is known. Hellmann showed that it was the relative reduction of destabilizing energy that contributed to the bond stability. He was, also, able to derive the variations of the energies as a function of the internuclear distance. In a way, Hellmann's work had a rather intriguing conceptual side to it: He attempted to *reformulate* the dominant view as to what caused the stability of bonds. It is interesting that the "physicist" Hellmann had adopted semiempirical methods in his theoretical calculations, where he made ample use of the experimentally known properties of the diatomic fragments occurring in a molecule.

Aware—as were most of the pioneers of quantum chemistry but not many of his German fellows—that results had to be made available to chemists in a special parlance, Hellmann together with W. Jost (a fellow assistant who was a physical chemist) published two papers discussing the "chemical forces" by using quantum mechanics (Hellmann and Jost 1934, Jost 1935). Later in the Soviet Union, he also wrote articles on quantum mechanics and chemical bonding, aiming primarily at chemistry-oriented audiences. He also delved into the intense debate around the issues of quantum measurement and, generally, of the possibilities offered by dialectical materialism for the further understanding of nature, without succumbing to the extreme ideological views of some of the popularizers of science in the Soviet Union.

Hellmann's leftist political views did not help him during the first months of Nazi rule in Germany. Things became even worse when in his application for the submission of the habilitation, as stipulated by the Reich Law of June 30, 1933, he had to include information about his wife being Jewish. He accepted the offer in 1934 of the post of "head of the theory group" at the prestigious Karpov Institute in the Soviet Union. The institute was in the Ukraine, which was also the place of his wife's origin. The institute had existed since 1918 and was dedicated to research in chemistry, and it had developed into a world class center with very important work in physical chemistry and quantum mechanics. By the time Hellmann arrived, there were already 150 scientists and 250 technical staff members working at the institute.

There he was associated with Ya. Syrkin, who later wrote one of the standard textbooks on quantum chemistry and who was to oppose his colleagues' critical stand against Pauling's ideas in quantum chemistry (see chapter 2). By the end of 1936, he had been granted a doctoral degree that made him eligible for a lectureship at universities in the Soviet Union. He was given various prizes for his research, was invited to present his research to the Academy of Sciences, and on January 1, 1937, he was promoted to a full member of the institute, a position that corresponded with a professorship.

The work he undertook at the Karpov Institute dealt with the investigation of the ways to derive the affinity relations between the various chemical elements. Hellmann put forth the notion that came to be known as pseudopotential. When atoms approach each other, valence electrons are coupled because of the Pauli principle. This creates a repulsive effect between the occupied shells, because as a result of the Pauli principle, there is an increase in the electrons' kinetic energy. There is, thus, in Hellmann's words an "additional potential," later to be called pseudopotential: It is the potential that compensates for the electrostatic-nuclear attraction. Thus, the effective potential that now appears in the Schrödinger equation is the sum of the two. This expressed, again, his characteristic methodology: Hellmann proceeded to the determination of the value of this potential not only through approximate calculations when they were possible, but also through fitting to "experimental energies of suitable atomic states" (Schwarz et al. 1999, 16).

Despite his past as a lecturer in a chair of theoretical physics and applied mathematics, Hellmann was not among those who sought the strict application of formal mathematical methods while discussing issues in quantum chemistry. His was a semiempirical approach, followed by a careful analysis of some of the intricate problems in the development of quantum mechanics aiming basically at a chemistry-oriented audience. He was advocating the validity of physical laws for chemical phenomena, but at the same time he tried to avoid very strict and formal presentations. He made ample use of graphical methods, attempting, in a way, to circumvent the difficult problem of visualization in quantum mechanics generally, and quantum chemistry in particular. Hellmann the physicist, when it came to dealing with problems of (quantum) chemistry, was rather receptive to the possibilities offered through the semiempirical approach.

He claimed that "a purely theoretical derivation of properties of materials always means returning from quality to quantity," by which he meant that what one knows qualitatively can be physically explained if there is a (proper) theoretical derivation. The perception of quantum chemistry by chemists in the Soviet Union did not appear to have been any different from that of their colleagues in Germany. He wondered whether "organic and inorganic chemistry are engaged in organic and inorganic substances, which substances are the topic of theoretical chemistry? Purely theoretical substances? Does quantum theory have any useful role in chemistry at all?" (Schwarz et al. 1999, 15).

Hellmann believed that he had enough material to write a book. In fact, the manuscript in German was with Jost, who was seeking a publisher after Hellmann's forced emigration in 1933—to no avail, of course. The translated book was eventually published in the Soviet Union, appearing at the beginning of 1937 with the title *Quantum Chemistry*. An abridged German edition appeared the same year (Hellmann 1937). Hellmann had a tragic end. He became one of the countless victims of the purges of 1937. He was denounced by two colleagues (both having served as the Communist Party's local secretaries) at the Karpov Institute. He was arrested in March 1938 and executed on May 29, 1938. The only surviving document bearing his signature is his "confession," which "admits" deeds leading to espionage.

Friedrich Hund: Foundations of Molecular Spectroscopy in Quantum Mechanics

Among those who initiated research in quantum chemistry in Germany, Friedrich Hund followed a rather different path, inaugurating what came to be known as the molecular orbital approach. His contributions were remarkably close to Mulliken's, and both had a very cordial relationship. Born in Karlsruhe, Germany, Friedrich Hund (1896–1997) did not attend the Humanistische Gymnasium but instead the Oberrealschule and the Realgymnasium. He had a fair training in experimental physics but, since his high school days, he had shown a keen interest in mathematics, which he learned by himself, and thought for a long time of becoming a high school mathematics teacher. During the formative period before attending college, Hund equated physics with experimental physics and only later realized that the flourishing area of theoretical physics suited his interests better by fusing a solid mathematical methodology with a physical conceptual structure.

Hund attended the universities at Göttingen and Marburg and took courses in mathematics, physics, geology, and geography. In Göttingen, he took a course on quantum theory given by Debye and studied partial differential equations with Richard Courant. When Hund decided to become a theoretical physicist, he started to work with Born on the physics of crystals (*Gittertheorie*), as many of Born's students did, but he finally wrote his dissertation on the Ramsauer effect.

After completing his Ph.D. degree in Göttingen in 1922, he became Born's assistant and helped him write the book *Atommechanik* (Born 1960). Then he became privatdozent in the University of Göttingen in 1925, extraordinary professor in Rostock University in 1927, and professor at the same university 1 year later. From 1929 to 1946, he held a professorship of mathematical physics in Leipzig, after which he became successively a professor of theoretical physics in the universities of Jena, Frankfurt, and Göttingen.[27]

Hund's interest shifted to spectroscopy in the aftermath of Bohr's visit to Born's institute in June 1922 for "Bohr's festival," during which Bohr delivered a major series of lectures on quantum theory and atomic physics that were said to have "revolutionized" physics at Göttingen. He avidly discussed spectroscopy with James Franck, Hertha Sponer, Jordan, and Heisenberg (figure 1.2) and started to work on the interpretation of complex atomic spectra in terms of the Russell–Saunders vector model. The first to show how the notion of spin and the Pauli exclusion principle could be used to explain the periodic system of the elements, Hund's book *Linienspektren und Periodisches System der Elemente* contributed greatly to familiarize scientists with these two rules (Hund 1927).

In his last year as privatdozent at Göttingen, Hund divided his time between the study of complex atomic spectra and the spectra of molecules. The paper "Zur Deutung einiger Erscheinungen in den Molekelspektren" (Hund 1926) marked Hund's debut in

Figure 1.2
Group picture at the University of Chicago. Front row: Werner Heisenberg, Paul Dirac, Henry Gale, and Friedrich Hund. Back row: Arthur Compton, George S. Monk, Carl Eckart, Robert S. Mulliken, and Frank Hoyt.
Source: Max-Planck-Institute, courtesy AIP Emilio Segre Visual Archives. Gift of Max-Planck-Institute via David C. Cassidy.

the field of molecular spectroscopy. There, he discussed the nature of molecular electronic states from a theoretical point of view, introducing electron spin into band structure. He further suggested that in a diatomic molecule, the interaction of the two atoms produced a deformation of the spherical symmetry of the atomic field force into an axial symmetry around the internuclear axis, a perturbation that bore a strong analogy to the Stark effect. Based on this assumption, he suggested a molecular vector model and analyzed the different cases of band structure corresponding with different types of coupling between the electronic orbital angular momentum (specifically, its projection along the internuclear axis), the spin angular momentum, and the nuclear motion.

Later on, Hund used the new quantum mechanics to show that, in opposition to the old quantum theory, one could conceive an adiabatic transition from the states of two separated atoms to the states of a diatomic molecule, and then to the states of an atom obtained from the hypothetical union of the two atomic nuclei.[28] This fact allowed him to interpolate the electronic quantum states of a diatomic molecule between two limiting cases: the situation where the two atoms were separated (separated atom case) and the opposite situation where the two nuclei were thought to be united into one (united atom case). The idea was that one could imagine the molecule already latent in the separated atoms, so that the molecular quantum numbers existed already before the atoms come together, but started to play a dominant role (relative to the atomic quantum numbers) only in the situation where the two atoms were already at molecular distances from each other (Hund 1927a, 1927b).

Hund's contributions to elucidate the nature of the electronic states were almost "duplicated" by Mulliken, so that it becomes virtually impossible to analyze Hund's contributions to quantum chemistry without discussing simultaneously the approach of Mulliken. This we do analytically in chapter 2. As it often happened with approaches put forward initially by German and American scientists, they were frequently attempting to answer similar questions starting from different theoretical assumptions and developing complementary or even opposing methodologies. In chapter 2, we will discuss such issues when we deal with Heitler and London's later attempts to re-enter the field after the incursions of Pauling, Slater, and Van Vleck.

The case of the duo Hund–Mulliken is exceptional. After their first encounter and talks in 1927 when Mulliken visited Europe, the two became friends, discussing topics of common interest and complementing each other in their approaches to molecular spectroscopy and valence related questions. Their friendship, which grew stronger with time, seems to have facilitated their scientific dialogues and their symbiotic participation in building quantum chemistry.

Born in the same year, and revealing intertwined scientific trajectories up to 1937, Hund and Mulliken had very different backgrounds, yet very similar interests at the beginning of their careers, both being interested in band spectra.[29] In 1928, Hund had completed a paper where he discussed various points concerning molecular orbitals. Just before sending his paper for publication, he was sent a preprint by Mulliken who had essentially done the same calculations. But Hund decided to go ahead and publish his paper because "Mulliken's paper is rather American, e.g. he proceeds by groping in an uncertain manner, where one can say theoretically the cases for which a particular claim is valid."[30]

Some Further Remarks

Almost all the people involved in this first phase of the development of quantum chemistry—Heitler, London, Hund, Hückel, Hellmann—went through a university education where the mathematical training reigned supreme. Though nearly none was an expert in chemistry, almost all were highly sophisticated mathematically. Dirac's claim may have gone unnoticed by the chemists and was thought of as something obvious by physicists, but what followed the Heitler–London paper was, in effect, an attempt at circumventing the catastrophic state of affairs prophesied by Dirac. A generation later, chemists would be in a position to articulate a set of sophisticated theoretical schemata with impressive empirical confirmations and claim that there was a new culture joined by all.

Nevertheless, historically the Heitler–London paper set the stage for the (uneasy) coexistence of chemists with quantum mechanics—definitely a physicists' area of jurisdiction. For ages, mathematics and chemists did not make an agreeable contact. Hence, it was not all that welcome to have realized that one of the mysterious forces of chemistry—that of the homopolar bond—could be understood *only* in terms of quantum mechanics, bringing out at the same time the extreme significance of the exclusion principle. This principle acquired the status of a basic principle for chemistry. As Van Vleck and Sherman (1935, 173) aptly noted, "the Pauli exclusion principle is the cornerstone of the entire science of chemistry." The Heitler–London paper made the community of physicists as well as chemists aware of the spectrum of possibilities of the newly formulated quantum mechanics. Though the possibilities covered a wide area, no one really knew how to realize them. The program was there, its promises loosely defined, but the attempts to get specific results were bogged down in almost insurmountable technical difficulties.

For a short period, group theory appeared to be doing the trick—a trick that could not be brought about by the use of the Schrödinger equation alone. Heitler and London demanded too much from the chemists: to accept the new mechanics and change their theoretical outlook. Using group theoretical calculations, one could hope to articulate new concepts, and London attempted to give the valence numbers of the homopolar combinations an appropriate interpretation that "rests on the conceptual representations" of wave mechanics. Furthermore, the use of group theory brought about the realization that the "uniqueness of the chemical symbolism is actually a consequence of the most fundamental theorems of the theory of the representations of the symmetric group" (London 1928b, 48). But soon this approach was again at a dead end. And this dead end was of a technical as well as an epistemological character: Any theoretical attempt could not go any further than actually explaining what chemists already knew experimentally. There was no prediction. To use the theoretical apparatus of the physicists was one thing. To have theories with no predictive power was another, and particularly embarrassing, thing. The danger for the chemists was not to become physicists, but to become the physicists' poor relatives, with theories that lacked one of the cardinal characteristics of the physicists' theories—their predictive character. Examining the possibilities of group theory brought out the issue of theory *versus* rules. For a short period, it looked probable that group theory would lead to rules, something so dear to the chemists' culture. And in this respect, one can sense questions related to contingency. The use of group theory delineated a totally different direction, where it all depended not on the empirically more satisfactory schemata of the rival approaches but on the consensus of the community as to what constitutes a more "proper/scientific/strict" theoretical schema.

2 Quantum Chemistry *qua* Chemistry: Rules and More Rules

After publication of the papers of Heitler and London as well as those of Hund, a new approach—less intimidating to the chemists—started developing. It was an approach mostly developed in the United States, with an intense pragmatic streak, and which, in a few years, was almost universally accepted by the chemical audiences (Simões 1993, 2003; Gavroglu and Simões 1994, Simões and Gavroglu 1997). The main protagonists of such an approach were Mulliken in the Department of Physics at the University of Chicago and Pauling in the Department of Chemistry at the California Institute of Technology (Caltech). Their (semi)phenomenological approach, where strict matters of quantum mechanics were not high on their agenda, created a theoretical framework quite attractive for the chemists. Mulliken's strategy aimed at devising an *aufbau* procedure for molecules, after the amazing success of Bohr's use of this principle for atoms. Pauling's approach, though less intuitive, appeared to be building on an idea articulated some years back by a chemist, Gilbert Newton Lewis, who had a rather idiosyncratic research agenda.

The Young Mulliken: Hinting at Molecular Orbitals

In 1923, Kemble and Raymond Thayer Birge, together with Walter F. Colby, Francis Wheeler Loomis, and Leigh Page, started preparing a comprehensive report for the National Research Council on the spectra of molecular diatomic gases (Kemble et al. 1926). Kemble, Birge, and Colby represented the three major American centers where research in molecular spectroscopy took place—Harvard University, the University of California at Berkeley, and the University of Michigan, respectively. Loomis, who had discovered the isotope effect in molecular spectra, was then at New York University, and Page was the senior professor of theoretical physics at the prestigious Yale University. The report's framework was old quantum theory, and the attempt of the authors to be quite thorough delayed its publication so much that when it finally appeared in print, after the advent of quantum mechanics, it was a bit outmoded but still very informative. Molecular spectroscopists had been able to obtain information on the structures of molecules through the analysis and interpretation of the spectra of molecules that was based on the assumption of three different types of contributions: rotational, vibrational, and electronic. An increasingly more sophisticated model of the rotational and vibrational nuclear motions of diatomic molecules guided them through the maze of band spectra to offer in the end a detailed knowledge of molecular structural features (Assmus 1991, 1992, 1993).

The person who turned out to be the best "reader" of molecular (band) spectra was Robert Sanderson Mulliken (1896–1986) (Mulliken 1989; Simões 1999, 2008). The son of Samuel Parsons Mulliken, a renowned organic chemist, Mulliken studied chemistry at the Massachusetts Institute of Technology from which he graduated in 1917. After graduation, he accepted a wartime job as a junior chemical engineer for the U.S. Bureau of Mines and conducted research on poison gases at the American University in Washington, D.C. After the end of the First World War, he worked as a chemist for the New Jersey Zinc Company, and in 1919, attracted by the work on separation of isotopes of the physical chemist William Draper Harkins, Mulliken entered the graduate program in chemistry at the University of Chicago. There he earned his Ph.D. in 1921 with a dissertation on the partial separation of mercury isotopes by evaporation and other processes. He stayed one more year as a National Research Council postdoctoral fellow extending his former research to obtain bigger isotope separations with mercury by using improved equipment and methods. In the process, he built the first "isotope factory," an apparatus that was based on the different behaviors of isotopes under the processes of evaporation and diffusion through a membrane.

Still a fellow, Mulliken moved to the Jefferson Physical Laboratory at Harvard in 1923. Helped by F. A. Saunders and Kemble, he started working on molecular spectroscopy and became deeply involved in the preparation of the National Research Council report on the spectra of diatomic molecules mentioned earlier. By 1926, Mulliken became assistant professor of physics at New York University, already recognized as an expert on band spectra.

While investigating isotope effects in the spectra of diatomic molecules such as boron nitride, Mulliken's attention was caught by the electronic distribution in molecules (Ramsay and Hinze 1975).[1] By 1925, several electronic levels had already been identified in very simple molecules and molecular fragments, such as CO (five electronic levels), N_2 and NO (four levels), and BO, CN, CO^+, and O_2 (three levels) (Kemble et al. 1926, 238). As their number grew steadily, the need for a classification increased concomitantly. The search for analogies in the spectroscopic behavior of different compounds soon became the yardstick used to guide spectroscopists into unknown territory. Following earlier suggestions on the similarities between certain molecular and atomic spectra (Sommerfeld 1923) and on the physical similarities of isosteric molecules (compounds with the same number of elements and the same total number of electrons) (Langmuir 1919), Mulliken decided to look for similarities in spectroscopic behavior in isosteric molecules. He found that the spectroscopic analogy between isosteric molecules could be extended to the chemical element with the same number of electrons.

The parallels between molecular and atomic spectra (similar values for the energies of electronic levels and their multiplicities, values of other molecular constants, etc.) served as the basis for the classification of diatomic molecules into different families and suggested that similar electronic structures were responsible for corresponding systems of energy levels. Although "pretty speculative,"[2] the analogy had been used in one form or another by several scientists, like Rudolf Mecke and Sponer in Germany and Birge in the United States (Mecke 1925; Sponer 1925; Birge 1926, 1926a). It became a recurring theme in the extensive correspondence between Birge and Mulliken.[3] Together with evidence that the electronic levels of CO, N_2, and H_2 could be arranged in series fitting approximately the formulas known to hold for line spectra (such as the Rydberg or Ritz formulas), these analogies led Birge, in a letter to *Nature*, to a bold generalization postulating that "the energy levels associated with the valence electrons of molecules agree in all essential aspects with those associated with the valence electrons of atoms" (Birge 1926b, 301).

In case Birge's challenge could be accepted, one could classify electronic states in diatomic molecules by means of the same nomenclature (Russell–Saunders notation) used for atomic states to represent term symbols (S^2, P^2, S^3, P^1). Mulliken decided immediately to look for corroborative evidence (Mulliken 1926), and going one step further, he introduced three postulates that accounted for the band spectra structure of known band spectra and enabled one to make structure predictions of yet unanalyzed band spectra (Mulliken 1926a).[4] In a short while, Mulliken addressed the question of molecule formation and molecular structure and for the first time hinted at what he later called "electron promotion," a concept essential to his theory of chemical binding: In the formation of molecules, a radical rearrangement of some electrons may take place, corresponding with their "promotion" to orbits with a higher n quantum number (Mulliken 1926b).

When Mulliken read Hund's discussion of the nature of the electronic states (see chapter 1) (Hund 1926), he immediately recognized its importance and excitedly confided to Birge that Hund had included *everything* in his paper. "Almost all of my conclusions seem to agree with his theory."[5]

Mulliken (1927) went on to publish a summary of Hund's theory and to provide an extensive discussion of the empirical evidence for it, which relied heavily on his own work.[6] Hund's quantum mechanical approach to molecules found corroboration in the evidence largely gathered by Mulliken in his work in the systematization of band spectra, and Mulliken's phenomenological theory gained a legitimizing framework it did not possess before. He thought that Hund's first papers explained with remarkable success "why, and how, the electron levels in molecules resemble those of atoms, and why and how they differ."[7] He also noted that Hund used the new quantum mechanics to help explain how various atoms could be united to form a molecule. However, Mulliken's approach preceded and in large measure was independent of quantum mechanics. The following gives the climate of relief when phenomenological approaches were successful even though there was the awareness that they were lacking a proper theoretical treatment: "If there were then some feelings or misgivings that perhaps this [the interpretation of band spectra structure] ought to be done differently and that we now knew how to do it, there was also the feeling 'well, here it is.' Of course there was some kind of awareness that the new quantum mechanics might improve things but we had the feeling, 'well, here it is; this is probably worthwhile and good as far as it goes.'"[8] Much of what characterized Mulliken's conceptual scheme of molecular orbitals as the foundation for a radically new approach to valence theory was not really dependent on the new quantum mechanics. The Pauli exclusion principle, whose empirical origin as well as its independence from Schrödinger's formulation was rather convenient for Mulliken, played a crucial role in the genesis and development of the phenomenological approach to molecular structure and chemical bonding.

Mulliken went to Europe in summer 1927. He visited Göttingen, Zürich, and Geneva, meeting with Hund, Schrödinger, and Heitler and London, among others, and ended the summer with a hiking trip to the Black Forest with Hund and some friends. His aim was to discuss the problems of molecular structure and spectra with several spectroscopists, but especially to discuss with Hund the latter's new contributions to a quantum mechanical theory of molecular structure. Mulliken's attitude toward the new quantum mechanics was rather pragmatic. He was satisfied with a general knowledge of quantum mechanical methods and principles in order to understand particular molecules or types of molecules, their properties, and, especially, their spectra. "I was more interested in getting better acquainted with molecules than with abstract theory about them" (Mulliken 1989, 59).

While working on the assignment of quantum numbers to electrons in molecules, Mulliken came to realize that he had found something truly important. He communicated his preliminary findings during the 1928 February meeting of the American Physical Society,[9] and, when completed, circulated the draft of the paper among colleagues.[10] Mulliken first presented the aims of his program and, then, an explanation of the methods used. He formulated a set of working rules with the purpose of assigning quantum numbers to electron states in actual molecules and gave examples of their application. He cautiously noted that the method developed had so far been applied exclusively to diatomic molecules made of atoms of the first row of the periodic table. Only a few of the molecular states discussed were the unstable states of chemically stable molecules. Always with an eye to future chemical applications,

Mulliken remarked that, besides their purely theoretical interest, a knowledge of the numerous excited states and chemically unstable molecules would prove to be indispensable in deducing electron configurations for those special cases that correspond with stable molecules, and also for understanding the intermediate steps in various chemical reactions. Although the "essential ideas and methods" were those introduced by Hund, the paper's great novelty consisted "in the attempt to assign individual electronic quantum numbers" (Mulliken 1928, 190) and obtain a knowledge of the energies of individual electrons in molecules, in an analogous way to that possessed already in the case of atoms. Though they were not yet named as such, Mulliken, in a way, attempted to assign electronic configurations to experimentally observed molecular orbitals.

Hund's proof that an adiabatic transition could connect the separated atoms to the diatomic molecule and then the united atom gave theoretical support to Mulliken's former hypothesis that electronic quantum numbers could change drastically in the process of molecule formation (Mulliken 1926b).[11] Simultaneously, it gave a theoretical justification for the "marked analogies," found by Mulliken, between the spectra of certain groups of diatomic molecules (Mulliken's "octet" molecules) and certain associated atoms—which were Hund's united atom (Na, Be, and Al, respectively).

Besides the theoretical support given by Hund's work, there was not much in the paper to imply that Mulliken was thinking more along the lines of the new quantum mechanics rather than the old quantum theory. Schrödinger's equation was not used, and the language employed was not that of quantum mechanics: Mulliken's highly "visual" spectroscopic experience seemed to be consistent with the existence of orbits. The only paragraph where Mulliken addressed "the meaning of quantum states of electrons in the new mechanics" functioned rather as a cosmetic appendage to his largely "pre-quantum mechanical" language.

By analogy to what Bohr had done in his "grand synthesis,"[12] Mulliken pictured the molecules as being formed by feeding electrons into orbits that encircled two or all nuclei. As he later recalled: "Bohr's *Aufbauprinzip* for atoms made a very great impression on me and so I thought something similar for molecules would be nice. If you translate orbits into orbitals for atoms, then for molecules it is molecular orbitals; it is something that goes around all the atoms or however many atoms there are and the *Aufbauprinzip* transferred to molecules simply means molecular orbitals."[13]

To apply the *aufbauprinzip* to molecules, two sorts of questions called for clarification. The first concerned the nature of quantum numbers appropriate to characterize electrons in molecules and the nature of closed shells, molecular states, and multiplets. The second concerned binding energies and the type of energy relations resulting from the union of two atoms. To address the first set of questions, the relation between a *molecule* and a *molecule-as-united-atoms* was emphasized. To address the second set of questions, the relation between a *molecule* and the *separated atoms* was all important.

To find out the possible quantum numbers for each electron in the molecule, Mulliken suggested that they were obtained from those of the associated united atoms by placing them in a strong axially symmetrical electric field, so that the two resulting nuclei were fixed. This simplification was justified because "we are not directly interested here in the effects of nuclear rotation and vibration" (Mulliken 1928, 191). Several coupling schemes could be applied and, contrary to what happened in the atomic case, in molecules there was no limiting case, and "the actual condition usually lies more or less in the midst of a region between several limiting cases" (Mulliken 1928, 191–192).

The relation to the separated atoms enabled Mulliken to discuss the energy conditions favorable to the formation of molecules. It was noted that often, in order to obey the Pauli principle for a molecule as united atoms, some electrons could have their n value increased in the process of molecule formation. These electrons were called "promoted electrons," and the associated energy increase was called "energy of promotion." To analyze the nature of the energetic conditions necessary for the formation of the molecule, Mulliken considered the total energy divided into two components: the positive potential energy of nuclear repulsion and the negative binding energies of each electron in the field of the nuclei and the other electrons. In order for a molecule to be formed, the following conditions should be satisfied: for $r > r_0$ (r_0 = internuclear equilibrium distance), the electronic binding energy had to increase more rapidly than the nuclear repulsion energy. For $r = r_0$, the two types of energy should increase at the same rate, so that the total energy of the molecule attained a minimum. For $r < r_0$, the nuclear repulsion had to increase faster than the electronic binding energies. When (reasonably) stable molecules were formed, the binding energy had to increase considerably faster than the nuclear energy over a considerable range of r values as r decreases toward r_0 (Mulliken 1928, 194).

As the nuclear distance diminished, the binding energy of the *unpromoted electrons* should be expected to increase steadily, because the electron comes into the influence of the two nuclei, reaching a maximum when the molecule is formed. In the case of *promoted electrons*, the binding energy may either increase or decrease, because the increase in effective nuclear charge is, at least partially, and often more than not, outweighed by the effects of the increase in energy associated with the increase in n. This qualitative analysis pointed already to the reformulation of some of the most cherished concepts in chemistry. Although, following Lewis's views, electrons were usually divided into "bonding" (the paired electrons that hold the molecule together) and "nonbonding," Mulliken concluded that it was possible to assign various degrees of "bonding power" for various orbit types. Electrons could be regarded as having positive bonding power if their presence in a molecule tended to make the dissociation energy large or the equilibrium internuclear distance small (Mulliken 1928, 196). The converse was also assumed to be true. There were then two possible definitions of bonding power, *energy-bonding power* or *distance-bonding power,* arising either by the application of the energy criterion or the distance criterion, and a set of rules to be used in the analysis of spectroscopic data and in the assignment of quantum numbers to electron states of actual molecules (Mulliken 1928, 201). As we will discuss later, these considerations were the basis of Mulliken's criticism of Heitler and London's valence theory.

The completion of this phase of Mulliken's work was accompanied by his move, in 1928, to the University of Chicago as an associate professor in the Department of Physics. As a recognition of his outstanding contributions, while still at New York University, Mulliken was offered several jobs, all of them related to the creation or implementation of research programs on molecular structure: to succeed Loomis as head of the Department of Physics at New York University and continue developing the program he helped to start; to accept an offer made by R. W. Wood for Johns Hopkins University and work toward the inauguration of a research program on the study of molecules; to go to Harvard and help Kemble and John Clarke Slater in developing a molecular research program; or to follow the invitation of Arthur H. Compton and accept the offer at the Department of Physics of the University of Chicago.

Mulliken opted for Chicago. Besides sentimental reasons—according to Slater, "he liked everything about the great city, even its gangsters" (Slater 1964, 19)—the physicists at Chicago, especially Compton and the spectroscopist H. A. Gale, who was also the head of the Department of Physics, were the most persuasive in arguing for their molecular research program. Chicago already possessed a good spectroscopic laboratory, Eckart Hall, designed by the spectroscopist G. S. Monk, and Gale had promised Mulliken a new high-resolution grating. Besides, conditions seemed to be propitious to the expansion of their program with the transformation of the Ryerson Laboratory into a sort of molecular research center. Endowments were to be used in the acquisition of new equipment, and the university had always been willing to hire research assistants and pay visiting professors. In 1929, Hund, Heisenberg, and Dirac were to spend the summer in Chicago. Slater was, also, being pressed to join the department.

For Mulliken, it was essential to develop a molecular program both along theoretical as well as experimental lines. He did not consider himself a theoretical physicist but a sort of middleman between experiment and theory, so that the interaction with theoretical physicists was considered crucial for "stimulus and cooperation." It was suggested that Mulliken could give an advanced undergraduate course and a graduate course that boiled down to the supervision of three or four graduate students. This was exactly what Mulliken was looking for: minimum teaching load, just for "stimulus," in order that his creative energy could be channeled into scientific research.[14] Mulliken himself, owing to a delay in getting the high-resolution spectrograph, shifted into more theoretical matters (Mulliken 1989, 67). Perhaps this delay was decisive to get him into writing the review articles on "The Interpretation of Band Spectra" (Mulliken 1930, 1931, 1932), a series of articles that never turned into a book as first hoped by Mulliken. It was in this series that the correlation diagram for homonuclear diatomics appeared for the first time (figure 2.1). These diagrams provided a comparative representation of the electronic energy levels of diatomic molecules (relative to the energy levels of its separated atoms and the corresponding united atom) as a function of internuclear separation. They contained all the information necessary to describe the electronic structure of diatomic molecules made up of identical atoms, represented visually in a mode appealing to chemists (Park 2001). The chemical importance of the correlation diagram was such that Van Vleck and Albert Sherman, in their influential review article of 1935, proposed that: "[The correlation diagram] might well be on the walls of chemistry buildings, being almost worthy to occupy a position beside the Mendeleev periodic table so frequently found thereon. Just as the latter affords an understanding of the structure of *atoms*, so does the former afford an under-

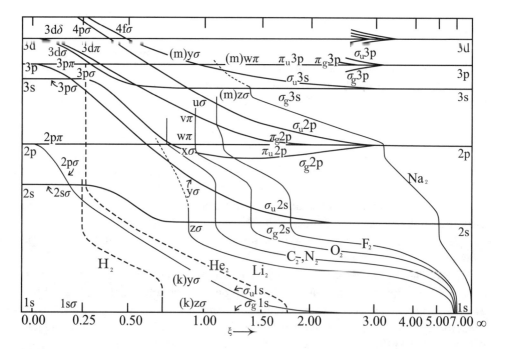

Figure 2.1
Mulliken's 1932 correlation diagram.
Source: Reprinted with permission from Robert Sanderson Mulliken, "The Interpretation of band spectra, III," *Reviews of Modern Physics* 1932;4:1–86 (on p. 40). Copyright © 1932 by the American Physical Society.

standing of the structure of *molecules*, with which the chemist is often concerned" (Van Vleck and Sherman 1935, 175, emphasis in original).

The review articles on the interpretation of band spectra and the agreement on notation for diatomic molecules in which Mulliken was actively involved marked the end of the period of Mulliken's scientific life in which he successfully worked out a systematization of the data on the spectra of diatomic molecules and a concomitant understanding of their structure.[15] He then shifted to the study of polyatomic molecules and to valence-related problems. The transition was accompanied by an increasing awareness of the necessity to propagandize among chemists his work on band spectra, his preliminary ideas on the chemical bond, and his criticism of Heitler and London's suggestions.

Gilbert Newton Lewis: A Precursor

No other person was as close to the heart of chemists in the 1930s as Linus Pauling, comforting many members of the chemical community that quantum mechanics could indeed be used in chemistry in ways that were relevant to them, and not only to physicists. Counterintuitive yet convincing, well versed in the physicists' trade yet squarely within the chemists' culture, Pauling managed to form a rather effective theoretical framework by articulating his views about the nature of the chemical bond. As Pauling (1926, 1926a) acknowledged,[16] he was following Lewis's steps.

Coming of age at the turn of the century, when 19th century science and technology were undergoing deep changes, Lewis was an attentive witness and active participant in disciplinary readjustments and innovations. His work on the chemical bond was but a piece of a lifelong effort to explore the frontiers of chemistry and physics. One might even claim that Lewis was as much a physicist as he was a chemist. This hybrid outlook that was shared by many American scientists of the "lucky generation"[17] he helped to mold is of key importance in understanding the context that favored the genesis and development of quantum chemistry in the United States. The versatility revealed by Lewis enabled him to cross disciplinary boundaries with extreme ease, to be sensitive to problems of articulation of neighboring disciplines or of specialties within disciplines, and to use his scientific contributions as a starting point for a philosophical reflection on the methods, structure, and unity of science. He became the author of the first paper on relativity to be published in the United States (Lewis 1908) and one of its most outspoken advocates,[18] and he paid much attention to facets of science other than strict scientific production. Eager to build around him a group whose organization mirrored his own views about chemistry and science, his impact extended to the educational and popularization realms.

In what follows, we look at the rather convoluted discovery process that gave birth to the concept of the shared electron-pair bond as developed by Lewis, to be subsequently appropriated by the American founders of quantum chemistry, and highlight the complex relations between conceptual development and the different contexts in which ideas are created and presented.[19] This concept continued to be (re)formulated throughout a 20-year period, while Lewis was not only trying to extend its applications to ever more chemical phenomena but also to investigate its epistemological status within the newly formulated quantum theory of Bohr. He first used it for teaching classes (1902), then, after Bohr's papers, he proceeded to publish his results (Lewis 1913, 1916), and, eventually, he analytically presented them in *Valence and the Structure of Atoms and Molecules*, a quasi-textbook written in 1923.

An Atomic Model Conceived for Teaching Purposes

Gilbert Newton Lewis (1875–1946) attended the University of Nebraska and Harvard University, receiving a B.Sc. in 1896 and then a Ph.D. in 1899 with a dissertation on electrochemical potentials supervised by the physical chemist Theodore William Richards (1868–1928), the first American Nobel laureate in chemistry (Hildebrand 1958; Kohler 1973; E. Lewis 1998). He stayed at Harvard until 1905 but spent a year (1900) in Germany at the two top laboratories for physical chemistry, working first with Walther Nernst in Göttingen and then with Wilhelm Ostwald in Leipzig. In 1905, he joined Arthur Amos Noyes and his team of physical chemists at the Massachusetts Institute of Technology, where he stayed for 7 years. There, he laid the foundations for his important work on thermodynamics based on the systematic measurements of free energies. In 1912, he accepted the offer of President Benjamin Ide Wheeler and became dean and chairman of the College of Chemistry at the University of California at Berkeley. This move was part of Wheeler's renewed attempt to revitalize chemistry through the promotion of physical chemistry (Servos 1990, 240–249).

Lewis moved west with a group of able young chemists including William C. Bray and Merle Randall (Servos 1990, 153), aiming at reforming teaching and research in chemistry. Lewis proceeded to reduce the number of basic courses in the undergraduate curriculum and encouraged the development of a critical spirit even at the freshman level. At the research level, everybody was supposed to be conversant about any chemical specialty, discussions were encouraged, and cooperation among researchers was fostered.

Lewis's scientific interests covered subjects as disparate as foundational issues in thermodynamics, valence theory, and theory of radiation and relativity.[20] In the last decade of his career, Lewis tried to devise a new chemistry of deuterium compounds, a field he abandoned for research on photochemistry in 1938. He died in 1946 in the laboratory while performing an experiment on fluorescence.

Lewis's extensive work in thermodynamics, including his sophisticated treatment of foundational issues and the textbook *Thermodynamics and the Free Energy of Chemical Substances* written together with Randall, played an important role among the chemi cal community, which from the very beginning was rather averse to accepting the intrusion of mathematics into chemistry. Without snubbing or ignoring the close attachment of the chemists to laboratory practice, Lewis's work insisted on the significance of mathematical treatment, thus familiarizing many members of the chemical community with the indispensable role of mathematics in chemistry.

According to Lewis's recollections, offered in *Valence* 20 years after the event, the cubic atom emerged while attempting to explain to an elementary class in chemistry the polarity and periodicity properties of valence (Palmer 1959, 1965; Ihde 1964; Partington 1964; Lagowski 1966; van Spronsen 1969; Cassebaum and Kauffman 1971; Russell 1971; Stranges 1982). Joseph John Thomson's experiments with cathode rays, the subsequent discovery of electrons, the discovery of the inert gases, and the identification of electrons in radioactivity composed the context for Lewis's proposal. It was founded on a number of hypotheses: that the electrons in an atom are arranged in concentric cubes; that a neutral atom of each element contains one more electron than a neutral atom of the element next preceding; that the cube of eight electrons is reached in the atoms of the rare gases, and this cube becomes in some sense the kernel about which the larger cube of electrons of the next period is built; and that the electrons of an outer incomplete cube may be given to another atom (Lewis 1966/1923, 29–30).

Chemical and physical considerations were at the origin of the cubic atom. Chemical and, specifically, valence considerations enforced a valence shell completed with eight electrons, and the cubical configuration was the result of assuming that electrons obeyed Coulomb's repulsion law and therefore tended to be as far apart as possible. The model expressed in electronic terms the "empirical" rule of eight according to which rare gases show no chemical reactivity. For Lewis, in rare gases, including helium, chemical inertness meant that their outer cube was completely filled with electrons, while chemical reactivity signified that reacting atoms had incompletely filled outer cubes. The cubic structure was the most symmetrical arrangement of eight electrons that ensured that they were the farthest apart. Sensing the limited applicability of laws, such as Coulomb's, at the atomic level, he guessed that "it seems inherently probable that in elements of large atomic shells (large atomic volume) the electrons are sufficiently far from one another so that Coulomb's law of inverse squares is approximately valid, and in such cases it would seem probable that the mutual repulsion of the eight electrons would force them into the cubical structure" (Lewis 1916, 780).

Lewis's 1902 theory offered a "remarkably simple and satisfactory" explanation of the formation of polar compounds such as sodium chloride (NaCl). It fitted nicely the old electrochemical theory by specifying what was meant by the transfer of electricity from one part of the molecule to another in a chemical union. The explicit statement of the transfer of an electron from one atom to another as the paradigm for chemical bonds appeared in print, shortly after Lewis had his first thoughts about such a mechanism. In the framework of Richard Abegg's theory of electrovalence (1904), each element had two kinds of valences—normal valence and contra-valence[21]—whose arithmetical sum was eight. In 1907, the nature of the chemical bond was addressed by the physicist J. J. Thomson in the framework of his "plum-pudding" theory of atomic structure, first proposed in the Silliman Lecture delivered at Yale University. It became the dominant atomic model until Ernest Rutherford's suggestion of the planetary atom. The chemical bond resulted also from a transfer of electrons from an atom to another, interpreted as the production of a "unit tube of electric force between the two atoms." Furthermore, it provided a physical interpretation to the lines by which chemists represent bonds in graphical formulas: They represented "the tubes of force which stretch between the atoms connected by the bond" (Thomson 1907, 138) and should be replaced by vectors symbolizing these tubes.

The Shared Electron Pair: Inventing Quantum Effects with Classical Entities
But dissenting voices tarnished the period of hegemony of polar theory. In 1913, the year of the publication of Bohr's model of the dynamic planetary atom, a few criticisms resulted in the adoption, by Lewis and others, of a dualistic view, according to which the usual polar bonds should be complemented by nonpolar bonds. This is especially clear in Lewis's paper "Valence and Tautomerism," in which he reviewed the chemical properties of both polar and nonpolar compounds and represented their opposite characteristics in a table, side by side. At this time, opposing properties forced Lewis to "recognize the existence of two types of chemical combination which differ, not merely in degree, but in kind" (Lewis 1913, 1448). Still unsure about how to accommodate nonpolar bonds in the framework of the cubic atom, Lewis assumed that "upon each atom there are definite regions, or points, at which direct connection to similar points on other atoms may be made, and that the number of occupied regions on a given atom is the valence number of that atom" (Lewis 1913, 1451).

In the following years, different dualistic theories came to the chemical fore (Kohler 1971, 1975). After having been an important advocate of the polar theory of valence, J. J. Thomson (1914) changed his mind and defended the existence of two sorts of chemical bonds. Polar bonds were formed by the transfer of electrons and were represented by a single vector bond as in his polar theory, whereas nonpolar bonds were associated with two physical bonds, two tubes of force connecting two electrons, one from each of the interacting atoms. Thus, for Thomson, the number of bonds in structural formulas should be doubled whenever bonds were nonpolar. For example, a single bond in a structural formula of a nonpolar compound was represented by two vectors with opposite directions symbolizing that two electrons, one coming from each atom, were involved in each bond; in the case of double bonds in nonpolar compounds, four electrons were involved, and the bond was represented by two pairs of opposite vectors.

For William C. Arsem (1914), working at the Research Laboratory of General Electric in Schenectady, New York, it was obvious that molecules such as H_2 should have bonds that did not involve an actual electron transfer.[22] He imagined that one single electron was responsible for both sorts of bonds, but in nonpolar bonds the electron, instead of being transferred from one atom to another, remained in oscillation between the two atoms, becoming simultaneously part of both atoms. Surprisingly, Arsem assumed that the molecule of H_2 possessed just one electron. Because of the oscillatory movement, the same electron was shared by both atoms.

Alfred L. Parson, a visiting graduate student from the United Kingdom, first at Harvard and then at the University of California at Berkeley, proposed an atomic theory that caught Lewis's attention. Parson (1915) introduced both a two-electron bond and a cubic octet and conceived the physical origin of the chemical bond as magnetic. The electrons should be represented by circular currents (magnetons), arranged at the corners of cubic octets. The magnetic moment generated by them was the source of chemical bonding and the stability of the octet. For Parson, there were three different kinds of chemical union associated with three different kinds of bonds (bonds such as in H–H, H–Cl, and Cl–Cl). In the case of bonds such as in Cl–Cl, and in order for both atoms to have complete octets, Parson proposed that magnetons from both atoms formed a "mobile group" that oscillated between the atoms, forming a full octet in one atom, and then in the other.

All three models introduced the idea of electrons shared by two atoms, a fact acknowledged by Lewis in the case of Thomson's and Parson's models (Lewis 1916, 763, 773–774; Lewis 1966/1923, 79).[23] It is quite probable that the shared pair bond was the outcome of appropriating the former ideas, exploring and translating them into the framework of Lewis's own picture of the cubic atom. For example, if one took the step of representing two atomic cubes with an edge shared, Thomson's positive and negative bonds coalesced in one single type of nonpolar bond, and the need for vectors and tubes of force disappeared.[24] This was the revolutionary idea of the 1916 paper, in which Lewis further explored the cubic atom, presenting his results in axiomatic form and using a formal deductive style of presentation:

1. In every atom is an essential *kernel*, which remains unaltered in all ordinary chemical changes and which possesses an excess of positive charges corresponding in number to the ordinal number of the group in the periodic table to which the element belongs.
2. The atom is composed of the kernel and an *outer atom or shell*, which in the case of the neutral atom contains negative electrons equal in number to the excess of positive charges of the kernel, but the number of electrons in the shell may vary during chemical change between 0 and 8.
3. The atom tends to hold an even number of electrons in the shell and especially to hold eight electrons, which are normally arranged symmetrically at the eight corners of a cube.

4. Two atomic shells are mutually interpenetrable.
5. Electrons may ordinarily pass with readiness from one position in the outer shell to another. Nevertheless, they are held in position by more or less rigid constraints, and these positions and the magnitude of the constraints are determined by the nature of the atoms and of such other atoms as are combined with it.
6. Electric forces between particles which are very close together do not obey the simple law of inverse squares, which holds at greater distances (Lewis 1916, 768).

While in the first and second postulates Lewis merely restated the basic assumptions of his unpublished 1902 theory, in the third postulate Lewis added to the former "rule of eight" a new "rule of two." This rule embodied two new sorts of experimental data, one provided by chemical, the other by physical methods, both pointing to the fact that almost all compounds contain an even number of electrons.[25] The "rule of two" was the property that in electronic terms corresponded with the novel concept of shared electron pairs introduced in the fourth postulate. Interpreted jointly with the third postulate, the fourth postulate meant that a chemical bond may occur due to the interpenetration of cubic atomic shells, and this may happen precisely by the sharing of a pair of electrons. The second and fourth postulates taken together meant that in interpenetrable atomic shells, an electron does not belong exclusively to one single atom. The electron is shared by two distinct cubic atomic shells, and hence neither atom loses or gains an electron. Reassessing the interconnections between both types of bonds, Lewis now claimed that whereas only in purely polar molecules is the electron transfer complete, in the framework of the new theory "it is not necessary to consider the two extreme types of chemical combination, corresponding to the very polar and very nonpolar compounds as different in kind, but only as different in degree" (Lewis 1916, 771–772).

Further clarification of the specific way in which the interpenetration of atomic shells occurred was provided by the visual representation of the chemical union of two cubic atoms. Single bonds such as in F_2 were represented by the sharing of an edge. Double bonds such as in O_2 were represented by the sharing of a face. The cubic atom could also account for intermediate states of valence, in which an electron pair was unequally shared between atoms.

The two last postulates addressed some physical implications of an atomic model conceived mainly to answer chemical questions and supported both by chemical and physical data. In the fifth postulate, Lewis seemed to be struggling once more with the idea of "indistinguishability." This postulate arose as a possible way of interpreting the nondetection of the so-called "intra-atomic" isomers, which differ in the positions occupied by the electrons of the outer cube.[26]

Finally, in the last postulate, Lewis imposed a restriction on the applicability of Coulomb's law, limited to distances greater than the atomic separation, in order to justify his idea of shared electron pairs as the mechanism for bond formation. In this move, Lewis was possibly helped by the awareness that Bohr's atomic model also violated the classical laws of electromagnetism. If the atomic world behaved differently than the macroscopic realm, to limit the validity of Coulomb's law was not as big a heresy. In any case, Lewis tried to explain the attraction between two electrons in physical terms. The justification stemmed from their magnetic properties. Properly oriented magnetic electrons accounted for the stability of paired electrons.

In the new model, and contrary to Lewis's former assumptions, both types of valence had a common cause. They were different in degree but not in kind, so that molecules could pass "from the extreme polar to the extreme nonpolar form, not *per saltum* but by imperceptible gradations" (Lewis 1916, 775, emphasis in original). This statement expressed a breakthrough in Lewis's understanding of the nature of the chemical bond. As of 1902, the cubic atom merely represented polar bonds. In 1913, Lewis advocated a dualistic theory that admitted two different kinds of bonds. But he still had no way to represent nonpolar bonds. In 1916, the novelty was the accommodation of the two types of bonds into a single framework, with the straightforward representation of nonpolar bonds by the sharing of an edge or a face. Lewis's choice of representing the atom as a succession of concentric cubes played a crucial role in the suggestion of the shared electron-pair bond.[27] This novel idea, which grew out of the exploration of a pictorial representation in the context of suggestions by other scientists and Lewis's own musings over the matter, introduced a new theoretical entity—the shared electron pair—into chemistry and entailed a reappraisal of the notion of individuality. Furthermore, it enabled Lewis to extend the scope of his 1902 theory in order to accommodate in one single framework the mechanism of bonding in polar and nonpolar substances, and by that token to unify inorganic chemistry and structural organic chemistry. In 1919, the cubic atom and the shared pair bond were taken up by Irving Langmuir (Kohler 1974). Langmuir elaborated Lewis's theory and did such a good job in popularizing it that the Lewis–Langmuir theory (as it came to be known) was widely discussed and accepted. However, in 1921 Langmuir abruptly stopped publishing on valence, apparently convinced of the supremacy of Bohr's atomic model.

During the period from 1916 to 1923, Lewis kept thinking about how to harmonize his model with Bohr's model of atomic structure (Arabatzis and Gavroglu 1997; Gavroglu 2001; Arabatzis 2006). The existence of two different atomic models—the chemists' and the physicists'—was a problem for a chemist so physically minded as Lewis, a problem ever more pressing in view of the increasing sophistication of Bohr's model, but which Lewis believed could be solved trivially. But by the time of the publication of *Valence*, Lewis had already convinced himself, and tried to convince his readers, that the contradictions between the two models could be easily removed, championing the view that the chemical and the physical atom could be merged into one unified description of atomic structure.

Believing that a scientist well informed in physics and chemistry had a crucial role to play in the study of molecular structure, Lewis's study of Bohr's atomic model assured him that "while the orbit of one electron may as a whole affect the orbit of another electron, we should look for no effects which depend upon the momentary position of any electron in its orbit" (Lewis 1966/1923, 31–32). In that case, it was possible to translate the positions of the electrons in the static model into the average positions of more or less mobile electrons in the atomic model of the old quantum theory, and this equivalence would be fundamental in bringing together the chemical and physical evidence into a unified theory of atomic structure.

By 1923, Lewis added to the chemical evidence he had previously gathered in favor of electron pairing some new physical evidence pertaining to ionization potentials and spectroscopy (Lewis 1966/1923, 1924). Going one step further, he suggested a physical explanation for the mechanism of pairing. His magnetochemical theory of chemical affinity was based on the assumption that in an atom or molecule, electronic orbits act as magnets in such a way that two orbits conjugate with one another so as to eliminate magnetic moment; and that the condition of maximum chemical stability for atoms, except for hydrogen and helium, corresponded with a valence shell completed with four pairs of electrons situated at the corners of a tetrahedron.

An obvious consequence of these rules was the prediction that odd molecules would reveal magnetic moment, so that chemical unsaturation went hand in hand with magnetic unsaturation, in the sense that every condition increasing unsaturation, or residual affinity, made the substance more paramagnetic. On the contrary, diamagnetism would go hand in hand with chemical saturation. Owing to the incipient stage of development of a physical theory of magnetism, the physical basis of the magnetochemical theory remained no more than a mere conjecture, which Lewis predicted could be clarified by "experiments of the type of Stern and Gerlach" (Lewis 1966/1923, 152). Two years later, in 1925, the mysterious phenomenon of electron pairing was connected with Pauli's exclusion principle and electron spin. A particular spin configuration (purely quantum phenomenon) accounted for the existence of attractive (exchange) energy and gave way to a classical picture of the *sharing* of two electrons.

The utmost plasticity and potential for appropriation of Lewis's model made it survive the transition from classical science to quantum mechanics. The model's openness offered its creator and newcomers the possibility of reinvention of some of its constitutive features together with new perspectives for future developments. Born when the role of electrons was still a mystery, when the idea of quantum particles and of properties such as spin or indistinguishability were still ahead, the model was developed in different contexts and articulated empirical evidence of different provenances, undergoing in the process modifications, from a three- to a two-dimensional model, amenable in any case to pictorial representations dear to chemists, to be finally incorporated in different ways in the framework of the work of the first generation of the American founders of quantum chemistry. The initial shortcomings of Lewis's novel idea of the pairing of electrons, pointed out by Lewis himself, became in the end the basis for the success of its future appropriation, on account of its potential to foster and incorporate novel adjustments and readjustments, to outrun and outlast its own multiform contexts of production.

Linus Pauling: Exploring Different Possibilities for a Quantum Mechanical Theory of Valence

Very soon, the young post-doc Linus Pauling (1901–1994) was to make the most out of Lewis's notion of shared electron bonds, first in the framework of the old quantum theory, then accommodating it in the context of quantum mechanics. Pauling attended the California Institute of Technology from 1922 to 1925 working toward his Ph.D. He was supervised by Roscoe G. Dickinson and worked on the determination of the structure of several crystals by means of X-ray diffraction, a technique that enabled one to "see" structural arrangements. He then stayed at Pasadena one more year as a National Research Council postdoctoral fellow. Pauling's first contributions to the subject of the chemical bond were made in this period.[28]

These were within the framework of the old quantum theory but shared with future contributions the methodological interplay of theoretical considerations and empirical evidence that so distinctively characterized Pauling's successful approach to the explanation of the chemical bond. In one of his first papers, Pauling used the information gathered on crystal structures together with Lewis's shared-electron bond to suggest a new guiding principle to analyze the relative stability of groups of molecules composed by the same atoms and having the same total number of electrons (Pauling 1926). And in another paper, Pauling (1926a) represented Lewis's shared electrons by means of binuclear orbits and together with experimental evidence stemming from his crystal structure work suggested dynamic models for the ammonium ion (NH_4^+), benzene (C_6H_6), and other aromatic molecules.[29]

Following the advice of his mentor Arthur Amos Noyes, Pauling applied for a John Simon Guggenheim Memorial Foundation Fellowship. He planned to stay in Munich at Sommerfeld's Institute for Theoretical Physics for a year and to visit Bohr in Copenhagen. He also planned to pay brief visits to other centers where work on crystal structure, either theoretical or experimental, was being carried out, and in fact Pauling made sure to stop at Born's Institute for Theoretical Physics in Göttingen and at the Braggs' laboratory in the University of Manchester. Although not mentioned in the plan, upon his arrival in Europe, Pauling was able to arrange for a stay with Schrödinger in Zürich.

Pauling arrived at Munich with his wife in April 1926. His adaptation to Munich's scientific lifestyle went smoothly even though things were very different from what he was accustomed to at Pasadena. He attended Sommerfeld's lectures on the new wave mechanics, which had just been formulated by Schrödinger.[30] He realized how much he missed the "invigorating discussions," now that he attended seminars devoted to "formal presentations" of new papers and not to presentations of students' research problems.[31] Pauling also realized that, in his first meeting with Sommerfeld, he had undiplomatically mentioned a research topic he would like to work at, instead of following the rules and waiting for Sommerfeld's advice.[32] However, he believed that there was a high probability of being assigned a problem of his own interest,[33] as he knew that the study of electronic motions in binuclear orbits was a topic of research at the institute. Sommerfeld suggested to Pauling to work on the spinning electron,[34] but in the meantime Pauling mentioned to Sommerfeld work written before leaving Pasadena on the effect of electric and magnetic fields on the dielectric constant of hydrogen chloride (Pauling 1926b, 1926c) and convinced Sommerfeld to allow him to extend that work to fields of arbitrary strength.[35] Pauling compared the results obtained by use of the old quantum theory with those obtained by use of the new mechanics and proved that the latter gave values of the dielectric constants in good agreement with experiment. It was this result more than anything else that convinced Pauling that quantum mechanics was necessary for the solution of chemical problems. He wrote to Noyes announcing: "I am now working on the new quantum mechanics, for I think that atomic and *molecular chemistry* will require it. I am hoping to learn something regarding the distribution of electron-orbits in atoms and molecules."[36]

When Pauling was in Munich, a new line of research emerged as a result of a paper published by Gregor Wentzel, a former doctoral student of Sommerfeld and a privatdozent at the institute. Taking into account the electron spin and the new quantum mechanics, Wentzel calculated the screening constants for electrons in complex atoms, but there was still poor agreement between the values obtained and the experimental results. The concept of screening constants, the ability of the innermost electrons to partially shield the outer electrons from the positive nucleus, had been introduced by Sommerfeld to explain the fine-structure of X-ray levels. Pauling decided to examine this question in a systematic manner. He noticed that some of the approximations used previously were not correct and when properly refined led to a good agreement with experiment (Pauling 1970, 993). Although Noyes advised Pauling to keep on publishing in American journals while in Europe, this paper, which was a reply to Wentzel, came out in the *Zeitschrift für Physik*.[37] Happy with this successful result, Pauling (1927a) continued to work on screening constants, quickly realizing the possibilities offered by the quantum mechanical treatment of screening constants for the study of the electronic structure and the prediction of the physical properties of complex atoms and ions. Encouraged by Sommerfeld, Pauling continued to develop his ideas and at Sommerfeld's suggestion published his results in the *Proceedings of the Royal Society of London* (Pauling 1927b).[38] Pauling's own assessment of this paper was that it was the "start of quantum mechanics of polyelectronic atoms."[39]

After staying with Sommerfeld in Munich for approximately 13 months, Pauling paid a visit to Bohr in Copenhagen (where he stayed less than a month). There he worked with Samuel Abraham Goudsmidt on the hyperfine structure of bismuth without much success. He then moved to Zürich, where he stayed for about 2 months. On the whole, he "rather regretted" the time spent in Zürich. He saw Schrödinger only at the weekly seminars, never found out what he was working at, and did not manage to interest him in his own work: "I offered to make any calculation interesting to him since he was not interested in my work; but without success."[40] He met Heitler and London, who, shortly before his arrival, had finished writing their paper on the formation of the hydrogen molecule. He talked with them a little but was not invited to collaborate with them. Thus, he continued alone to muse over the nature of the chemical bond by attempting to treat the interaction between two hydrogen atoms as well as between two helium atoms.

Pauling's notebooks from that period are especially helpful in tracing some of his early thoughts about the nature of the chemical bond.[41] The first set of notes are dated from 1926, and Pauling wrote later on the first page that they were written before arriving in Zürich, when he was still in Munich. During the period immediately preceding his trip to Europe, Pauling started to think about the electron-pair bond in terms of electrons in binuclear orbits (Pauling 1926a),[42] a hypothesis Lewis initially explored himself. Both were eager to take full advantage of visual representations as constitutive elements of theory building.

At the same time, and in order to study diatomic molecules of identical atoms, Pauling tried to determine the form of molecular (binuclear) orbitals in the case the internuclear potential was approximated by two identical unidimensional square-well potentials. He then integrated Schrödinger's wave equation for the electron in the different regions associated with this simplified potential. He used the boundary conditions to determine some of the constants of integration and analyzed the extreme situations corresponding with the separated atoms (infinite distance between the atoms) and the united atom (zero distance between the atoms). The eigenfunctions obtained were either symmetrical or antisymmetrical in the position coordinate, and that meant that there existed either symmetrical or antisymmetrical electron orbits. Then, by assuming that shared electrons were necessarily symmetrical electrons in two-center orbitals, whereas unshared electrons were half symmetrical and half antisymmetrical, Pauling tried to write down the Lewis formulas for several diatomic molecules, namely F_2, O_2, and N_2. For those cases where more than one formula could be written, he used all the empirical information available together with the results of his theoretical treatment to devise rules to decide among them. In these notes, he hinted at the role of the spinning electron in bond formation and jotted that "the spinning electron accounts for electron pairs." He also referred to resonance energy: "There is, of course, a continual interchange of energy among the various electrons, so that one electron cannot be assigned to a given state; and there is the corresponding resonance effect on the energy of the system."[43] Finally, he tried to extend his square-well potential model to the analysis of the case of a diatomic molecule composed of two different atoms and of polyatomic molecules composed of two large and two small atoms.

Since his application for the Guggenheim Fellowship, Pauling had already proposed to apply quantum theory to the study of the motion of electrons in Burrau type of orbits (see chapter 1) but could not derive any results he considered worth publishing. Furthermore, he recalled having discussions with several people "about the explanation of chemical bonding in terms of the Burrau paper on the hydrogen molecule ion and the Pauli principle."[44] Although never explicitly stated in the notes, it is not altogether improbable that he might have been already thinking of the electron-pair bond in terms of two electrons with opposite spins in a two-center molecular orbital.

While still in Munich, Pauling learned about the "empirical method" developed by the atomic physicist Edward U. Condon in 1927 for the hydrogen molecule and was immediately struck by "Condon's ingenuity." One of the early American physicists to embark on the use of quantum mechanics to understand the atom and its nucleus, Condon took the interaction of the two electrons in the hydrogen molecule to be small in comparison with the electron–nuclei attraction, representing each electron by an H_2^+ eigenfunction. To estimate the value for the perturbation energy, he reasoned by analogy and considered the energy of the hydrogen molecule to be twice that calculated by Burrau accrued by the interelectronic energy taken as a perturbation. He assumed that the relation between interelectronic energy and total energy was the same for the hydrogen molecule and for its united atom—the helium atom—and used the known value of the latter to estimate the value of the former. In this way, Condon "got results as good as Heitler and London got later."[45]

Although Pauling kept on thinking about the chemical bond while in Copenhagen, the next set of important notes were written while he was in Zürich.[46] There, Pauling discussed the chemical bond with Heitler and London. He did not agree entirely with their results, an understandable reaction for he had been thinking about the electron-pair bond as a pair of electrons in molecular (binuclear) orbits.[47] In Munich, he had attempted to make the first calculations using a very rough approximation to the molecular potential. Calculations with molecular orbitals were by no means easy to carry out. Burrau and Condon used molecular orbitals for the hydrogen molecule ion and the hydrogen molecule, respectively, but even in these cases the integrations were not straightforward. In the case of the hydrogen molecule, Condon's ingenuity in developing an "empirical method" enabled him to circumvent some mathematical obstacles and arrive at an estimate for the interaction energy as good as that obtained later by Heitler and London. Burrau's and Condon's successes undoubtedly reassured Pauling about the potentialities of a method alternative to that of Heitler and London. In Zürich, he worked hard to develop such an alternative computational method. Curiously, in doing so he used the idea of resonance and the Pauli principle, which he later classified as the two fundamental factors influencing chemical valence (Pauling 1928a).

The idea was the following.[48] Taking two hydrogen atoms at a distance d apart, representing each electron by an H_2^+ wave function (as Condon did), which, according to the conclusions of former calculations, could be either symmetric (Ψ) or antisymmetric (Φ) relative to the nuclei, and neglecting at first the repulsion between the two electrons, the wave function for the hydrogen molecule could be approximated by the product of the two electronic wave functions, that is, $\Psi(H_2) = \Psi_1\Phi_2$ (or $\Psi_2\Phi_1$). However, due to the indistinguishability of the electrons, a resonance or exchange phenomenon occurs, so that two different complete eigenfunctions can be formed for the hydrogen molecule:

$$\Psi^S = \Psi_1\Phi_2 + \Psi_2\Phi_1$$

$$\Psi^A = \Psi_1\Phi_2 - \Psi_2\Phi_1.$$

Taking next the Coulomb interaction between the two electrons as the perturbation force, represented by

$$f = \frac{e^2}{r_{12}},$$

one can use perturbation theory to find an expression for the perturbation energy due to the electrons:

$$W^A = \int \frac{e^2}{r_{12}} (\Psi_1^2\Phi_2^2 - \Psi_1\Phi_1\Psi_2\Phi_2) dr_1 dr_2.$$

Representing the two-center wave functions in terms of one-center wave function ψ and ϕ,

$$\Psi = \psi + \phi$$

$$\Phi = \psi - \phi,$$

and substituting in the expression for the perturbation energy and making the necessary simplifications, Pauling obtained

$$W^A = \iint \frac{e^2}{r_{12}} (\psi_1^2\phi_2^2 - \psi_1\phi_1\psi_2\phi_2) dr_1 dr_2.$$

The first term represents the electrostatic repulsion between the electrons and the other term the resonance between the electrons.

Next, Pauling considered the electrons in the two hydrogen atoms as spinning electrons. Six different cases were possible: one with both electrons represented by the symmetric H_2^+ eigenfunctions; another with both electrons represented by the antisymmetric H_2^+ eigenfunctions; and four with one symmetric and another antisymmetric H_2^+ eigenfunction. Pauling's analysis led him to the conclusion that of these six situations, four corresponded with the formation of the hydrogen molecule from neutral atoms, whereas the other two corresponded with the formation of the hydrogen molecule from the ions H^+ and H^-. By representing the two-center wave functions (molecular orbitals) in terms of one-center wave functions (atomic orbitals), Pauling could compare his result with that obtained by Heitler and London in order to relate his six cases (with two-center wave functions) with the four cases considered by Heitler and London (with one-center wave functions). As we saw previously, Heitler and London obtained three cases where the electrons attracted each other and one in which they repelled. Pauling proved that if one started from the consideration of two separated like nuclei (one-center wave functions) and two electrons, then by taking into account the degeneration due to similar nuclei, and then the indistinguishability of the electrons, one would arrive at the six cases he obtained previously with the two-center wave functions. Heitler and London did not consider the formation of the hydrogen molecule from two ions, and this is the reason why they had two solutions less than Pauling who stated his conclusions analytically in the notebooks[49]:

London and Heitler have a different method of treatment, starting with separate atoms (or ions). Their eigenfunctions are not those obtained by starting from H_2^+ eigenfunctions. I suppose, though, that they are roughly correct for large distances, and that in some way there is transition to the H_2^+ eigenfunctions, probably through degeneration. Thus if it could be shown that their $H^+ + H^-$ and $H + H$ eigenvalues approach each other; it would be satisfactory. There would have to be triple degeneracy—both $H^+ + H^-$ and one $H + H$ coming together.

I suppose that this is what happens, and that neither my treatment nor London's and Heitler's is correct, but that the correct treatment of the secular problem would give London and Heitler's results for large distances, and mine for small, with intermediate ones in between, and with degeneracy at points, so that all the transitions indicated are possible adiabatically (switchings being called adiabatic). Thus both $H^+ + H^-$ and $H + H$ could go adiabatically to H_2.

Besides working on the interaction between two hydrogen atoms, Pauling spent most of his time in Zürich trying to treat the interaction between two helium atoms, but he stumbled upon some integrals for which he was not able to find good approximate values. But he kept on trying as he was convinced that "if I worked in this field I probably would find something, make some discovery, and that the probability was high enough to justify my working in the field. Of course, it led to hybridization and all of this stuff."[50]

1931: The Annus Mirabilis for Quantum Chemistry

Pauling's Valence Bond Theory

In 1931, Pauling returned to Berkeley to deliver a series of lectures on "The Nature of the Chemical Bond," which spanned almost a month, from March 23 to April 20. The lectures were a big success. Birge enthusiastically confided to Mulliken that Pauling had just finished a series of 12 lectures, which he classed as the most exciting he ever attended. Birge further confided that Pauling was duplicating Slater's work, but Pauling denied it, having already clarified it to Slater. Above all, Birge was certain that "we are on the threshold of a very large new development."[51]

Confident of the importance of the results that eluded him for nearly 3 years, Pauling did not miss the opportunity to publicize them immediately. Even before their publication, he presented and explained them in detail in the lecture series at Berkeley. But, beforehand, Pauling wrote a letter to A. B. Lamb, the editor of the *Journal of the American Chemical Society*, urging him to publish the paper as soon as possible. The letter leaves no doubt about Pauling's self-perception of the importance of his work and of the way he was planning to establish his hegemony. He acknowledged that the paper was longer than most of the papers published in the journal. A shorter version "would render it more difficult for chemists to understand what has been done," especially as he thought that the paper was primarily of chemical interest. If prompt publication was "prevented through time-consuming consideration by referees," then he was planning to send it to *Physical Review* or to a European journal.[52]

Pauling's wish was fulfilled. The first of what became the famous series on "The Nature of the Chemical Bond" was soon published by Lamb (Pauling 1931). Pauling had also sent a letter to the editor of *Physical Review* in which he called the attention of physicists to the *Journal of the American Chemical Society* article (Pauling 1931a), outlined its main conclusions, and emphasized differences between his paper and an earlier short paper by Slater published there in 1931.

John Clarke Slater (1900–1976) attended the University of Rochester and then graduate school at Harvard from 1920 to 1923 (Schweber 1990). Supervised by Percy W. Bridgman, his Ph.D. dealt with the compressibility of the alkali halides. Receiving the Sheldon traveling fellowship, Slater visited Europe during the academic year 1923–1924, staying at the Cavendish Laboratory in Cambridge, and then at Bohr's institute in Copenhagen where he had an unpleasant experience while working on the famous joint paper on the nonconservation of energy with Hendrik A. Kramers and Bohr (Jammer 1966; Klein 1970; Stuewer 1975; Hendry 1981; Konno 1983; Schweber 1990). Finding himself working on the same problems as Dirac, Slater decided to quit the field because Dirac "got ahead of me each time."[53] Slater stayed at Harvard until 1930 working actively with Theodore Lyman, Kemble, and Bridgman to upgrade the Department of Physics. In the meantime, he refused offers from

Princeton (1927, 1929)—the "moving spirit behind the offer"[54] was Compton—from Stanford and from Chicago (1928). In 1931, he finally accepted the position of full professor and head of the Department of Physics at the Massachusetts Institute of Technology (MIT). He joined forces with Compton, then MIT president, and worked toward the revitalization of the physical sciences at the institute. Slater's friends were overjoyed. Pauling, who first met Slater at Harvard, wrote: "I must say that I'm probably much more glad than if I were at Harvard—for even though MIT is not far away, it is far enough so that the change would be noticed by one at Mallinckrodt."[55] Van Vleck sent a postcard from Leipzig congratulating him for being "simultaneously a youthful and venerable departmental chairman."[56]

The general ideas described in Slater's brief paper were presented at the meeting of the American Physical Society in Washington in April 1930. Later, Slater recalled that after his presentation, Pauling and Condon participated in the discussion.[57] And afterwards, he worked out the results on hybridization that were published in the 1931 paper. Slater acknowledged the influence of B. E. Warren, who worked on X-ray crystallography and had studied with W. L. Bragg: "Through him I learned of the angular arrangements of bonds in such crystals as the silicates. This set me to thinking about the directional properties of atomic bonds and at molecular orbitals."[58] He paid attention to atoms of elements F, O, N, and C where the valence electrons are p electrons. He discussed qualitatively directional effects and illustrated his ideas by many examples. The two valences of O and the three valences of N tend to be at right angles from each other, whereas the four valences of C have tetrahedral symmetry. Examples were amply illustrated by pictures of charge distributions and molecular models. Tentative explanations were put forward without theoretical justification. In a letter to Pauling, Slater expressed his pleasure for their simultaneous discovery: "I am glad things worked out as they did, we both deciding simultaneously to write up our ideas. I haven't had a chance to read yours in detail yet, but it looked right as if we were in good agreement in general. I'm glad you have thought some about d valences too. Incidentally, our general points of view seem so similar that we shall want to compare notes . . . ."[59] The mathematical justification of his preliminary ideas was postponed to a paper that came out later in the year (Slater 1931a). Contrary to the first paper in which particular examples illustrated general properties, in this one Slater used a method more congenial to his general approach toward science. He progressed from general principles to particular cases. He outlined a general method to solve Schrödinger's equation for molecules, using perturbation theory in order to guide the search for approximations, determining the matrix components of energy, and solving the secular equation. Each step was discussed in detail, starting from unperturbed wave functions, obtained from atomic orbitals with his determinantal method. It was followed by an analysis of real problems treated sequentially including "two atoms, each with one s electron," "two atoms, one s and one p," and the case of methane treated in the end under the heading "five atoms, one with four s and p electrons, the others with an s each." Pauling replied enthusiastically to this paper: "your paper in the Physical Review is just what was needed to uphold the ideas of directed valence."[60]

In the first lecture at Berkeley, Pauling (figure 2.2) pointed out the extent to which his ideas went much beyond Slater's. In fact, Slater had used the criterion of maximum overlapping of wave functions to derive the directional properties of the tetrahedral carbon atom and to suggest that *p*-bonds were formed at right angles from each other. But Slater did not find a criterion for the change in quantization as did Pauling, who offered in addition many more interesting results.[61] Pauling denied that Slater's paper acted in any way as a sort of catalyst to him. The 3-year lag resulted from his dissatisfaction with the initial mathematical treatment of the problem, which he considered

Figure 2.2
Linus Pauling at the blackboard.
Source: From Ava Helen and Linus Pauling Papers, Special Collections, Oregon State University.

"too complicated to be convincing or reliable." By the end of 1930, he "had figured out a remarkable simplification in the equations,"[62] and then, most of the calculations for the paper were done in one single evening.[63]

In the same lecture, the intellectual debt to Lewis was, also, evident. Pauling came to realize that the notion of the electron-pair bond might hold the clue to the understanding of the properties of chemical substances. He considered himself "not a stranger" who was bringing something new, but he felt, in fact, like a student of Lewis who since 1916 had been working on similar problems. "For ever since I first learned of the electron-pair bond, in 1920, I devoted my efforts to attempting to understand the properties of substances from this viewpoint."[64] In fact, Pauling interpreted the Heitler–London paper as providing the quantum mechanical justification for Lewis's empirical theory of the electron-pair bond. Besides, it was while studying one of London's papers (London 1928a) that Pauling hit upon the idea of changed quantization, which he later called "hybridization."[65]

Pauling presented his fully developed theory in a series of seven papers spanning a 3-year period. The first paper on the nature of the chemical bond and the two papers published three years before on the same subject—the *Chemical Reviews* article and a brief communication to the *Proceedings of the National Academy of Sciences of the United States of America* (Pauling 1928, 1928a)—revealed already several of the important characteristics of Pauling's style. Believing that quantum mechanics provided an indispensable tool to attack the problem of the chemical bond, he prepared his papers having in mind a chemical audience largely ignorant of physical theory. The fluidity of his style and the clarity of explanations made it relatively easy for the interested chemists to take the first steps into a territory largely unknown to them.

In the opening paragraph of the first paper of "The Nature of the Chemical Bond" series, Pauling assessed the situation concerning work on the chemical bond and expressed in an unambiguous way the general guidelines of his program:

During the last four years the problem of the nature of the chemical bond has been attacked by theoretical physicists, especially Heitler and London, by the application of quantum mechanics. This work has led to an approximate theoretical calculation of the energy of formation and of other properties of simple molecules . . . and has also provided a formal justification of the rules set up in 1916 by G. N. Lewis for his electron bond. In [this] paper it will be shown that many more results of chemical significance can be obtained from the quantum mechanical equations, permitting the formulation of an extensive and powerful set of rules for the electron-pair bond supplementing those of Lewis. These rules provide information regarding the relative strengths of bonds formed by different atoms, the angles between bonds, free rotation or lack of free rotation about bond axes, the relation between the quantum numbers of bonding electrons and the number and spatial arrangement of the bonds. (Pauling 1931, 1367)

Texts such as this are, in a way, pace-making texts. This particular series of papers exerted a powerful influence in the way chemists started to deal with the problems of valence and contributed to the further articulation of the chemists' culture in order to accommodate the newly emerging quantum chemistry. Pauling conceded that it was the theoretical physicists who first *applied* quantum mechanics to a chemical problem. At the same time that he considered his own work as an extension of their program, he pointed out how his own approach differed from that of the theoretical physicists. His applications provided "many more" results that could be obtained in the *form of rules* supplementing other rules. In the paper, a sketch of the reasoning leading to the rules was presented. Many approximations and arbitrary assumptions, which might have inhibited a theoretical physicist, were made in the process. They had been obtained as generalizations of the quantum mechanical rigorous treatments of simple systems such as the hydrogen molecule, the helium atom, and the lithium atom, together with generalizations supported by Pauling's systematic study of the behavior and properties of chemical compounds, for which rigorous quantum mechanical proofs were inaccessible. In the lectures, the quantum mechanical justification of the rules was postponed to the end but was never given for lack of time. Even if the rules were not derivable from quantum mechanical principles, their usefulness in providing information regarding bond types in chemical substances was enough as an empirical criterion for their acceptance.[66] Pauling's pragmatism left no doubts as to the ways he was planning to tackle the problem of the chemical bond.

Besides the rules for the electron-pair bond, in the second part of the paper and in the last part of the lectures, Pauling introduced a group of rules for the determination of the magnetic moments of complex ions, which provided "little more than the justification and unification of previously developed rules" (Pauling 1931, 1391). This set of rules permitted him to correlate magnetic moments and types of bonds in molecules and complex ions: They enabled him to get information on bond types in molecules or ions from knowledge of magnetic moments, and conversely, they enabled prediction of magnetic moments from knowledge of bond types and atomic arrangements.

The rules for the electron-pair bond were divided in two classes. In the first, Pauling summarized the conclusions of Heitler and London's work in three postulates, which he considered to express essentially the quantum mechanical underpinning of Lewis's 1916 results: that the formation of an electron-pair bond results from the interaction of an unpaired electron coming from each of two atoms (as mentioned previously, Lewis objected to the specification of the electron's provenance); that in bond formation, the spins of the interacting electrons are antiparallel so that they do not contribute to the paramagnetic susceptibility of the compound[67]; and that two electrons forming a shared bond cannot participate in the formation of additional pairs.

The second class embodied Pauling's own conclusions as to the main factors influencing bond formation, and it is for this group that Pauling provided a sketchy justification. The rules were as follows:

1. In the formation of an electron-pair bond, the main resonance terms are those involving just one eigenfunction from each atom.
2. For eigenfunctions with the same radial dependence, the eigenfunction that will form the stronger bond is the one with the largest value in the bond direction. For a given eigenfunction, the bond will tend to be formed in the direction for which the eigenfunction has the largest value.
3. For eigenfunctions with the same angular dependence, the eigenfunction that will form the stronger bond is the one corresponding with the lower energy value for the atom.

An implicit assumption underlying Pauling's considerations is that the conditions for bond formation are not influenced much by the presence of other atoms in the molecule. The bond direction is the direction where the concentration of individual bond eigenfunctions is highest. The strength and directional character of bonds is thus explained as a result of the overlapping of individual bond eigenfunctions, which is itself a reflection of a greater density of charge concentrated along that particular direction. The "remarkable simplification" in the equations Pauling had been searching for consisted in working just with the angular component of the atomic wave functions, that is, in taking $\Psi = R(r) \cdot Y(\theta, \varphi)$ instead of working with the complete atomic wave functions $\Psi = R(r) \cdot Y(\theta, \phi)$, where r, θ, ϕ are spherical coordinates. For s and p eigenfunctions with the same total quantum number, n, this approximation seemed to be generally valid.

In all the cases where he used his theory to explain various combinations of molecules—among them the successful explanation of the bent nature of the water molecule—Pauling assumed implicitly that s and p eigenfunctions were not altered during bond formation. But, according to Pauling, there were cases when a change in quantization occurred, and this new phenomenon led to novel results. The criterion for the change of quantization was defined quite loosely: when the bond energy is greater than the s–p separation, then s–p quantization is broken. Carbon was an example of an element that satisfied the criterion. When the s–p quantization was broken as a result of the interaction of carbon with other atoms (molecule formation), the s and p eigenfunctions must be combined together to form the original ground state. The new eigenfunctions (obtained as linear combinations of the s and p functions) were particularly suited to bond formation. Thermochemical and band spectral data enabled one to calculate values of the s–p energy differences and of bond energies. Pauling, then, proved that the best set of four bond eigenfunctions that could be formed had a maximum value of 2 along lines making tetrahedral angles of 109° with each other.

By the application of the rules for the electron-pair bond, Pauling was able to remove the apparent incompatibility between chemical and quantum theory. The equivalence of the four bonds formed by a carbon atom was not at first easily reconciled with the quantum description of the electronic structure of carbon as having one $2s$ orbital and three $2p$ orbitals in the valence shell. The reconciliation between these two sets of information was attained when it was recognized that as a result of resonance, a tetrahedral arrangement of the four bonds was achieved. One more step had been taken in the reconciliation between the physicist's and the chemist's conception of the atom, and this time the merging of the two pictures had been effected by the modification of the physicist's picture with the introduction of the new idea that quantization could be broken under certain conditions.

In cases where the criterion for s–p changed quantization was not clearly satisfied, Pauling suggested that the tetrahedral bond angles set an upper limit to the angle between bonds, which could vary between 90°, when the s–p quantization was not broken, and 109°, when it was broken. Pauling's results were checked against crystal structure data on bond angles in non-ionic crystals. In many instances, Pauling had gathered himself the information he needed while he was working on crystal structure.

Although the conclusions concerning the tetrahedral carbon atom were of the utmost importance for understanding organic chemistry by means of quantum mechanical notions, a host of other results on directed valences was also obtained involving s, p, and d electrons. In this way, Pauling extended his analysis to compounds of elements in rows other than the first. When d electrons exist besides s and p electrons and quantization is broken, the number and type of bonds increased, and they were classified by applying an analogous reasoning used for the tetrahedral carbon atom.

In 1892, Alfred Werner introduced a new type of chemical bond, which he named coordination link, to account for the fact that many apparently saturated inorganic molecules showed a tendency to combine and form more complex aggregates. His subsequent studies of the composition and properties of inorganic complexes led him to suggest that in these complexes, a central atom of a metal coordinated around itself four atoms at the corners of a square or tetrahedron or eight atoms at the corners of an octahedron. Thirty years after Werner's ideas about the geometry of inorganic complexes, Pauling proved that the square, tetrahedral, or octahedral eigenfunctions gave strong bonds; yet, he was unable to prove that they gave simultaneously the strongest bonds possible.[68]

Five months after the appearance of Pauling's paper, Slater (1931a) published a detailed quantum mechanical justification of the results he had already presented qualitatively. As referred to at the beginning of this section, Pauling thought that Slater's paper was just "what was needed to uphold the ideas of directed valence."[69] Pauling's results on directed valence had been obtained from a set of postulated quantum mechanical principles that Pauling never attempted to derive rigorously. His approximations were immediately criticized by Slater and Mulliken. They argued that it was a very poor approximation to assume that the overlap integrals are the same for an *s* function and a *p* function and that Pauling's definition of bond strength, which involved exclusively the angular part of the wave functions, was not proportional to the overlap integral, and therefore was not a satisfactory approximation.[70] But Pauling argued for the postulates' *usefulness*. They could be accepted on an empirical basis, on the grounds of their practical implications. This approach was especially tailored to chemists who could follow his reasoning without having to face the details of a quantum mechanical presentation that would possibly have frightened them away. The success of Pauling's method depended largely on its ability to legitimize the use of empirically found quantities as a heuristic for the formulation of chemical rules. The information on bond angles and interatomic distances gathered through his work on crystal structure determinations served to control Pauling's quantum mechanical speculations and functioned in the end as an extra asset to gain the chemists' support.

This methodology gave especially good insights in the case of polyatomic molecules, which were systems too complex to be treated on a rigorous quantum mechanical basis, but which were, nonetheless, the systems of real interest to chemists. It was also reassuring to chemists that Pauling's valence theory resonated with former chemical ideas about bonds and valence. Pauling gave the quantum mechanical justification for the chemical viewpoint according to which molecules are made up of atoms, which interact with each other along privileged directions. Starting by accepting Heitler and London's quantum mechanical justification for the electron-pair bond, Pauling's semiempirical method accounted for the directional properties of chemical bonds through the introduction of new ideas such as changed quantization and new principles such as the criterion for maximum overlapping of bond eigenfunctions.

Mulliken's Molecular Orbital Theory

Mulliken's papers rarely shared the same clarity of presentation as those of Pauling. They were usually very long and dense, and, often, crucial points and details were both given the same emphasis. Although Mulliken was fully aware of the importance of his work for chemists,[71] only by 1929 did he start to publish in chemical journals. He had previously published in journals read mainly by physicists, and it took him some years to realize the importance of writing for a chemical audience. The year after Pauling's 1928 publications, Mulliken presented at the spring meeting of the American Chemical Society a review about band spectra and their significance for chemistry (Mulliken 1929). There, the theory of band spectra was summarized with special emphasis on the applications to chemistry, and notably on its usefulness to valence theory. Mulliken's ideas on molecule formation were explained together with the different considerations by Heitler and London on the same problem. When the paper appeared, it was a model of clarity.

Nothing was fundamentally new in the paper. The basic ideas of Mulliken's new valence theory had been introduced already in some of his earlier papers. But now, for the first time, the core of his new conceptual scheme was spelled out, and chemists could assess its significance. Mulliken argued convincingly that the interpretation of band spectra offered refreshing new means to address the problem of chemical bonding. And the strength of his approach became evident when he extended his theory to more complicated molecules.

Birge was overjoyed by the paper. It proved to him that Mulliken could indeed write a "simple and clear" article if he wanted to. He classified it as the "best simple, yet comprehensive, introduction to band spectra that has yet been written by anyone, in any language."[72] This was precisely the type of article that was tailored for a chemical audience: It contained all the information that a chemist, not working on band spectra, needed to know, but omitted all unnecessary details and complications. It was an excellent paper to use in lectures and introductory talks. With articles such as this one, Mulliken would reach an enlarged audience, composed of specialists as well as nonspecialists on the subject. For Birge, writing this sort of papers was fundamental and perhaps the most important task Mulliken should consider at this stage of his career[73]:

You certainly know more about band spectra than anyone else in the world, and *now the important thing is for you to give out your knowledge in a way that will enable others also to understand it.* Your Chemical Review article shows that you can do this better than I can, while your Supplement article shows that you do not try to make things clear, and in part, non-mathematical, when you are trying to write a serious article. If you write a book, I hope you will write a simple, more or less non-mathematical account of each part, and then put all the details and abstruse things into appendices, even if the latter occupy more than half the book. Only in that way will there be any sale, I think.

In his enthusiastic remarks, Birge did not mention Mulliken's views on valence theory. This is understandable: They were not the main topic of the review article, and Birge, as a more conventional physicist, was not as deeply interested in chemical problems as Mulliken. By 1931, when Pauling's first paper on "The Nature of the Chemical Bond" came out, Mulliken recalled Birge's friendly advice. He realized that his preliminary ideas on valence theory were buried amid all sorts of information on band spectra, in journals seldom read by chemists such as *Physical Review* and *Reviews of Modern Physics*. A presentation especially devoted to them was called forth, now that another approach to valence theory appeared in America. The paper on the "Bonding Power of Electrons and the Theory of Valence" (Mulliken 1931a) was intended to fulfill that aim. Its contents were first presented at the end of March 1931 at the meeting of the American Chemical Society at Indianapolis, Indiana. Mulliken confided to Birge that he was preparing himself for "an attack on the Heitler and London valence theory" and reiterated his belief in understanding better chemical bonding by means of molecular electron configurations than by Heitler and London's method.[74] Mulliken and Birge had previously discussed certain aspects of Heitler and London's theory, and Birge realized that certain of their statements disagreed with spectroscopic evidence (like the impossibility of having HeH or He_2^+). He worried about the compatibility between the spectroscopic evidence for the analogies between atoms and molecules and chemical knowledge of saturated and unsaturated bonds and electron pairs.[75] But Mulliken had come to the conclusion that there were important flaws in Heitler and London's theory[76]:

> It is becoming clear that the London valence theory is really not so good, just happens to agree with chemical theory. I mean, the ⇔ [pairing] of electron spins is not the real essential thing. It is that the molecular electron configuration of lowest energy quite usually (not always) happens to have the electrons all ⇔. . . . His idea of ⇔ being all-important was a far-too-large generalization.

He was thus preparing himself for a public criticism of their theories and a restatement of his own bold considerations on valence theory. Mulliken (1931a, 384, 369) was now convinced of the "arbitrariness" and "superfluity" of concepts such as valence, valence bonds, and bonding electrons. He intended to prove in this paper that there were serious reasons to believe that these concepts "should not be held too sacred" (Mulliken 1931a, 347). To show the fragility of one of the stronger pillars of chemical science was already a brave act. Yet simultaneously, Mulliken made a constructive proposal: He outlined the guiding assumptions and basic concepts of a new valence theory, which dispensed with the cherished tenets of classical valence theory. To accomplish this aim, the reader was guided through a careful analysis and criticism of the most recent additions to valence theory.

In Lewis's theory, the valence bond was defined as a pair of electrons held jointly by two atoms. This definition implied, according to Mulliken, that pairs of electrons acted as a sort of interatomic glue binding the atoms together and thus could be called *bonding electrons*, whereas the remaining unpaired electrons played a relatively minor role in bonding atoms together and thus could be called *nonbonding electrons*. On the contrary, Mulliken's work on band spectra led him to conclude that, besides bonding and nonbonding electrons, there were electrons that actively opposed bonding. Mulliken called these electrons *antibonding electrons*. They were the electrons that, during the process of molecule formation and in order to satisfy the Pauli exclusion principle, had to be promoted to orbits of a larger n quantum number. This tripartite classification of electrons implied that the adequate concept was that of a continuously varying *bonding power of electrons*, which Mulliken had previously defined in terms of energy relations (or alternatively in terms of distance relations). Then, one could see that the "problem of valence is really one of energy relations" (Mulliken 1931a, 350). If there were rules to calculate the energies of the different kinds of possible molecular orbits and to correlate them with the energies in the associated atoms, the "rules of valence should follow automatically" (Mulliken 1931a, 351). Their absence limited the scientist to qualitative reasoning and approximate calculations. Mulliken relied largely on empirical spectroscopic data to calculate and predict energies of atomic interactions and molecular constants.

Mulliken presented a qualitative analysis of the formation of the H_2^+ ion and the H_2 molecule in terms of energy considerations. When the H atom and the H^+ ion approached each other, the electron orbit was continuously deformed and tended to surround both nuclei. In the process, the 1s atomic electron became more firmly bound because it came under the attraction of the two nuclei. The formation of the ion may be represented by

$$H(1s) + H^+ \rightarrow H_2^+(1s\sigma, \Sigma_g^2),$$

which means that the atomic orbit 1s was changed into a molecular orbital $1s\sigma$. In this case, the energy of binding in the molecular orbit outweighed the energy of repulsion of the two nuclei, and the electron acted as a bonding electron. There was, however, another possibility corresponding with repulsion. It was represented by

$$H(1s) + H^+ \rightarrow H_2^+(2p\sigma, \Sigma_u^2),$$

and in this case the energy of promotion, together with the nuclear repulsion, acted in such a way as to cause the atom and ion to repel each other at all distances. The electron was an antibonding electron.

Two similar situations occur when two atoms of hydrogen approach each other. They were represented as follows:

$$H(1s) + H(1s) \rightarrow H_2(1s\sigma^2, \Sigma_g^1)$$

$$H(1s) + H(1s) \rightarrow H_2(1s\sigma 2p\sigma, \Sigma_u^3).$$

In the first mode of interaction, the two electrons became $1s\sigma$ electrons and acted as bonding electrons, forming a stable hydrogen molecule. In the second mode of interaction, one of the electrons was promoted to a $2p\sigma$ orbit, and the energy of promotion was so large that it outweighed the increased energy of binding of the 1s electron, and no molecule was usually formed. However, Mulliken once again noted that, contrary to Heitler and London's claim, there might be a considerable attraction at large distances, leading to the formation of a relatively stable molecule, even though at smaller distances the large energy of promotion led to a strong repulsion between the two atoms.

At this point, Mulliken made a crucial distinction in his criticism of Heitler and London's contributions. He distinguished between their *method* and their *valence theory*. The method had great potentialities at least in principle, if not always in practice. Yet, for Mulliken it presented a serious drawback: it "fails to give a detailed insight into the nature of the changes which take place in the electron orbits when the atoms come together" (Mulliken 1931a, 354). But, above all, Mulliken really objected to their *valence* theory, not so much to their *method*.

Heitler and London's *spin theory of valence* postulated as the paradigm of a valence bond the establishment of a "symmetrical" relation in the position coordinates and an antisymmetrical relation in the spin coordinates between two electrons belonging to two separate atoms. Mulliken could not disagree more with their explanation of chemical bonding. He emphasized that the pairing of spins, or for that matter the pairing of electrons, was misleading because it pointed to a phenomenon that is purely incidental and concealed something much more fundamental. For example, the case of the ion, with only one electron, illustrated that the phenomenon of pairing could not be of any relevance to the analysis of molecule formation and further suggested that "we should regard a *single bonding electron as the natural unit of bonding*, an antibonding electron as a negative unit" (Mulliken 1931a, 383).

The effect of a symmetrical relation in the positions of two electrons, with the concomitant antisymmetrical relation in their spins, is "to make the electrons keep on the average closer together than they otherwise would . . . hence, unless other indirect effects are important, a symmetrical relation increases the energy of repulsion of the electrons and so the total energy, while an antisymmetrical relation decreases it" (Mulliken 1931a, 357). For Mulliken, the presence of unpaired electrons with antiparallel spins and their subsequent pairing in molecule formation acted just as a "convenient indicator" of valence and the formation of valence bonds but did not "hit the nail on the head" (Mulliken 1931a, 359).

In Heitler and London's theory, the valence V of an atom was therefore defined as $V = 2S$, with S the resultant spin of the atom. The valence of an atom was equal to the number of unpaired electrons, and as each electron had a 1/2 spin quantum number, if there were n unpaired electrons, the resultant spin would be $S = n/2$. This meant that $n = 2S = V$. The theory further implied that when an atom did not have any unpaired electrons, $n = S = V = 0$, the atom could not form any bonds. This rule was bluntly contradicted by evidence from band spectra, where it was found that molecules such as HeH and He_2^+ could be formed under certain conditions. According to Heitler and London, the helium atom could not bond to a hydrogen atom because the two 1s electrons were already paired in the helium atom. However, when the hydrogen atom approached the helium atom, nothing forbade the 1s electron of the hydrogen atom to be promoted to a $2p\sigma$ orbit. What happens in general is that the promotion energy is so large that the two atoms repel each other. However, in analogous cases, such as when a helium atom and a helium ion approach, a physically stable molecule could be formed:

$He(1s^2) + He^+(1s) \rightarrow He_2^+(1s\sigma^2 2p\sigma)$.

In this process, which was confirmed by spectroscopic evidence, the two $1s\sigma$ bonding electrons outweighed the antibonding $2p\sigma$ electron. The existence of molecules such as HeH or He$_2^+$ illustrated the arbitrariness of the concept of valence and "the impossibility of accepting it as corresponding to an always sharply definable, whole number property of atoms" (Mulliken 1931a, 384–385). Taking advantage of his command and deep understanding of vast spectroscopic data, Mulliken clarified the meaning of the concept of stability by making a distinction between *chemical* and *physical* stability. HeH and He$_2^+$ are not chemically stable but are physically stable.

Mulliken suggested that chemical binding could best be understood in terms of quantum numbers of individual electrons and their changes when one goes from the united atom to the molecule, as well as in terms of the type of energy relations resulting when one goes from the separated atoms to the molecule. The strength of the chemical bond was related to the electronic quantum numbers. Mulliken was making a very bold suggestion. His radical proposal amounted to dispensing altogether with classical valence theory, which he called the "atomic point of view," and to adopt instead a new "molecular point of view," where the molecule as a distinct individual built up of nuclei and electrons is emphasized, as opposed to the atomic point of view where a molecule is regarded as composed of atoms. From the molecular point of view, "it is a matter of secondary importance to determine through what intermediate mechanism (union of atoms or ions) the finished molecule is most conveniently reached. It is really not necessary to think of valence bonds as existing in the molecule" (Mulliken 1931a, 369).

Two Parallel Research Agendas

The Development of Pauling's Program: The Nature of the Chemical Bond
By 1931, when Mulliken was extending his program to polyatomic molecules, Pauling was hard at work developing his own program, which appeared in the sequel to the first paper on "The Nature of the Chemical Bond." The next six papers in the series came out in the following 2 years. A sense of urgency emerges from the correspondence between Pauling and Lamb.[77] Pauling had mixed feelings concerning publication in the *Journal of the American Chemical Society*. On the one hand, he believed that it was the best place to publish articles of chemistry content such as his, and he wanted to secure from the start the chemists' attention. On the other hand, he worried constantly about possible delays in publication in view of the "rapid developments in this field and the speed with which other contributions are published in other journals."[78] He was certain of the novelty and originality of his results, and he did not want to be outrun by anyone else.

In the formation of an electron-pair bond, the resonance phenomenon of quantum mechanics accounted for the energy of the covalent bond. Resonance occurred with like or unlike atoms in view of the indistinguishability of electrons. The same resonance phenomenon was responsible for the formation of the one-electron and the three-electron bond. (Pauling hinted at the latter in the first paper of the series as a possible way to explain the magnetic behavior of the oxygen molecule.) However, in these cases certain conditions had to be met for the stabilization of the bond by resonance energy. In the case of two identical or nearly identical atoms, a one-electron bond might be formed if the two configurations A. B and A .B (where in the first/second the unpaired electron is attached to A/B), have "essentially the same energy." In the same way, a three-electron bond might be formed if the two configurations A: .B and A. :B (in one of which the atom A contains an electron pair and the atom B an unpaired electron, and in the other of which the reverse occurs) have "essentially the same energy." The meaning of "essentially the same energy" was defined loosely as the situation when the energies of the two configurations differed by an amount less than the possible resonance energy (Pauling 1931b).

Pauling used the one-electron bond to explain the properties of molecules such as H_2^+, H_3^+, and the boron hydrides (B_2H_6). He used the three-electron bond to explain the properties of molecules such as the He_2^+, NO, and O_2. Some of these molecules or ions, such as H_2^+, NO, and He_2^+, possess an odd number of electrons. The designation "odd molecules" was introduced by Lewis in his 1916 paper as part of an argument to justify the "rule of two," that is, the tendency of atoms to hold an even number of electrons in the outer shell. The relatively few molecules with an odd number of electrons known to chemists were generally unstable, showing a great tendency to react and form compounds by pairing. The molecule of nitric oxide, NO, was quite exceptional in that it did not associate into pairs. If for some reason the odd electron in NO was more tightly bound than in other cases, then one could explain why the molecule behaved as if saturated. Lewis had not found any mechanism to explain the exceptional behavior of NO. The three-electron bond was intended as a solution to such a puzzle, even though it did not meet with the agreement of Lewis, who insisted that a bond between two atoms could not be formed by more than two electrons.[79] As nitrogen and oxygen have nearly the same nuclear charge, the two configurations : Ṅ :: Ö : and : N̈ :: Ȯ : have nearly the same energy, so that through resonance a more stable configuration was formed involving a double bond and a three-electron bond, $:N\!\stackrel{..}{:}\!O:$.

The three-electron bond was used to explain the properties of the oxygen molecule. Although in order to conform to the "rule of eight," Lewis suggested the formula : Ö :: Ö : for oxygen, another formula, : Ȯ : Ȯ :, would be more appropriate to account for its paramagnetism, which Lewis associated with the presence of unpaired electrons. In his notebooks, Pauling discussed several of Lewis's formulas for the oxygen molecule,[80] but by 1931 he suggested that two three-electron bonds together with an electron-pair bond, : O ⋮⋮ O :, would account for its magnetic and chemical properties. Yet, Pauling's structure for the oxygen molecule looked quite farfetched. Besides, to explain the cases for which no single acceptable formula could be written in terms of Lewis's electron-pair bonds by the introduction of two distinct types of bonds seemed too high a price to pay, especially as the molecular orbital theory did not have to introduce any ad hoc explanations to account for these cases.

Pauling, who viewed his work as an "extension and refinement of Lewis's conceptions,"[81] considered the introduction of the one-electron and the three-electron bond as a generalization, rather than a violation of Lewis's electron-pair bond. The new definition of a shared-electron bond was now presented as a "complex between two atoms involving one eigenfunction for each atom and one, two, or three electrons, and giving attraction."[82] All three types of bonds, different as they might look at first, were united by the same underlying physical mechanism. It was the quantum mechanical resonance phenomenon that, in all cases, accounted for the energy of the bond and provided the extra stability of the molecule. It became the fundamental aspect of Pauling's theory of valence.

Lewis's 1916 theory of the electron-pair bond was introduced in the context of a discussion as to whether or not there existed two different types of valences to account for the two classes of compounds (polar and nonpolar) known to chemists. Lewis accommodated the two types of bonds in a unified framework through the idea that a pair of electrons might be equally or unequally shared by two atoms. For him, there existed just one chemical bond, and the differences between polar (or ionic) and nonpolar (or covalent) compounds were quantitative rather than qualitative or, as he put it, were differences in degree, not in kind. Although Lewis advocated that molecules could pass continuously from one extreme form to another, the question as to whether an ionic or a covalent structure should be assigned to a given molecule, and whether or not a continuous transition from one extreme type to the other could occur, was a point of contention among chemists throughout the 1920s. Once more it was the concept of resonance that enabled Pauling to answer these questions (Pauling 1932).

In case the electronic structures corresponding with the two extreme types (ionic and covalent) had nearly the same energy and the same number of unshared electrons, a continuous transition between the two types could take place, and the actual state of the molecule might be anywhere between the two extremes. In practice, however, it was difficult to do the calculations. To circumvent this obstacle, Pauling developed an alternative semiempirical method based on the comparison between the potential energy curves associated with the electronic structures of the extreme ionic and covalent states of molecules. If the curves do not cross or cross in a region where there is small overlapping, then the bond is either essentially ionic or essentially covalent, depending on the relative position of both curves; if the curves cross in a region of considerable overlapping, then there is mixing between the two states in the region of overlapping.[83]

In the fourth paper of the series, Pauling (1932a) went a step further in the discussion of the partial ionic character of bonds. To the qualitative criterion for bond character presented before, he added a quantitative semiempirical criterion that enabled him to determine the approximate percentages of ionic and covalent character of bonds and then to map atoms in a scale of relative electronegativities. Pauling was thus able to suggest a viable alternative to the quantum mechanical treatment outlined in the former paper, which, as he pointed out before, was impossible to carry out except in the simplest cases.

Pauling's earlier crystallographic work on the determination of atomic sizes and interatomic distances came in handy. The sizes of atoms and ions determined how close they may come together, and their closeness in turn provided an indication of the energy involved in a bond. It was empirical evidence of this sort, not rigorous theoretical justification, that led Pauling and one of his collaborators to the formulation of the additivity principle, according to which the energies of normal covalent bonds were additive, if one assumed that bonds acted independently from each other in a molecule (Pauling and Yost 1932).

Since the first paper on "The Nature of the Chemical Bond," Pauling had implicitly assumed that a bond between two atoms was not strongly affected by the presence of other atoms in a molecule. This assumption was behind the rule according to which in bond formation the main resonance terms involved just one single-electron orbital wave function from each atom. Pauling now claimed that the success of the criterion of maximum overlapping as well as the data collected on interatomic distances supported the assumption of bond independence. In that case, the total energy of formation of a molecule from separated atoms would be equal to the sum of the energies of the individual bonds, and hence the additivity postulate was justified.

This was the starting point of a program for determining bond energies from the experimental values of heats of combustion and heats of formation of gaseous molecules. Pauling compiled 21 values of single bond energies and used this database to generate an electronegativity scale. For each experimentally determined value of the bond energies, he found the corresponding value of the covalent bond energy given by the application of the additivity postulate. He then subtracted the covalent bond energy from the experimental value for the bond energy and found that in almost all cases, the difference was positive and it increased with the ionic character of the bond. This meant that the ionic contributions should be taken into consideration whenever calculating the actual bond energy. Introducing the designation Δ for the difference between actual bond energy and covalent bond energy, Pauling suggested that the value of Δ could be used to position atoms on a map according to their relative electronegativities. By assigning positions to several atoms in the electronegativity scale, Pauling confirmed that there was an increase of electronegativity for successive elements in each row of the periodic table and a decrease of electronegativity for successive elements in the same column of the periodic table. Besides its mere confirmatory value, the new electronegativity scale was used to predict the bond energies of 20 elements for which no experimental data were available.

For the first time, a numerical value was associated with the qualitative concept of electronegativity loosely defined as the tendency of atoms in molecules to attract electrons. To avoid any misunderstandings, Pauling pointed out in the closing paragraph of the fourth paper in the series that his definition of electronegativity was not equivalent to the electron affinity of atoms (the energy released when an atom or molecule in the gaseous state gains an electron to form a negative ion) although it bore a strong similarity with what he called the chemists' "intuitive conception" of electronegativity. Pauling had been able to give approximate values for the relative electronegativities of elements in case certain assumptions and approximations were valid. He was never inhibited in formulating rules that could be abstracted from the empirical evidence concerning molecular structure but had no "proper" theoretical justification.

Pauling published the first four papers of "The Nature of the Chemical Bond" series in the *Journal of the American Chemical Society*. Although he made extensive use of quantum mechanics, the introduction of the quantum mechanical concepts of resonance and hybridization of bond orbitals served to give a proper underpinning to chemical facts long known to chemists. Such was the case of the directional character of bonds, bond angles, and bond strengths. Furthermore, he was able to give answers to questions debated among chemists such as whether a certain bond was ionic or covalent or how to decide among alternative Lewis structures. The last three papers of the series were published in the first volume of the new *Journal of Chemical Physics*, which was created in 1933 with the purpose of housing a growing number of papers that were "too mathematical for the *Journal of Physical Chemistry*, too physical for the *Journal of the American Chemical Society* or too chemical for the *Physical Review*."[84] The appearance of a new journal expressed the feelings of a great number of chemists that a new area of problems had emerged out of the application of quantum mechanics to chemistry, an area sufficiently autonomous relative to physical chemistry to deserve a distinct name and a distinct outlet for its papers. Although Pauling was not involved in the decision to create the journal, he participated to some extent in its formation, served as associate editor through 1937, and published there regularly.

In a period of readjustments of subdisciplinary boundaries, Pauling did not think of himself as working in the field of physical chemistry, but instead he strove to lay the foundations for what he called "modern structural chemistry." Pauling's agenda was, in a way, at odds with the program of physical chemists. For him, the study of how structure determined the behavior and properties of molecules was the dominant issue. He made it a priority all across chemistry, from inorganic to organic chemistry, and in his later papers about the chemical bond he was ready to address some old structural puzzles of organic chemistry. With them, Pauling concluded the discussion of what he called, back in 1928, "the big problem which yet remains."[85]

The lack of criteria to choose among alternative Lewis structures had been instrumental to Pauling's introduction into valence theory of the one-electron and the three-electron bond, as well as the notion of the partial ionic character of substances. A similar uncertainty had existed in organic chemistry for more than half a century. There was no single structural formula consistent with the observed properties that could be assigned to several organic compounds. The structure of benzene (C_6H_6), and of other aromatic compounds in general, had been a constant source of annoyance to organic chemists. Consider for example the case of benzene. Shortly after Friedrich A. Kekulé's pioneering work of 1866, several structural formulas had been introduced as alternatives to Kekulé's hexagon structure with alternating single and double bonds. But without exception, all models for the structure of benzene, be it Kekulé's hexagon structure, James Dewar's hexagon structure with a diagonal bond (1866/1867), A. Ladenburg's prism structure (1869), or the centric formulas of A. Claus (1867) and H. E. Armstrong and Adolph von Bayer (1887, 1888), were subjected to serious objections. In 1872, in order to make all bonds equivalent in his former benzene model, Kekulé introduced the oscillation hypothesis according to which the double bonds were in rapid oscillation in such a way that two adjacent carbon atoms were connected part of the time by a double bond and part of the time by a single bond (Russell 1971).

In the years that followed, a number of chemists, including F. G. Arndt and B. Eistert in Germany, Thomas M. Lowry, Arthur Lapworth, Robert Robinson, and C. K. Ingold in Britain, and H. Lucas in the United States, became increasingly aware that the properties of aromatic and conjugated compounds could not be correlated with a single valence bond structure[86] and instead should be described by several structures. In the theory of intermediate states, Arndt suggested that if two different formulas could be assigned to an organic compound, then the molecule was not in either of the states associated with each of the valence bond structures but was instead in a new "intermediate" state involving both classical formulations. Ingold's theory of mesomerism assumed that the actual structure of the molecules was "in between" those described by the different valence bond formulas. The new word "mesomerism" was put forward to emphasize that this phenomenon was quite different from the classical phenomenon of tautomerism in which two isomers did indeed change into each other. Tautomeric molecules existed in two or more different forms, whereas in the new situation there was just one type of molecule.

The development of quantum mechanics facilitated further understanding of these ideas. The structure of the actual molecule was thought of as the resonance of various individual valence bond structures, and its wave function was taken to be a linear combination of the wave functions of each of the valence bond structures. In the paper in *Physical Review*, Slater applied these considerations to the structure of benzene (Slater 1931). The wave function representing the structure of benzene was considered to be a linear combination of the wave functions associated with the two Kekulé structures. The equivalence of all the carbons in the ring followed immediately, and the extra stability of the molecule was accounted for as a result of the quantum mechanical resonance effect on the energy of the molecule. At the same time, as we already saw, Hückel (1931, 1931a, 1932) was developing a quantum theory of the benzene molecule in the framework of molecular orbital theory.

In the fifth paper of the series on "The Nature of the Chemical Bond" and the first written in collaboration—inaugurating Pauling's strong association with George W. Wheland, then a National Research Fellow at Pasadena and the person whose future work would be decisive in the further establishment of the theory of resonance—Pauling and Wheland suggested an alternative method to study the benzene molecule. Slater's contribution is referred to only in a note, and in the paper there is just a general reference to those who hinted at resonance to explain the benzene structure. In the same way as Pauling contrasted his approach to that of Heitler and London in the first papers of the series, Pauling and Wheland now compared the methodological premises of their approach with Hückel's. They thought his approach was too cumbersome, and even though they acknowledged that both papers reached the same results, they believed their method was easily applied to other molecules. Furthermore, they claimed that they could extend their treatment to the problem of free radicals, where they obtained "surprisingly good qualitative agreement with experiment" (Pauling and Wheland 1933, 363).

Pauling and Wheland did not hesitate to make bold simplifications and approximations as long as this enabled them to extend the scope of the theory. Simultaneously, this quantum mechanical treatment was complemented and supported by an extensive amount of empirical information on resonance energies. The interplay between theoretical and empirical considerations had become crucial for the development and extension of a program that, in order to be of any help to chemists, had to show its usefulness in dealing with a great variety of molecules, from the simplest to the more complex molecules of organic chemistry.

Pauling and Wheland used perturbation theory to calculate the wave function representing the normal state of the benzene molecule and to find an expression for its resonance energy. They assumed that the wave function could be written as a linear combination of five wave functions that represented the five independent canonical structures contributing to the normal state of the molecule. The principal contributions were made by the two Kekulé structures and the rest by the three different forms assumed by the Dewar structure. But it was clear that "in a sense all structures based on a plane hexagonal arrangement of the atoms Kekulé, Dewar, Claus, etc. play a part, with the Kekulé structures most important. It is the resonance among these structures which imparts to the molecule its peculiar aromatic properties" (Pauling and Wheland 1933, 365). The extra energy of the molecule due to resonance among the five canonical structures was calculated as a function of the exchange integrals involving adjacent atoms. To find a numerical value for the exchange integral, Pauling and Wheland used the empirical values obtained by thermochemical methods for the resonance energy of the benzene molecule (Pauling and Wheland 1933). In this way, they were able to calculate the percentages of the total resonance energy contributed by the two groups—Kékulé and Dewar—of independent structures.

In the sixth paper of the series, Pauling and J. Sherman continued the calculation of bond energies carried out in the former paper and complemented them by the empirical determinations of bond energies from thermochemical data already in existence for a long time (Pauling and Sherman 1933). Previous attempts to extract from such data information on bond energies had failed. Pauling's former work had convinced him that there were essentially two distinct classes of compounds: those for which one single electronic structure of the Lewis type was consistent with the observed properties, and those for which no single valence bond structure was entirely compatible with the properties and behavior of the compound. For the first class of compounds, Pauling assumed that the energy of formation of the molecule could be calculated as the sum of the energies of each bond. But for the latter class of molecules, the energy of formation of the molecule from the separated atoms had to be greater than the energy associated with each individual bond structure, the extra stability of the molecule resulting from the resonance among several valence bond structures.[87] The same treatment was applied to naphthalene ($C_{10}H_8$), a benzene-like structure with two hexagon rings fused together, and extended to other conjugated systems in the last paper of the series (Pauling and Sherman 1933a). In the case of naphthalene, the calculations were much more involved as the number of canonical structures increased to 42. In fact, Pauling's method became increasingly more "cumbersome" as it was extended to more complex molecules.

In a letter to Pauling, Hückel emphasized the advantages of his approach whose results had been in good agreement with experiment and where he could derive the stability of benzene with six electrons on the ring and had "found a number of systematic discrepancies" in their calculations.[88] In a paper published much later, he was still unconvinced: "from the methodological point of view, the work by Pauling and his school signified a step forward—although, from the substantial point of view, initially a step backwards as well" (Hückel 1937, 759). In what became a typical manifestation of his style, Pauling did not budge, though acknowledging that they should have mentioned Hückel's work. He defended his own method and criticized Hückel's approach on various counts. He did not feel comfortable about the fact that for a hydrogen molecule, Hückel's method led to a near equality of the ionic and covalent parts in the eigenfunction, something that had been shown not to be the case as the covalent part was dominant. Pauling thought that this result was also valid for more complicated molecules. But more importantly, Pauling pointed out a serious shortcoming in Hückel's approach in which he thought that no account whatever of the exclusion principle was taken, feeling that all of the electrons in the molecule could be piled up on the same atom. "I feel that this makes [your results] very unreliable."[89]

The Development of Mulliken's Program: What Are Electrons Really Doing in Molecules?

The extension of Mulliken's program to polyatomic molecules started with a series of 14 papers titled "The Electronic Structure of Polyatomic Molecules and Valence," which spanned the years 1932–1935. The series appeared in *Physical Review* and, after the fourth paper, in the newly created *Journal of Chemical Physics*, which some claimed was founded just to house Mulliken's papers (Platt 1966, 746)! In this way, foundational papers for the subdiscipline of quantum chemistry written by two of its founders were included in the first and subsequent volumes of the *Journal of Chemical Physics*, which thereafter functioned as a privileged outlet for the new subdiscipline.

The first paper contained a brief outline of Mulliken's proposal (Mulliken 1932a; figure 2.3). Originally intended as a rather harsh critique of the Slater–Pauling approach, the letter-turned-paper underwent considerable modifications as a result of discussions with Van Vleck, and in the final version, a soften criticism of alternative approaches was relegated to a long "added in proof."[90] He pointed out to Van Vleck that

$$\frac{\text{One-electron } \Psi\text{'s in Molecule}}{\text{Localized binding of Pauling and Slater}} \equiv \frac{\text{Deep understanding of molecular Structure and Valence}}{\text{Conventional valence theory with electron-pair bond}}.$$

He was convinced that the method he was developing brought one much closer to acquiring an *insight* into the electronic structures of molecules. He thought that the "Slater-Pauling theory" gave a *poorer* approximation that only fit the phenomena of homopolar valence. "The Slater-Pauling theory, using the Heitler-London methods, offers a *method of calculation*, which is important." But he insisted that it was as important to get a deeper qualitative insight into molecular structure by visualizing polar and nonpolar intermediate cases, not as linear combinations, but directly. He acknowledged that he had assumed the tetrahedral structure, and they, through the use of directed valence, were able to get good results for water, yet, together with Hund, he disagreed with their treatment of the double bond.

Mulliken started using, in an uninhibited manner, the language of the new quantum mechanics in the papers dealing with the extension of his theory to polyatomic

Figure 2.3
Robert Sanderson Mulliken outdoors (photograph by Samuel Goudsmit).
Source: Courtesy of AIP Emilio Segre Visual Archives, Goudsmit Collection.

molecules. An atomic orbital was associated with the motion of an electron in the field of a single nucleus and of other electrons, whereas a molecular orbital was associated with the motion of an electron in the field of two or more nuclei and that of other electrons. His initial idea was to use many-center nonlocalized molecular orbitals, even though other types of orbitals could be used, like two-center ("localized") orbitals as suggested previously by Hund (Hund 1931). In fact, the correlation diagram implied that for large internuclear separations, one could approximate a molecular orbital by a linear combination of atomic orbitals. Although considered as a partial departure from the original formulation of Mulliken's program, John Edward Lennard-Jones's earlier suggestion for diatomic molecules was followed and extended to the case of polyatomic molecules (Lennard-Jones 1929). Unshared electrons (nonbonding electrons) were represented by atomic orbitals, and shared electrons (bonding or antibonding electrons) were represented by molecular orbitals considered to be entities quite independent from atomic orbitals and represented by linear combinations of atomic orbitals (LCAO) for reasons of calculational expediency: "The present method of thinking in terms of the finished molecule, used already by Lewis in his valence theory, avoids the disputes and ambiguities, or the necessity of using complicated linear combinations, which arise if one thinks of molecules as composed of definite atoms or ions" (Mulliken 1932b, 51).

After the definition of the terms to be used, a survey of some of the most recent valence theories was presented with the aim of showing that there were neither empirical nor theoretical reasons "for placing primary emphasis on electron pairs in constructing theories of valence" (Mulliken 1932b, 52). He then proceeded to compare the valence theories of Lewis, Heitler and London, and Slater and Pauling by assessing the different weight attributed in each theory to the following three theoretical components: (A) Each nucleus tends to become surrounded by a set of closed shells of electrons (group of eight, or octet). (B) Shared electrons in covalent bonds are localized between the nucleus they glue together. (C) A chemical bond usually consists of a pair of electrons relatively tied to each other (group of two).

In Lewis's theory, component (C) was emphasized relative to components (A) and (B). In Heitler and London's spin theories of valence, component (C) had priority over (B) and (A). In Slater and Pauling's theory, component (B) was emphasized relative to (C). Relative to Lewis's, the Heitler-London-Pauling-Slater (H-L-P-S) theory was more limited in the sense of assuming that in a bond, each electron comes from each atom. To stress this difference, Mulliken suggested calling the H-L-P-S theories *electron-pairing* theories in contrast to Lewis's theory, which he called *electron-pair* theory.

Contrary to all former theories, in the theory of molecular orbitals the emphasis was on component (A), component (B) was adopted in a generalized form (many-center nonlocalized molecular orbitals), and component (C) was considered to be a mere incidental characteristic of chemical combination. Mulliken believed that the

H-L-P-S approach put an unjustified emphasis on the role of electron pairs and electron pairing. He further argued that relative to the electron-pair bond of H-L-P-S, the concept of molecular orbitals presented a number of advantages: it did not assume that two electrons were necessary to form a chemical bond; bonding molecular orbitals may have any degree of polarity corresponding with different degrees of sharing of electrons between the nuclei; and bonding molecular orbitals are not restricted, like electron-pair bonds, to hold just two electrons together. And he claimed that the concept of bonding molecular orbitals was more general, more flexible, and undoubtedly more natural than the concept of electron-pair bonds, even though "the electron-pair bond method may for many problems be more adapted to quantitative calculations than the present method" (Mulliken 1932b, 60).

Despite such a thorough (re)statement of the guiding assumptions of his program, three years later, in 1935, Mulliken still believed that the theoretical framework of his unconventional valence theory was not understood clearly, and there was yet another attempt to clarify the situation. Mulliken regretted that his radical proposal was meeting with a generalized incomprehension but admitted that he might not have stated precisely his original intentions. Mulliken sent a copy of the manuscript to Van Vleck and wrote[91]:

I have always had in mind the idea of a "conceptual scheme" to be compared with empirical data, but seem never to have stated this very clearly. The conceptual scheme using "natural" or "real" or "best" (even though of not known exact form) molecular orbitals must represent a better approximation than the use of rough LCAO orbitals.

The situation was the following: Two different descriptive chemical theories provided alternative methods for the assignment of molecular electron configurations, one of them relying entirely on the use of atomic orbitals, whereas in the other molecular orbitals of some sort were used to describe shared electrons. The first method followed "the ideology of chemistry" and treated each molecule as composed of definite atoms or ions. The electron configuration became the sum of the configurations of these atoms or ions. He noted that this method's success resulted from the qualitative conceptual scheme it provided for "interpreting and explaining empirical rules of valence and in semiquantitative, mostly semiempirical calculations." The second method was "departing from chemical ideology" because it treated each molecule as a unit. "It is the writer's belief that the present . . . [method of nonlocalized molecular orbitals] may be the best adapted to the construction of an exploratory *conceptual scheme* within whose framework may be fitted both chemical knowledge and data on electron levels from molecular spectra." He considered his work as a preliminary stage from which quantitative calculations would "logically follow later" (Mulliken 1935, 376, emphasis in original), even though the values obtained by the molecular orbital method for dissociation energies were not as good as the ones obtained by the valence bond method. The molecular orbital method gave also an excessive ionic character to wave functions.

Through the systematic presentation of Hund's contributions to the quantum theoretical foundation of band spectra, which started with Mulliken's fourth paper of the "Electronic States" series, Mulliken played the role of a translator of Hund's very abstract suggestions into a language more appropriate to an American audience (Schweber 1986; Cartwright 1987; Holton 1988).[92] From the start, the differences between Mulliken's and Hund's approaches were so obvious that Van Vleck called attention to them in his correspondence with Mulliken. For Van Vleck, Hund made a "lot of abstract suggestions without working over the experimental data carefully," whereas Mulliken "as a man thoroughly conversant with the experimental data" was able "to bridge the gap between theory and experiment." Hund's work provided a legitimizing framework for Mulliken's phenomenological approach.[93] Mulliken's work provided a justification for Hund's very abstract suggestions, translated them into a language understood by the Americans, and went further than Hund's through a systematic study of the relevant empirical data guided by theory.

According to Mulliken, it was utterly unfair to criticize the molecular orbital theory on the basis of the poor results obtained by the use of a rough approximation such as the LCAO representation of a molecular orbital: "While the LCAO type of approximation is very simple and convenient as a qualitative guide, it is in no way an essential part of the present method. Especially it is not essential to the qualitative conceptual scheme of the latter."[94] The molecular orbital theory was also criticized on the grounds that it neglected the interactions between electrons. Mulliken argued that "their qualitative inclusion has always formed a vital part of the method of molecular orbitals used as a *conceptual scheme* for the interpretation of empirical data on electronic states of molecules." Part of its success was the "qualitative explanation of the paramagnetism of the oxygen gas" (Mulliken 1935, 378). In the evaluation of his valence theory, Mulliken challenged people to distinguish between its conceptual scheme and the methods available for computational calculations. The LCAO simplification gave a useful method for getting quantitative results by approximating molecular orbitals by linear combination of atomic orbitals. But this approximation should not be taken as the representation of the "real" or "best" orbitals. Thus far, it had been impossible to express them mathematically, except in the simplest case of the hydrogen molecule ion. Yet, Mulliken had in mind these "real" or "best" molecular orbitals while suggesting the conceptual scheme of his valence theory. Mulliken had attempted to construct "an exploratory conceptual scheme" whose assumptions were at the antipodes of Pauling's valence bond theory. He continued to argue for the establishment of the molecular point of view, which is of the view that molecules should be regarded as made up of nuclei and electrons, and not as aggregates of atoms or ions. The chemical molecules were those aggregates whose lowest states happen to be particularly stable in free competition with other possible aggregates. Nevertheless, "in pursuing this point of view it is advisable, however, at the outset to make a partial concession to the chemical point of view, so as to regard molecules as *aggregates of atom-cores or atom-ions and electrons*, rather than just of bare nuclei and electrons. In other words, a certain number of electrons, in general, *most* of the electrons, may for all practical purposes be assigned definitely to specific nuclei so as to form atom cores or ions, leaving only a generally smaller number to belong to the molecule as a whole."[95]

He objected to the oversimplification he saw in a title such as "The Nature of the Chemical Bond." For him there was not one nature of the chemical bond. There were many. Besides, he did not want to commit himself to the conventional view according to which "atoms were still atoms when they have formed molecules." Echoing Gertrude Stein, whom he recalled, he would rather claim "a molecule is a molecule is a molecule." The young man whose high school graduation speech was on the electron had come a long way to understanding what "the electron . . . is and what it does" (Mulliken 1975, 13).[96]

In two of his interviews, Thomas Kuhn discussed the issue of difference in styles concerning the different approaches to chemical bonding. He asked Mulliken whether there were a series of schools—a Heitler–London school, a molecular orbital school—with some geographical localization and whether in different places different approaches were being used "so that people were somewhat past each other?"[97] Mulliken was noncommittal in his answer: "I do not know. The way I was thinking was not in such terms as to notice things quite in that framework. I would say there were some people who were stronger for one thing than for another, but whether they were more abundant in one particular place I do not know."[98] Wigner, in contrast, stated that he never felt this opposition, as it was very clear to him from the very beginning that these approaches had different objectives. For example, the molecular orbital method does not speak about the bond, but rather it has molecular orbitals that extend over the whole molecule: "This is too far away from the very useful and very fruitful chemical concepts."[99]

Despite participants' assessments as stated above, differences in styles played a role in the early developments of quantum chemistry not only in what was referred to as differences between the Heitler–London school and the molecular orbital school but, more dramatically, in what related to the contrast between physically oriented approaches of German physicists and chemically oriented approaches of American scientists. Geographical differences were after all a subsidiary manifestation of strong differences in the institutional settings that provided the educational backgrounds of members of future quantum chemical communities.

In Germany, there was a sharp division between the chemical and the physical communities, which hardly if ever communicated. And German chemists were in general ill prepared to cope with the challenges of quantum mechanics. One example was Kasimir Fajans, a leading professor of physical chemistry in Munich when the young Pauling was there in 1926–1927. Many years later, in 1987, Pauling remembered that Fajans's inability to understand quantum mechanics was a problem that bothered him for the rest of his life.[100] In contrast, there were people like Heitler and London who showed a marked disrespect for chemists' opinions, which they considered as rather "soft" when they assessed anything related to quantum mechanics.

By the early 1930s in Germany, chemistry and physics entertained few disciplinary, methodological, or institutional ties to each other. Therefore, scientists whose profile could favor an attack on chemical problems using the tools of the newly developed quantum mechanics were hard to find. Several German physicists, not chemists, were interested in applications to chemistry and contributed initially to the field but were unable in the long run to carry out their research programs. Such were the cases of Heitler, London, Hund, Hückel, and Born.

The situation was very different in the United States. Interdisciplinarity between chemistry and physics, pragmatic and instrumental outlooks became common to many American scientists who delved into quantum chemistry. A particular kind of institutional atmosphere accounted for the appearance of a new type of scientist, whose definition as a chemist or physicist was in many instances a matter of chance, personal preference, or of institutional affiliation. The institutional ties between chemistry and physics were stronger in the United States than in Europe. In universities like Berkeley and Caltech, chemistry students were often learning as much physics as chemistry, and thus were more apt to learn and accept quantum mechanics than their European counterparts. Pauling's knowledge of physics was impressive, and Mulliken was an expert on the quantum theory of molecules. Besides Berkeley, Caltech, Harvard, and MIT, more universities were promoting cooperation between physics and chemistry departments. Examples were Princeton, Chicago, Michigan, Minnesota, and Wisconsin. But before Mulliken, Pauling, Slater, and Van Vleck, the preceding generation of chemists and physicists—chemists like Lewis and Noyes, Richard Tolman, and Harkins, and physicists like Kemble and Birge—planted the seeds that blossomed into quantum chemistry.

Playing the Devil's Advocate

The Odd Man Out: John Clarke Slater's Determinants and Mathematical Equivalence

It was two American physicists—Slater and Van Vleck—whose work convinced many that the approaches of Mulliken and Pauling were, in fact, complementary.

Van Vleck, who was Kemble's first Ph.D. student, considered Slater and Mulliken to be Kemble's "illegitimate children" as both were considerably influenced by Kemble but neither was supervised by him. Striking parallels unite Slater and Van Vleck's careers: They both worked in atomic physics, although their youthful attraction for atoms soon gave way to a lifelong effort to understand the properties of molecules. They were both eager to contribute to the education of their peers in the new quantum mechanics. In 1928, Van Vleck's article on "The New Quantum Mechanics" appeared in *Chemical Reviews* (see chapter 1). That same year, Slater organized a symposium on "Quantum Mechanics" at the New York joint meeting of the American Physical Society and the American Mathematical Society, in which Slater, Van Vleck, Hermann Weyl, and Norbert Wiener spoke. In their lectures, Slater and Van Vleck decided to "hammer on" the inherently statistical character of quantum mechanics.[101]

In the 1970s, they committed themselves to "take a crack at the idea that American physicists prior to World War II were no good except for European training and influence."[102] The accomplishments of American quantum physics in the 1920s were such that they defined this period as the coming of age of American physics, a time when American physicists were undeniably charting waters independently from their European colleagues (Van Vleck 1964, 1971; Slater 1967, 1967a). Their successful careers proved, like those of Mulliken and Pauling, how far Americans could go. Once, Van Vleck expressed his admiration for Slater's contributions to science in the following manner: "Slater is in the very front ranks of American physicists of the century. This is true even when one includes the distinguished émigrés from Europe in the 30's . . . Luiz Alvarez once said 'physicists are like baseball players . . . when they approach forty they are pretty much washed up as far as their old skills are concerned. Baseball players open bowling alleys or become managers. Physicists turn into deans or college presidents or become managers of teams of younger physicists.' However, there are exceptions to what Alvarez said. Slater's productive research spans half a century and is remarkable."[103]

After a brief flirtation with foundational issues, Slater switched to the application of quantum mechanics to atoms and molecules. Several questions concerning the properties of the alkali halides had been puzzling him since his dissertation time. The fact that a molecule of an alkali halide had a valence shell completely filled with electrons in analogy with the inert gases suggested to Slater that the quantum mechanical study of the helium atom was the convenient place to start. He tried to carry out more accurate calculations than those of Heisenberg for the helium atom and spent a considerable time during 1927–1928 working out these ideas (Slater 1927, 1928) and then applying them to the understanding of the repulsion between ions in an alkali halide crystal.

In this period, two different works exerted a considerable influence on him. Egil A. Hylleraas, a Norwegian working with Born's group in Göttingen, developed a more accurate method to treat the helium atom (Hylleraas 1928, 1929, 1963). And, also, Douglas Rayner Hartree's development of the self-consistent field method, which we address in detail in chapter 3, was also a significant factor in Slater's career: "My approach to physics has been more English than continental, if one wants to make such a distinction. I always felt much more sympathy with Hartree, for instance, than with most of the continental physicists."[104] Slater disliked group theory as much as the majority of physicists and was convinced that there was a way to circumvent the "elaborate complications"[105] of group theorists. His starting point was Heisenberg's and Dirac's independent discovery that in a two-electron system with parallel spins, the complete wave function, including the spin factors, must be antisymmetric and could be represented by a determinant (Heisenberg 1926, 1926a, 1927; Dirac 1926). Not long after, Pauli had shown how to interpret spin in terms of a two-valued factor in the wave function. However, it was not evident how to construct a determinantal antisymmetric wave function in the case where the electron spins were not all parallel.

Slater decided to start by studying the Li atom, the atom next to the two-electron case. His methodology was "to consider a special case, complicated enough to point the way to even more complicated cases, but simple enough so that it is easy to understand."[106] He represented the wave function for the three-electron case in terms of the Pauli products $u(1)\alpha(1)$ and $u(1)\beta(1)$, where u represented the orbital component and α or β the spin component of the wave function, and found that he could write the total wave function using antisymmetrized products of "spin-orbitals" rather than "orbitals." In fact, what Slater did was to generalize Pauling's four-electron wave function used in the study of the repulsion of two helium atoms to a system of an arbitrary number N of electrons (Pauling 1928). As Slater later recognized, in many cases more than one antisymmetrized product had to be used to describe the wave function, and he worked out ways for finding the linear combinations of these antisymmetrized products in order to get solutions for the multiplet energies. This led to a rather straightforward procedure compared with those of Wigner, Heitler, and the others.[107] Although in retrospect this seemed the obvious thing to do because the determinantal form including the spin factor automatically ensured the proper total antisymmetry, the fact of the matter was that, as Van Vleck recognized "almost two years elapsed between the appearance of Pauli's paper and Slater's. Even with all the galaxy of European stars, Bohr, Kramers, Heisenberg, Born, Dirac and all the rest, none of them had the imagination or simplistic insight to introduce this kind of determinant."[108] Slater circulated the manuscript among his friends. In the paper, which Slater later defined as "strictly USA without the slightest influence from Europe,"[109] the determinantal functions formed from spin-orbital functions were introduced for the first time as well the F and G parameters, integrals describing the energies of all states that arise from a given configuration (Slater 1929).

When the paper was completed, Slater left for Europe on a Guggenheim Fellowship for one-half year. He stayed in Leipzig most of the time. Debye, Heisenberg, and Hund belonged to the faculty of physics. This time, his European trip was a success. As Slater recalled, because many physicists strongly disliked group theory, they were quite overjoyed with his approach. "Remarks went around to the effect that 'Slater had slain the dragon of Group Theory,' and on the whole the paper made as much international sensation as the earlier Bohr, Kramers, Slater had done."[110]

An immediate consequence of Slater's paper was to get Condon—who belonged to the editorial staff of the *Physical Review* when the manuscript arrived—to start working on atomic spectra. Condon's fascination with atomic spectra finally led him to write *The Theory of Atomic Spectra* with G. H. Shortley. Van Vleck regarded the paper of 1929 as the greatest of Slater's papers.[111] Felix Bloch realized immediately that some of Slater's ideas could be of help for his magnetic theory.[112] Slater's complex spectra paper had a positive feedback even among the group theorists. Wigner, whom Slater met when he visited the Technische Hochschule in Berlin, wanted to know more about Slater's determinantal method, telling him that he had planned to work along the lines of Slater, but could not manage it.[113] Heitler called Slater's paper a "turning point" in the study of polyelectronic systems. He recalled that neither he nor London ever thought of including the spin in the wave function because they thought that the spin did not play any dynamic role, so they left it out altogether. "Then indeed this group theory treatment of the multi-body problem becomes very complicated, but if you include the spin wave function there is a certain simplification."[114] This simplification was introduced by Heitler in a paper with Rumer (referred in chapter 1) where they gave a general method for treating the interaction of several atoms (Heitler and Rumer 1931).[115]

In a letter to Born, Slater spelled out what was to follow. He informed Born that very interesting results were obtained regarding the general problem of valence in polyatomic molecules, wondering how closely these results resembled those of Heitler.[116] In their group theoretical papers, Heitler and London considered only those cases when all valence electrons were *s* electrons. This simplifying assumption enabled them to neglect the role of orbital orientation and therefore to analyze the valence behavior of substances exclusively in terms of spin. But even in this case, the mathematical secular problem associated with the coupling of spins proved to be hopelessly complicated if one used group theory as Heitler did. Born was able to prove that the calculations could be simplified by the use of Slater determinants (Born 1930, 1930a). Dirac (1929) was also able to "reinvent" group theory in a more physical way by showing that the exchange coupling of Heitler and London theory—the cement responsible for the chemical bond—is equivalent to a coupling of the spins obeying the relation:

$$V_{ij} = -\frac{1}{2}(1 + 4\vec{s}_i \cdot \vec{s}_j)J_{ij}$$

where $\vec{s}_i$ and $\vec{s}_j$ are the spin vectors of electrons *i* and *j*, and J_{ij} is the exchange integral. Soon afterwards, Van Vleck remarked that if one used Dirac's vector model "instead of recondite group theory or bulky determinants the entire interaction problem becomes essentially that of the vector model so dear to the heart of the atomic spectroscopists" (Van Vleck 1970, 243). In this paper, Dirac was first and foremost looking for approximations able to circumvent the complexities of atomic problems, but his paper became famous for his opening paragraph, charting a reductionist program in quantum chemistry (Simões 2003).

During his stay in Leipzig, Slater attempted to extend the determinantal method to the study of molecules and solids: "I needed to 'colonize' the field of molecules and solids instead of going on and working out all the details of the atoms . . . I felt I wanted to take a crack at each of these as soon as I could, so that at least I'd know the fundamentals."[117] As a result of his graduate work with Bridgman, Slater was more interested in solids than in molecules, but he believed that the understanding of molecules was a necessary preliminary step toward the understanding of solids. Slater chose to study the cohesive properties of monovalent metals such as lithium and sodium. Once more he used his method—to analyze in detail a simple case and then attempt to generalize the conclusions to more complex systems. It led him to start with the study of the interaction between two atoms before going on to more complicated cases, like the interaction of three atoms or a chain of atoms or finally a three-dimensional crystal. The simplest case of relevance to Slater's project was the interaction of two hydrogen atoms in the formation of the hydrogen molecule. This set him to think about the relation between localized and delocalized orbitals. Heitler and London had used localized orbitals, whereas Mulliken's proposal built on nonlocalized orbitals extending over the molecular framework. Slater noticed that a similar situation occurred in the case of solids. Heisenberg, in the theory of ferromagnetism, had started with separated atoms as Heitler and London did for the molecule, whereas Bloch, who took his degree with Heisenberg and Hund in 1928, used wave functions extending throughout the crystal. Slater used the determinantal method to explore the relationship between these methods, for molecules and for solids.

To establish the relationship between the two methods, Slater considered a configuration in which the electrons were assigned to molecular orbitals with the same symmetrical orbital function u but opposite spins and approximated the molecular orbital by a sum of atomic orbitals a and b (Slater 1965). In that case:

$$\begin{aligned}\psi_{MO} &= u(1)u(2) \\ &= [a(1)+b(1)][a(2)+b(2)] \\ &= a(1)b(2)+b(1)a(2)+a(1)a(2)+b(1)b(2) \\ &= \psi_{HL} + a(1)a(2)+b(1)b(2).\end{aligned}$$

The wave function obtained by the molecular orbital method contains, relative to the wave function obtained by the valence bond method, two extra terms that correspond with both electrons attached to the same nucleus; these are ionic terms, corresponding with the chemical structures H^-H^+ and H^+H^-. The covalent and ionic terms occur with equal weight, and that means that the molecular orbital function attributes equal probability to the dissociation of the hydrogen molecule into two neutral hydrogen atoms or two ions, a situation that is contradicted by the actual dissociation of the molecule in its ground state into neutral atoms. Thus, in the limit of large internuclear separations, the ionic contributions should not be present because the ionic state has a much higher energy than the state obtained from two neutral atoms. In this limiting situation, the molecular orbital function represents a poorer approximation to the true wave function than the Heitler–London wave function. In the molecular orbital method, the two electrons were treated as independent. The absence of correlation between the two gave too much weight to the unlikely situation of finding both electrons in the same atom. As soon as Slater recognized this fact, he was able to use his determinantal method to show how the two schemes could be brought together.

The correct wave function was a linear combination of two determinantal wave functions—one formed from the symmetric molecular orbital and the other from the antisymmetric one and gave a lower energy than the Heitler–London function alone by mixing in some of the ionic function. The procedure of using linear combinations of determinantal functions was called "configuration interaction." For Slater, configuration interaction not only modified the molecular orbital method so as to bring it into agreement with the Heitler–London method but also allowed a significant improvement on that method (Slater 1965). Configuration interaction was later to play a determining role in the development of computational methods in quantum chemistry (see chapters 3 and 4). Heisenberg, who used a method similar to Heitler and London in the case of solids, and Hund, who advocated the use of molecular orbitals in the study of molecules, showed a keen interest in Slater's work. At the end of his stay in Leipzig, Slater gave a colloquium where he presented his conclusions to an illustrious audience that included, besides Heisenberg and Hund, Pauli, his assistant Rudolf Peierls, Wigner, and London. The paper on "Cohesion in Monovalent Metals" was sent for publication upon Slater's arrival in the United States (Slater 1930).

Slater had been familiar with both approaches, from the early days of wave mechanics, partly as a result of his friendship with Mulliken. Although his name was associated with the Heitler–London–[Slater]–Pauling method after 1931, he was the first to point to the interrelationship of the two methods. In a short note, Slater came back to this topic and summarized his previous conclusions. He claimed once more that both methods were "complementary" rather than "antagonistic" and that the choice between the two should be made on the "basis of convenience rather than correctness" (Slater 1932, 255). The methods looked very different at first but when properly refined should come into closer agreement. Slater outlined the steps necessary for their refinement: in the case of the valence bond method, one should allow for possible ways of drawing valence bonds other than localized bonds between pairs of atoms, and for contributions of possible ionic states; in the case of the molecular orbital method, one should work with antisymmetrized combinations of molecular orbitals taking spin into account. Then he discussed their relative merits from the point of view of convenience. For a qualitative discussion both methods were equally convenient, and Slater argued that the best procedure to follow in this case was to discuss a problem qualitatively by both methods, in order to compare them and find their real relationship.

When carried out completely, either method should give the same integrals, and then it was irrelevant which scheme one was using. In the case of numerical applications, Slater pointed out that there was not yet any general method to treat molecular orbitals unless one represented them by linear combinations of atomic orbitals as suggested by Lennard-Jones, and this fact seemed to give an advantage to the valence bond method. In contrast, the fact that molecular orbitals are orthogonal and atomic orbitals are not made it easier to carry out calculations with the molecular orbital method. And Slater concluded: "The writer is quite sure that there are no facts explainable by the scheme of Mulliken and Hund which could not be equally well treated by his method, interpreting it in a suitable way. On the other hand, he would be the last to question the great power of the method of Mulliken and Hund, particularly for qualitative discussion" (Slater 1932, 257).

Slater had the impression that his papers up to 1934 went generally unnoticed, with the possible exception of the Bohr–Kramers–Slater paper and the complex spectra papers. He commented that people who were working on molecules were either interested in Heitler–London and doing fancy group theory or were interested in molecular orbitals and were not interested in carrying it any further than just to find the energy of the molecular orbitals. He did not seem to have a feeling that the community was paying any attention to his views: "I think that, in other words, people were all in their little lines of approach, and what I was doing was rather out of all of these lines."[118] Notwithstanding Slater's feelings, Hund and Mulliken certainly read his papers and were aware of the interrelationships between the two methods. Mulliken, for instance, immediately extended Slater's argument for the hydrogen molecule to the case of heteronuclear diatomic molecules (Mulliken 1932b). And, as we discuss in the next section, Van Vleck (1933, 1933a, 1935) compared the two methods in the case of methane and related molecules and, eventually, published with Sherman a review article on "The Quantum Theory of Valence," which became one of the hallmarks in the history of quantum chemistry (Van Vleck and Sherman 1935).

Van Vleck: The Optimists and the Pessimists Vis-à-Vis Quantum Chemist

John Hasbrouck Van Vleck (1899–1980) attended the University of Wisconsin from 1916 to 1920 and a summer school in Chicago in 1920 where he met Mulliken for the first time. Together they took a course on quantum theory given by Robert A. Millikan, which Van Vleck compared with Kemble's course in the following terms: "I cannot overemphasize Kemble's early role in training young theoretical physicists in quantum theoretical physics. For one thing, the lecture course which Kemble gave provided a systematic presentation of the mathematical basis of the old quantum theory, in sharp contrast to Millikan's course . . . which was a rather disconnected survey of topics in quantum theory as viewed by an experimentalist, with many of the lectures given by the students themselves, none too familiar with the subject."[119]

Van Vleck pursued graduate studies at Harvard. In the dissertation supervised by Kemble, he used the old quantum theory to compute a value for the energy of the helium atom using a crossed-orbit (nonplanar orbits) model that had been independently proposed by Kemble and Bohr. After completing his Ph.D. in 1922, he stayed at Harvard one more year as instructor. He spent summer 1923 in Europe. He met Pauling by accident on a train, and they became good friends (Van Vleck 1968). In fall 1923, he accepted the position of assistant professor at the Department of Physics of the University of Minnesota. He then moved to Madison, Wisconsin, which he considered his "alma mater," and stayed there until 1934 when he moved to Harvard. In the 1920s and early 1930s, the leading universities active in quantum theory included Harvard, Princeton, Chicago, Caltech, Michigan, Minnesota, and Wisconsin. During this period, the Department of Physics at the University of Wisconsin hosted as visiting professors Sommerfeld, Born, Schrödinger, Dirac, Heisenberg, Fowler, and Wentzel (Van Vleck 1971, 6).

During the 3 years after the doctorate, Van Vleck's research was on atomic physics. He worked on the theory of the specific heat of hydrogen and wrote a paper on the correspondence principle for absorption. His expertise in this area accounts for his selection to prepare the National Research Council report on the quantum theory of line spectra. After the advent of quantum mechanics, Van Vleck's research concentrated on the theory of magnetism. In 1932, he published the *Theory of Electric and Magnetic Susceptibilities* (Fellows 1985). His subsequent research opened the way to much of what is now called "magnetochemistry" and "ligand field chemistry."

While he was at Minnesota, the Department of Chemistry tried to hire Mulliken, and Van Vleck, eager to have his friend closer, argued for the move on the grounds that some of the chemistry faculty was also interested in "physics as well as chemistry."[120] Mulliken did not accept the offer, as he wanted to be in a physics department rather than a chemistry department.[121] Although geographically distant from each other, Mulliken and Van Vleck's overlapping interests resulted in a correspondence that extended throughout their lives. In 1927, Van Vleck found an explanation for the magnetic susceptibilities of the two common paramagnetic gases, NO and O_2. He recalled having benefited greatly from his rapport with Mulliken, who had been able to infer from spectroscopic data available that the ground state of the NO molecule was a doublet. Through discussions with Mulliken, Van Vleck became aware of all the data relevant to his theoretical work. When he sent the first theoretical formulas to Mulliken, Mulliken noted that Van Vleck's conclusions fitted the experimental results "almost like a glove."[122] The compliment could not have pleased Van Vleck more, who so much admired Mulliken's "bloodhound-like qualities" for the qualitative interpretation of band spectra (Van Vleck 1971, 8).[123]

In 1930 while on a Guggenheim Fellowship in Europe, Van Vleck met Kramers and went with him for a walk to the dunes of Holland. Then, by chance, Kramers called his attention to Hans Bethe's papers on group theory (Bethe 1929, 1930). Van Vleck remembered this walk as the highest point of his European trip. He considered that he "learned more in this walk than in the whole rest of my fellowship. One can never tell when a turning point will arise in one's career in research" (Van Vleck 1971, 12). This event also had important repercussions in Mulliken's work. It was Van Vleck that alerted him to Bethe's work, which Mulliken then successfully applied to the study and classification of polyatomic molecules. After Slater's effective proof that the use of mathematical group theoretical methods could be avoided in the analysis of complex spectra, Mulliken showed that when dealing with polyatomic molecules, group theory was not just another fancy invention of the mathematicians to make physicists and chemists feel miserable. It really made a difference in the classification of the electronic states of polyatomic molecules, as well as in the study of valence-related problems (Mulliken 1932a, 1933).

The manuscripts of Mulliken's first papers on the "Electronic Structure of Polyatomic Molecules and Valence" stimulated Van Vleck to contribute to the quantum theory of valence, and this happened in the period immediately preceding his move to Harvard in 1934, when his involvement with valence problems came to an end (Van Vleck, 1971). No longer close to Mulliken, lacking an appropriate chemical background, and overwhelmed by the labor involved in accurate calculations, he shifted to more familiar grounds: "I pulled out of the field because it seemed to me that you could make the thing quantitative only with very dinosauric calculations. It's been a revelation to me how seriously chemists in later years took these very skeletonized approximations because actually you've got an enormously complicated eigenvalue problem."[124]

In the three-paper series "On the Theory of the Structure of CH_4 and Related Molecules," Van Vleck contrasted advantages and disadvantages of both approaches and was able to prove that they led to similar predictions in respect to geometrical arrangements for methane and similar molecules (Van Vleck 1933, 1933a, 1934). Although in syntony with Mulliken on many grounds, his was not a partisan view of the field. As Slater, he demonstrated a similar theoretical attitude, both willing to compare the two methods and to discuss their equivalence without committing himself to either one.

More than anywhere else, this attitude was crystallized in a review paper published in 1935, which was even translated into Russian, and which offered the first overall assessment of what had been at the core of quantum chemistry during the first years of its development. Van Vleck and Sherman recalled what they dubbed as Dirac's "now classic sentences" in the 1929 paper where he claimed that the developments in physics guaranteed that all of the chemical problems were soluble. And they offered a *bilan* of the successes and the difficulties in attacking the complexities of the *n*-body problems faced in quantum chemistry. They recalled that "anyone is doomed to disappointment who is looking in Diogenes-like fashion for honest, straightforward calculations of heats of dissociation from the basic postulates of quantum mechanics." And they posed the question "how, then, can it be said that we have a quantum theory of valence?" Their answer left no doubts about the different possible attitudes to be taken in building quantum chemistry:

The answer is that to be satisfied one must adopt the mental attitude and procedure of an optimist rather than a pessimist. The latter demands a rigorous postulational theory, and calculations devoid of any questionable approximations or of empirical appeals to known facts. The optimist, on the other hand, is satisfied with approximate solutions of the wave equation. . . . The pessimist is eternally worried because the omitted terms in the approximations are usually rather large, so that any pretense of rigor should be lacking. The optimist replies that the approximate calculations do nevertheless give one an excellent "steer" and a very good idea of "how things go," permitting the systematization and understanding of what would otherwise be a maze of experimental data codified by purely empirical rules of valence. (Van Vleck and Sherman 1935, 168–169)

Van Vleck and Sherman proceeded to discuss the quantitative success already obtained in the study of the hydrogen molecule, to describe and compare the valence bond (VB) and the molecular orbital (MO) methods, to analyze their relative advantages and difficulties, their interrelationship in terms of group theory, and also Dirac's vector model with considerable detail. In the absence of rigorous calculations, they considered it advantageous to use as many methods of approximation as possible. Therefore, they emphasized the importance of comparing the predictions made by both VB and MO methods and gave as an example of their good agreement the calculation of resonance energies by Pauling and Hückel. They tried to be pragmatic about the two approaches because these represented two different types of approximation, neither one being particularly reliable. Hence "much more confidence can be placed in the results of the two methods when they agree than can otherwise. If certain properties are found to be true under these two different kinds of approximation, warranted under different idealized limiting conditions, it is natural to suppose that the same properties are also valid in the actual, more complicated intermediate case. It is therefore a comfort that both theories predict a nearly right-angled model for water, a tetrahedron for methane, etc. (Van Vleck and Sherman 1935, 214).

Van Vleck and Sherman did not name whom they considered to be the optimists or the pessimists, and in fact they claimed to have adopted a middle ground between the two opposite points of view. However, one cannot but wonder that there are striking similarities between the praxis of some like Mulliken and Pauling and the optimistic persona and the praxis of some like Heitler, London, or Hund and the pessimistic persona depicted in the review. In between both attitudes, Van Vleck and Sherman, as well as Slater, opted not to enroll in partisan stances, playing the necessary but difficult role of the devil's advocate in a time of stark disciplinary contrasts. Van Vleck's different style was even noticed by the angered Heitler in his correspondence with London.

Heitler and London: The Lost Battle

Apart from the letters Heitler and London exchanged in late 1927 about the possibilities offered by group theory, there was no contact between them until 1935, when they started frantically writing to each other. They were both in Britain having had to resign from their positions after the decrees by the Nazi government of April 1933.

In 1933, London was offered a research fellowship funded by the Imperial Chemical Industries at Oxford to collaborate with Frederick Lindemann, Viscount Cherwell, who had received his doctorate in Berlin with Nernst in 1910 and was elected as Dr. Lee's Professor of Experimental Philosophy in 1919. Lindemann was very keen to set up a group in low-temperature physics. He had brought to Oxford F. E. Simon, K. Mendelssohn, and N. Kurti, and in 1934 Fritz London's brother Heinz joined them. A couple of months after Heinz's arrival, the two brothers worked out the electrodynamics of the superconductors and offered a theoretical schema for the explanation of superconductivity—22 years after the phenomenon was first discovered. The research of the Londons was prompted by the discovery of W. Meissner and R. Ochsenfeld in 1933 that, in contrast to all expectations, superconductors were diamagnetic. In view of this result, the Londons considered the expulsion of the magnetic field rather than the infinite conductivity as the fundamental characteristic of superconductors and proceeded to formulate their equations. Hence, the vantage point for superconductivity shifted, and what was considered for more than 20 years to be a phenomenon of infinite conductivity came to be regarded, primarily, as a case of diamagnetism at very low temperatures. Their theory was not a microscopic theory explaining the phenomenon in terms of the dynamics of electrons. In 1935, during a discussion about low-temperature phenomena at the Royal Society of London, London formulated his most radical concept: The superconducting electrons acquired a rigid wave function with a wavelength the size of the superconductor and, hence, because of the uncertainty principle, superconductivity could be understood in terms of order in momentum rather than phase space. According to London, superconductivity became a macroscopic quantum phenomenon. During his stay in Oxford, London started thinking about the structure of helium at absolute zero (Gavroglu 1995).

In 1933, after the Nazis came to power, Heitler went to Bristol (figure 2.4). There was a standing arrangement between Göttingen and Bristol for one of Born's assistants to spend a year at Lennard-Jones's laboratory. Neville Mott had just become the professor of theoretical physics and was delighted with Heitler's going to Bristol. Heitler started working on quantum electrodynamics, and in 1936 he published his well-received *Quantum Theory of Radiation*, which was to become one of the standard books on the subject. By 1935, he had become a member of the staff of the Department of Physics at the University of Bristol.

The publication of the papers by the Americans and especially those by Slater, Pauling, Van Vleck, and Mulliken prompted a rather desperate exchange of letters between Heitler and London starting in the end of 1935. This correspondence is quite revealing. It shows the attitude of each about the possible development of the approach laid down in their common paper, the tension between them, as well as the search for a means to consolidate their theory at a time when the Americans appeared to be taking over the field of quantum chemistry. The correspondence reflects the different style of their respective environments. Faithful to the Göttingen spirit, Heitler was

Figure 2.4
Fritz London (left) and Walter Heitler (right) near Bristol, England, in 1934 or 1935.
Source: Courtesy of Edith London.

"more mathematical," London continued in the Berlin tradition of theoretical physics with its inclination to examine intuitive proposals. They discussed the possibility of writing an article in *Nature* to present their old results and include some new aspects that had not been emphasized properly in their earlier papers. These were the activation of spin valence and the possibility of a bond that will not depend on spin saturation. "That is what I meant in a past note—vaguely and wrongly—with the term orbital valence."[125]

Heitler's attitude concerning the approach by Slater and Pauling was that they were correct about the principles they adopted, and he was quite sympathetic about the direction of their research, even though a series of results did not follow exclusively from their theory. He thought a polemic against them to be quite unjustified: "I simply find that the importance of this theory has been monstrously overrated in America."[126]

For the first time, both Heitler and London expressed doubts about the character of the attractive forces. It was conceivable that these forces may not be only due to spin. There were other attractive forces of the same order of magnitude as the usual ones, and those did not follow from their original theory of spin valence. It was *wrong* to believe that these forces could originate only from the directional degeneracy of the ground state. It may be the case that these forces resulted in the formation of a molecule only if there were also spin valences. At this point, nothing much could be said about the claim that these forces did not have the characteristics of valence. They admitted that they did not know much about what happened when more than two atoms were near each other, because the mechanism of these forces differed from the mechanism of spin valences[127]:

The next question is whether one should consider these forces, that are added to our original ones, as *valence* forces. Well, the chemists undoubtedly do it, since they name, or, rather, they named in this way whatever gives molecules (in contrast to the v.d.Waals forces and the pure ionic molecules). This is exactly our job. To say *that* there are also other forces of molecule formation, beyond our old ones, and *which* phenomena of chemical valence depend on those, and that our old scheme can be extended.

Heitler's feeling was that there had been no attack against them by the Americans except for the case of the oxygen molecule whose supposed diamagnetism they could not explain[128]:

The nucleus of our theory is the spin valence and our theory is the only one that explains the mechanism of repulsion in a qualitatively exact manner. It is needless to write this since we surely agree on that. You could perhaps include the above discussion under the title: *Delineation of completeness* (so much of theory as well as of the chemical notion of valence that corresponds to theory). In any case, we should stress that the extension could be realized on the basis of our theory and, substantially, it includes whatever one could wish (this last thing only as a footnote for us). It is ridiculous even from a quantitative point of view.

London's answer was not exactly a eulogy to the chemical profession.[129]

The word "valence" means for the chemist *something more than simply forces of molecular formation*. For him it means a substitute for these forces whose aim is to free him from the necessity to proceed, in complicated cases, by calculations deep into the model. It is clear that this remains wishful thinking. Also the fact that it has certain heuristic successes. We can, also, show the quantum mechanical framework of this success . . . the chemist is made out of hard wood and he needs to have rules even if they are incomprehensible.

They were progressively realizing that part of the problem was their isolation, and this realization bred even more frustration. The fact that they had not even been attacked was not an indication of the acceptance of their theory. Their feeling was that their theory may have even been forgotten or that it "can be combated much more effectively by the conscious failure to appreciate and avoid mentioning it."[130] Heitler did not agree with London that their theory "is fought by the most unfair and secretive means."[131]

It may be true for some people in America. Not all people, however, are rascals (e.g. I would not believe it for Van Vleck), but only silly and lazy. And we should accept that our theory was quite complicated. I would gladly like to look at the books of Sidgwick and Pauling. I cannot get them here.

To write a book, as London suggested, did not find Heitler in full agreement, especially as he had just finished writing one (Heitler 1936). But his main worry was that they may not have many new things to include.[132] Clearly, Heitler thought that London was being too paranoid. To be more convincing, London asked him to look up the assessment of their work by two other American writers, Mulliken and R. Kronig, and implored him to "judge for yourself whether we are neglecting something or not, when we leave unanswered these kinds of distortions. And they are not at all isolated cases."[133]

Mulliken's main objection was methodological. The approach of Heitler and London required long calculations in order to make quantitative predictions, but "qualitative predictions can usually be made much more easily by a consideration of electron configurations of atoms and molecules" (Mulliken 1932, 30). It was these kinds of pronouncements that deeply angered London. He did not mind there being a theory "superior" to his own approach, but one had to play the game according to the rules and not devise new rules along the way. So much the worse when these rules were nothing but rationalizations of experimental data! Some years later he would be furious when he thought that Lev D. Landau was doing the same thing in superfluidity.

Still, Mulliken in 1928 had made some attempts to give due credit to the work of Heitler and London. He considered their joint work and the subsequent papers using group theory together with Hund's papers as promising. "at last a suitable theoretical foundation for an understanding of the problems of valence and of the structure and stability of molecules" (Mulliken 1928, 189). And in his next paper, he mentioned the agreement of some of his results with those of Heitler (Mulliken 1928a). A year later in a presentation of London's group theoretical approach, he thought that London's theory was a translation of Lewis's theory into quantum mechanical language (Mulliken 1929a). But such credit slowly waned. In 1933, he did not refer at all to the Heitler–London paper but rather to the theory of Slater and Pauling, which together with the molecular orbital approach was considered to illuminate Lewis's theory from more or less complementary directions (Mulliken 1933a). In his Nobel lecture of 1966, he referred to the paper as merely initiating an alternative approach to the molecular orbital method. He did not even recognize that it provided the quantum mechanical explanation of Lewis's schema, as the "electrons in the chemical molecular orbitals represent the closest possible quantum mechanical counterpart to Lewis's beautiful pre-quantum valence theory" (Mulliken 1966, 142). What was, however, rather odd was that London nowhere mentioned Mulliken's 1931 article in *Chemical Reviews* where Mulliken expressed in a detailed manner his objections to the Heitler–London method *and* theory.

The appearance of Kronig's book (1935) did nothing to alleviate the feelings of London that they should take a strong stand against distortions of their theory. Kronig's book written almost exclusively from a "chemist's viewpoint" was indeed quite harsh toward Heitler and London and welcomed the approach of Slater and Pauling. Kronig mentioned quite a few shortcomings of the Heitler–London approach. It could not deal successfully with atoms that were not in their ground state. It was not possible to explain the numerous compounds between oxygen, sulfur, and the halogens. The calculations were only in first approximation, and this made many of the results doubtful because near the normal states of the interacting atoms there were other states. Though it was recognized that the Slater–Pauling approach gave the same results as that of Heitler and London for atoms with only s electrons outside closed shells, its unquestionable advantage was the interpretation of the directed nature of valence bonds in the case of the p electrons. For the Slater–Pauling approach "the mathematical procedure is again a perturbation calculus starting from atoms in the limiting case of infinite separation, but the criticism of its applicability was not as severe as for the Heitler-London theory since all the low-lying atomic states are taken equally into account" (Kronig 1935, 201).

Heitler visited London at Oxford at the beginning of December 1935. Both were now fully aware that the Americans were starting to dominate the field. As soon as Heitler was back in Bristol, he read a paper by Wheland where the following passage exhausted the last vestiges of tolerance displayed by the more "objective" of the two:

The Heitler-London-Slater-Pauling (HLSP) method. This method, which was developed originally by Slater as a generalization of Heitler and London's treatment of the hydrogen molecule, was first applied to aromatic compounds by E. Hückel, but has since been greatly simplified and extended by Pauling and his co-workers. (Wheland 1934, 474)

Heitler was vitriolic in his response to London.[134]

I propose in the future to talk only about the theory of Slater-Pauling of the chemical bond, since, in the last analysis, the H_2, well now—what can this be compared with the feats of the Americans. I am afraid that the reading of the papers that we have voluntarily undertaken shall be the purgatorium of our souls. If you cannot restrain me, I think, I will write a very clear letter to this Pauling (he should give a better upbringing to his students). I think that you are right that we should publish in the blue journal. It would be really good to write something which will mostly have those things that are stealing in America. Do not think I am exasperated because (in the case of Pauling) it involves my paper with Rumer, but because of our common cause. Your achievements disappear equally in the lies . . . For Van Vleck I notice that his papers are more dignified than his report and does not thank Slater and Pauling for free.

They decided to find an excuse to write to Pauling admitting that the work of Pauling and Slater did, in fact, go beyond the version of their original theory. London took it upon himself to read carefully their papers: "We should find many points where it will be *evident* that the passages were written in bad faith . . . The best thing would be to have as an excuse a substantial question or a criticism to Pauling's papers."[135] Slater's "shameless behavior starts from 1931." In his fundamental work he claimed that the theory of Heitler and Rumer was valid only when the bond energy was small with respect to multiplet dissociation and, therefore, it had no physical meaning. This was not correct, and it was because Slater confused multiplet dissociation with the separation of terms via the Coulomb interaction. Heitler, then, made a specific proposal to London.[136]

The local chemists, in hordes, torment me with that wretched B_2H_6. It is a typical case where there should be special reasons for bonding. The examination of this reason would be useful for the following reasons: 1) the opportunity is given to underline your view that it is possible to exist special forces, but what is *generally valid* is the formation of pairs. 2) it would have impressed many chemists. 3) it would let the wind out of the sails of certain ill intentioned or silly people.

London's reaction was to clarify the situation with the oxygen bond, first, and they planned to meet in London in mid-February. They were both back to study chemistry. Heitler criticized Pauling's theory as follows[137]:

I was looking for ways to devour the so-called theory of Sl[ater]-P[auling]. These types are so proud about something which is not so bad, but which, under no circumstances, is so distinguished. It gives a *general formula* for the bond that corresponds to the pair bonding and the repulsion of the valence lines. The bonding energy is additive, and the directional properties are included. The approximation is as rough as in my semi-classical theory (without such mathematical gurus), but it is surpassed since it includes the directional properties. One, however, totally loses 1) the activation energy, 2) the non-additivity of the bond energy.

It is needless to say that it is fully based on our ideas . . . We should not, though, fall into the error and regard this work bad or insignificant (as *these* people do). It is a branching from our work, from about the point where we strictly suppose that the atoms are in only one state . . . Generally, I believe that we did the mistake not to give *more concrete* applications of the theory, it was a mistake to leave it to the chemists (who are nearer to this kind of work) or to types like Eyring. I do not find, though, that our direction is not being given any attention (apart from the details) in Europe.

There are not many places where we can read the opinions of either of them concerning the molecular orbital approach. Heitler thought that their basic objection with "Hund's People"—who both agree are not the bigger and most unpleasant enemies—was not related so much to the actual results derived by this method. Sufficient patience with the calculations and a lot of semiempirical considerations gave, in fact, correct results: "Nevertheless, no one could name this a general theory—much less a valence theory—since all the *general and substantive* points are forever lost."[138]

After Born had himself expressed the intention to write an article on the valence bond method, Heitler and London were seriously thinking of asking him to write it "since it is very difficult for us to correct the situation with the necessary emphasis." London reported to Born all the developments and wrote to Heitler details of who he informed and with whom he talked so that Heitler was sufficiently knowledgeable about everything, as he was the "sole representative of our enterprise in England"[139]—London, in the meantime, having moved to Paris.

Lennard-Jones in all his publications preferred the "one-electron-orbital-bonds" and presented the version by Heitler and London "as not so beautiful and as inadequate." London asked Born's advice on how to proceed and get out of the quagmire he feels they are in.[140]

Maybe it was a mistake that we never expressed the objections we had from the beginning on questions of principle concerning the approach of Lennard-Jones-Mulliken. Both of us thought it as superfluous, because we had both "transcended" this same phase of Lennard-Jones-Mulliken in the beginning of our observations in 1927, and we were very proud when we realized that we get the exchange degeneracy because of the similarity of the *electrons*. For this reason we thought as totally evident the self-destruction of the approach by Lennard-Jones-Mulliken, and maybe, for this reason we did not take it seriously. Recently, I talk very often with Heitler about this lost ground and repeatedly we tried to find a way to make up for it. We continuously fail and I think that we can do something more that has the weight that interests here. We have, undoubtedly, made a mistake by not taking seriously our competitors . . . The situation had become clear since 1932–33 when we should have thought to find new issues and not make enemies with our polemics.

Born did not wish to publish anything about the problems of valence, as he had not followed closely the developments. But he thought that it was absolutely necessary that London and Heitler take a position and publish something that would be accessible to chemists. He even promised that he would encourage his publisher to have a new series and that they be the first to write something about the chemical bond.[141] At long last they realized that "in the last analysis the pressure to do what is necessary falls on us. What is needed to keep the more dangerous of our colleagues, those, in other words, who work with our method, from falsifying history (Eyring, Pauling, etc.) in their place, is a good standard book. Would you not want to write it?"[142] Oxford University Press suggested that Heitler write another book, especially after the success of the book on radiation, and, now, he toyed with the idea of writing one with London about quantum mechanics and chemistry.

Then suddenly, as if by magic, there was no more talk about these issues—maybe all the reading and the discussions did indeed become the "purgatorium of our souls" as Heitler had suggested. London's move to Paris and the incomparably more pleasant atmosphere there in comparison with Oxford; Heitler's success with his book and his work in quantum electrodynamics; London's success with the theory of superconductivity . . . somehow, one cannot help but feel that both of them could now afford to be gracious.

Why is it that Heitler and London never managed to co-author an article or a book together? Obviously, their different scientific interests after 1933, their wanderings, and professional insecurity did play a role. Hard feelings and misunderstandings about the way each started publishing in group theory were overcome by 1933, so in 1935–1936 they were not a factor in undermining a possible collaboration. The main reason was a difference in their views about the role of physics in such an approach. Heitler's strong reductionism is evident in all his writings. The last paragraph of his book *Elementary Wave Mechanics* reads: "Reviewing the contents of the last three chapters it can be said that wave mechanics is the tool for a complete understanding, on a physical basis, of all the fundamental facts of chemistry" (Heitler 1956, 190). The understanding of chemistry commanded that its phenomena should be understood in terms of physical laws. "Thus the two sciences of physics and chemistry were amalgamated" (Heitler 1967, 14, 35).

But such an approach was particularly unappealing to London and not even its success could convince him. Despite the fact that their joint paper was a "classic" example of a reductionist approach, in London's subsequent work in quantum chemistry there is a confluence of timidly articulated trends expressing London's search for an alternative nonreductionist *approach*. One senses him trying to find the fringes of a net he had so successfully woven in his school essays and, especially, in his doctoral thesis. It is ironic that this first important paper of his was such a pronounced deviation from his grand schema to view theories as wholes! But, in the ensuing years after his correspondence with Heitler, London was immersed in the world of very-low-temperature physics dealing with two of the most intriguing phenomena: superconductivity and superfluidity. He was beginning to weave a new net to accommodate the notion of a *macroscopic quantum phenomenon* and move away from a reductionist view (Gavroglu 1995).

The epilogue is characteristically ironic. In a letter sent by Heitler in 1951 to thank London for sending him the first volume of his *Superfluids* (1954), he talked of life in Zürich after nearly 25 years since the time they were both there. "It may interest you that these days Cafe Globus has been demolished . . . I have not noticed whether they have put a plate with 'The chemical bond was born here.'"[143]

Legitimation through Pedagogical Considerations

Textbooks have always played a rather dominant role in the early stages of the development of subdisciplines: By formalizing the "principles" of a subdiscipline, making explicit the solutions to hitherto unsolved problems, reviewing the state of the field, codifying what there is to be taught, and giving background information for nonexperts to learn about the field, the early textbooks in a subdiscipline's history contribute to the legitimization and institutionalization of the field (Gavroglu and Simões 2000; Nye 2000a; Simões 2004, 2008a; Park 2005).

The development of quantum chemistry has been no exception. The contents of textbooks in general are—necessarily—ahistorical, and only in very few instances do we find a mention and, in even fewer cases, a discussion of some of the disputes in a discipline's early history. Notably, the early textbooks of quantum chemistry can also be read as polemical or partisan texts: By proposing and arguing in favor of particular (ontological) hypotheses and approximation methods, each one of them adopts a particular viewpoint on how the question of whether quantum chemistry is an application or use of quantum mechanics for chemical problems is to be answered.

Early textbooks in a discipline's history could also be viewed as a genre for consolidating a consensus as to the language to be used and the practice to be adopted. In the case of quantum chemistry, such an agenda revolved around the question of whether chemists should start diverging from the accepted norms of their disciplinary culture where chemistry is not thought of as a mathematical science or whether they should continue to be faithful to such a culture and appropriate the right dose of quantum mechanics for their own purposes. The dilemma, then, of whether chemists should apply quantum mechanics to chemical problems or use quantum mechanics in chemistry, and the ensuing issues as to the extent of mathematics to be introduced, was really a dilemma concerning the status of quantum chemistry, the question, that is, about the extent of its relative autonomy with respect to physics. In 1939, two important textbooks—Pauling's *The Nature of the Chemical Bond* and Slater's *Introduction to Chemical Physics*—came out. These two articulate writers who were among the founders of the new discipline aimed—by adopting different viewpoints—at educating an audience of students as well as professionals in the ways of the new discipline.

Pauling, the chemist, proceeded to a reform of the whole of chemistry from the standpoint of quantum chemistry. Slater, the physicist, saw the beginnings of quantum chemistry, which he christened chemical physics, as heralding the unification of physics and chemistry. Both books reflected the tendency to impose a new (sub)discipline by establishing a new language, a new practice, a new theoretical agenda, and a concomitant methodology, and finally by securing an audience. That was no longer the case with another textbook. The organic chemist Wheland, one of Pauling's former students and his long-time collaborator, contributed more than any other chemist toward the extension of the scope of the theory of resonance to organic chemistry. He adopted Pauling's research agenda and pushed it ahead by arguing that it is possible for organic chemists to use quantum chemistry without having to turn their discipline into a fully mathematized science. In fact, his textbook *The Theory of Resonance and its Application to Organic Chemistry* published in 1944 was to play a prominent role in the education of organic chemists. But it was a British chemist, Nevil Vincent Sidgwick, who wrote the first textbook to present systematically the virtues of resonance.

Sidgwick's Role in the Popularization of Resonance Theory

Together with Ralph Howard Fowler, Nevil Vincent Sidgwick (1873–1952) played a leading role in the emergence and consolidation of quantum chemistry both in the United Kingdom and the United States.

Sidgwick entered the University of Oxford in 1892. His tutor was Vernon Harcourt, a pioneer in the study of kinetic mechanisms in chemical reactions. In 1895, Sidgwick was placed in the First Class of the Honor School of Natural Sciences, and quite unusually he studied *literae humaniores*, in which he also gained another First Class, largely due to his performance in philosophy.[144] He then went to Germany and studied physical chemistry under Ostwald in Leipzig and organic chemistry under Hans von Pechmann at Tübingen. Upon his return to Oxford, he was elected a fellow of Lincoln College, where he went into residence in 1901 and remained there for the rest of his life, visiting many times the United States. Sidgwick always considered himself a chemist, and his academic positions were all in the field of chemistry. In 1922, he was elected a fellow of the Royal Society of London. He was president of the Faraday Society from 1932 to 1934 and president of the Chemical Society from 1935 to 1937. In 1935, he was appointed Commander of the Order of the British Empire (CBE) and was elected by the University of Oxford to a supernumerary Professorship of Chemistry.

Sidgwick's early scientific interests included the kinetics of organic reactions and the relation of solubility and chemical structure. His first book *Organic Chemistry of Nitrogen* was published in 1910. Here, as in future works, he applied the methods of physical chemistry to organic chemistry in a systematic and thorough way. In the preface, he warned readers that organic chemistry could no longer be treated satisfactorily without reference to the relevant questions related to physical chemistry. His innovative methodological approach accounted for the fact that many of his fellow chemists in Oxford dubbed him an "organic chemist gone wrong" (Tizard 1954, 242). In Oxford where the teaching and research in chemistry was divided among organic, inorganic, and physical chemistry, Sidgwick became the unifying agent among the different areas. His approach to chemistry was revealed in his publications, his colloquia, in the role he played at the Alembic Club, the university's chemical society, and, also, in his interventions at the Dyson Perrins tea club of the organic chemistry laboratory (Roche 1994). It was his friendship with Ernest Rutherford (whom he met in 1914) that awakened his interest in atomic and molecular structure. Later on, Bohr's book the *Theory of Spectra and Atomic Constitution* (1924), as well as Lewis' book *Valence and the Structure of Atoms and Molecules* (1923) set him to work on the electronic constitution of chemical compounds. He decided to bring together the ideas of the chemists and those of the new physics (Sidgwick 1923a).

In June 1923, Lewis visited Oxford, stayed with Sidgwick, and they both attended the meeting of the British Association for the Advancement of Science held in Liverpool. Sidgwick opened the discussion with a paper titled "The Bohr Atom and the Periodic Law." In July of the same year, they were both among the contributors to the 1923 Faraday Society Meeting held in Cambridge.

In 1927, Sidgwick published *The Electronic Theory of Valency*, a book whose organization and contents mirrored Lewis's *Valence*. A second part of this book was expected to come soon, but the intended scope was so large that the book appeared in 1950. It was called *Chemical Elements and their Compounds* and was a detailed compilation of all the evidence as to the properties of elements and their compounds.

When the *Electronic Theory of Valency* appeared, many chemists had become aware of the amazing explanatory power of the new quantum mechanics, yet it was difficult to see how this newly developing explanatory framework would be assimilated into the chemists' culture. Many feared that such assimilation might bring lasting changes to their culture and that there was a danger that chemistry might soon be reduced to physics. For Sidgwick, it was, nevertheless, a "risk" worth taking. He did not have any inhibitions about letting the new quantum mechanics invade the realm of chemistry. He expressed an unreserved enthusiasm about the new quantum mechanics and embraced wholeheartedly Lewis's theory of the nonpolar bond.

Confronting the developments of the new mechanics, but not yet its application to chemical problems, Sidgwick in the very first lines of the preface to his book attempted to clarify the methodological stumbling block that he sensed to be in the way of his fellow chemists. He considered that the chemist could follow two different courses of action in developing a theory of valence. He could use symbols with no definite physical connotation to express the combination of atoms in a molecule or he could adopt from start the concepts of atomic physics and use them in the explanation of known chemical properties. This latter option was, of course, the one advocated by Sidgwick, who, however, was eager to point out its full implications: "[the chemist] must not use the terminology of physics unless he is prepared to recognise its laws" (Sidgwick 1927, preface).

In 1931, Sidgwick joined the faculty of Cornell University as the George Fisher Baker Non-resident Lecturer. At Cornell, he decided to lecture on the covalent link in chemistry, focusing on the physical methods used to measure the more important properties of those bonds, such as heats of formation, dimensions, and electric dipole moments. His lectures were later published in the book *Some Physical Properties of the Covalent Link in Chemistry* (1933).

In this new textbook, Sidgwick made a very promising assessment of the methodological guidelines to be followed by the emerging discipline of quantum chemistry. Among the methods surveyed, Sidgwick discussed analytically the method of the electric dipole moments. More than one third of the book dealt with the different methods of measuring dipole moments, providing a survey of the literature on the subject. Lewis had considered that covalent bonds gave rise to nonpolar molecules, though Debye had shown that this was too narrow an interpretation of the chemical facts. It was, however, in the 1928 meeting of the Bunsen Gesselschaft in Munich, where Sidgwick played a leading role, that the significance of dipole moments for the clarification of molecular structure began to emerge. Debye wrote Sidgwick approvingly: "I have read the book from the beginning to the end and in doing so have learned a lot in a most delightful way."[145] In the book Sidgwick contended that there were basically two different types of bonds in molecules, which in practice could be sharply distinguished, and that bonds that are intermediate between electrovalent and covalent rarely exist. He lectured on the same topic in Philadelphia in 1931.[146]

The introductory lecture included in the book is in sharp contrast with the rest. Sidgwick discusses a number of quasi-philosophical questions he considered to be of utmost importance for the clarification of the relations of physics to chemistry, a topic occupying a central position in the assessment of the status of quantum chemistry in all future discussions over the extent of its relative autonomy with respect to both physics and chemistry. He specifically discussed what he considered to be the main difference between a chemical and a physical theory: "a chemical theory, dealing with more complicated phenomena, is less accessible to mechanical treatment. It takes account in the first instance of properties which can not be measured quantitatively, but which are clearly shown to exist" (Sidgwick 1933, 13). In the structure theory of organic chemistry, which he considered to be the paradigm of a chemical theory, chemists started by assuming the existence of chemical bonds, without any associated idea about their physical origin. By induction from the experimental data, they were able to correlate the structure of molecules with their properties, but up to the first decade of the 20th century, they had avoided any statements concerning the physical origin of valence forces. Things were changing. The chemist was being challenged to supplement his chemical information by considerations as to the physical mechanisms responsible for bonds—always obeying the constraints imposed by physics.

Sidgwick believed that chemists should use all the physical means available to them for the solution of chemical problems. They should collect and articulate chemical data in order to supply to the physicist "in a simplified form those questions arising out of our chemical experience which he is best able to solve." Physicists, in contrast, need the help of the chemists because the inherent complexity of most chemical problems forbade their solution by deductive reasoning alone. But "the chemist must not employ the language of physics unless he is willing to accept its laws. . . . The chemist must resist the temptation to make his own physics; if he does, it will be bad physics—just as the physicist has sometimes been tempted to make his own chemistry, and then it is bad chemistry" (Sidgwick 1933, 16).

In the future, so Sidgwick ventured, earlier chemical theories should incorporate the new discoveries concerning the physical structure and behavior of atoms, use the language of physics in translating former chemical concepts, and in the process reassess the question of the relative autonomy of chemistry vis-à-vis physics. Of course, for the older generation of organic chemists this was too daring a step to make. When Armstrong died, Rutherford wrote Sidgwick and said that he had been told that "Armstrong had never got beyond arithmetic, and that even algebraic symbols were Greek to him. This may account for his attitude to all mathematical theory."[147]

Sidgwick became Pauling's close friend and an immediate convert and enthusiastic follower of the resonance theory of atomic bonding and molecule formation. Although he did not make any original contributions of his own to the quantum mechanical resonance theory of valence, he played an active role in popularizing it, becoming one of the most effective expositors of the theory of resonance. Sidgwick viewed resonance theory as a translation of the older structure theory of organic chemistry into the language of the new quantum mechanics. He believed it was his task to communicate to the community of organic chemists the nuances of the resonance theory.

Another, related, problem was also worrying him. In a letter to *Nature* titled "Wave Mechanics and Structural Chemistry" (1934), he tried to reconcile the view of the molecular orbital approach with the prevalent views of the organic chemists. In the molecular orbital approach, the molecule is treated as a whole and not necessarily as the union of atoms. Organic chemists assumed that bonds are always localized so that their method of representation could, at times, provide two different formulas for one substance. Sidgwick tried to reconcile both viewpoints: "If these views are both true, it follows that if a molecule with one structural formula can have (in the sense of molecular orbital theory) more than one electronic constitution, these must be able to change into one another in less time than is required to isolate the substance" (Sidgwick 1934, 530).

Not long after the letter to *Nature*, Sidgwick asked Fritz London his view on a number of issues concerning resonance theory. London offered his own interpretation of the meaning of resonance, a concept appropriated from quantum mechanics often erroneously conflated with tautomerism.[148] He tried to show how misleading were the attempts to visualize a process behind the concept[149]:

Firstly there is a misunderstanding of Pauling's last papers, which it is true is suggested by Pauling's expressions. There is not the idea of a real periodical phenomenon, when he speaks of "resonance." It has directly nothing to do with the conception of moving atoms. Pauling wishes only to express that the stationary state of a configuration of electrons with fixed nuclei cannot always be represented by only one eigenfunction corresponding to an electronic structure of the Lewis type. In these cases—e.g. Benzene—the eigenfunction may be represented as a superposition of several Lewis structures, but that construction is only a matter of mathematical representation; the state is as stationary as any other stationary state. Important is that in this case the single Lewis structures are not stationary and, therefore, they cannot have sharp definite energy values; only a statistic of energy can be given of them; for instance the average of energy of your isomeric structures A and B can be spoken of.

It was not particularly easy to come to a consensus on what resonance is. Leslie E. Sutton, who was one of Sidgwick's students, spent an extended period with Pauling at the California Institute of Technology in Pasadena in 1933–1934. He did not understand exactly how resonance arose between structures that were not equivalent but had about the same energy. He was convinced, though, that Pauling's demonstration was strictly mathematical. In a letter to George Hampson, one of his Oxford colleagues and first collaborators, he confided[150]:

The non-mathematical chemist can develop a "feeling" for resonance, however, which will enable him to make intelligent use of it, and I hope that I am beginning to do this. I believe that what I said about the cause of resonance is correct, but I don't understand very clearly what it means. Nevertheless I was able to use the idea, as you saw. The various component structures can be worked out by the non-mathematical chemist if he follows certain simple, half-empirical rules which ensure that the structures he takes have somewhere about the same energy (the range can be two or three volt-electrons, I believe). We always consider the ground state in ordinary structural questions, so that in practice only simple non-excited structures and the lowest excited ones will be important . . .

Sidgwick considered it his main duty to address first and foremost his fellow organic chemists and to educate them in the ways of the new resonance theory. He did so in his many review papers, annual reports, talks, and in presidential addresses.[151] His unique ability to simplify many complex mathematical arguments for those chemists least receptive to mathematics as was the case with organic chemists, and his status in the chemical community, first as president of the Faraday Society and then as president of the Chemical Society, won many converts to the new theory. In the talk on "Some Problems in Molecular Structure," he attempted to reassess the classical structural theory of chemistry in light of recent knowledge.[152] He reviewed the successes and failures of the classical structure theory and recalled that the physical evidence for the mechanism of atomic bonding was derived from wave mechanics. Schrödinger's equation gave a complete account of the properties of each molecule, but for any but a few of the simplest molecules, its solution was so lengthy that it could not be carried out in practice. A variety of shortcuts and approximations enabled one to arrive at more than just a qualitative answer to the problem, and in fact one could get a general picture of the mechanism that is responsible for holding the atoms together in a molecule. However, from the modern physical point of view, the classical structure theory appeared to be very imperfect as it only took account of the interaction between one atom and its neighbor, whereas physics demanded that other atoms in the molecule had, also, a role to play. But in many other respects the structural theory was a good one, as, for example, when it came to the prediction of the number of isomers that can exist. The question Sidgwick wanted to clarify was to understand how an admittedly imperfect theory should be found perfect in this respect. In the course of the explanation, one was led to analyze the role of stereochemistry and the meaning of resonance.

In the 1936 presidential address to the Chemical Society, Sidgwick once again reviewed the major successes and failures of structure theory and described the ways in which recent physical methods had come to deepen the meaning of the former concept of a bond. He even talked about progress in science, which he considered to be the outcome of establishing "existing doctrines on a firmer foundation" and giving "them a deeper meaning" (Sidgwick 1936, 533).

Sidgwick then discussed how new principles of structure, such as resonance, enabled theorists to go much further. He stressed that resonance was a bad choice of name, but one had to comply with the "terminology of the physicists." This was exactly Pauling's attitude. He concluded by reassuring his fellow chemists that resonance theory "far from destroying the older doctrine [structural theory], has given it a longer and fuller life" (Sidgwick 1936, 538). Pauling could not be happier: "I have read your very interesting review of structural chemistry in your presidential address with great pleasure. I am indeed glad to know that you consider the ideas of resonance worthwhile and especially that the new methods of attack on resonance problems seem to you to be yielding good results."[153]

Sidgwick returned to the topic less than a year later in another presidential address to the Chemical Society to stress how "the general conclusions of the theory of resonance are of great practical importance, especially to the organic chemist" (Sidgwick 1937, 694). He did not expect in general organic chemists to have the time, patience, and ability nor the need to master the detailed mathematical operations of the theory, but he reassured them that it was not necessary to handle the technical language of wave mechanics in order to understand its conclusions.

Sidgwick's arguments to convince organic chemists to accept and adopt resonance theory emphasized again and again that resonance was a name detrimental to the aims of the theory, that the component structures were not real, and that resonance should not be confused with tautomerism. His task was facilitated by the existence of British research schools of organic chemists sympathetic to physical methods and techniques (Nye 1993). The explanation of several properties of organic compounds (occurrence of tautomeric forms, low dielectric constants, and low electric conductivities) was an important stimulus for the analysis of different models of chemical bonding, and later on, it led to the proposal of reaction mechanisms in carbon compounds. If physics and, especially, quantum mechanics enabled researchers to understand many of the properties and behavior of carbon compounds, organic chemistry was to constitute a sort of "epistemological laboratory" for quantum chemists.

Slater: A Project of Bringing Physics and Chemistry Together

Slater's contributions to quantum chemistry ran parallel with the implementation of a theoretical agenda that aimed at bringing physics and chemistry together, and which is nowhere expressed more coherently than in his textbook *Introduction to Chemical Physics* (1939).

Two episodes in Slater's career are quite illuminating. One was the move to MIT as head of the Department of Physics, the other his attempt to lure Pauling to join him at MIT. When Compton decided to leave Princeton and become president of MIT, he believed that conditions were favorable to strengthen the fundamental sciences at the institute,[154] an enterprise he considered to depend heavily on the total reorganization of physics. He chose Slater to head that process, a "gamble" that Slater accepted after some hesitation and attempts from Bridgman to deter him.[155] In 1931, Slater became a very young full professor and head of the Department of Physics at MIT.

In a couple of months, he prepared a project for a "Research Program in Physics."[156] The first priority was to strengthen theoretical physics, a very recent area in the American context, for "the new field of atomic physics is one with a difficult mathematical background requiring much theoretical work to tie it up with experiment." Then, build a strong experimental group—with lines of research in spectroscopy, discharges and arcs, X-ray crystallography and properties of dielectrics—able to cooperate effectively with the theoretical group. Also, cooperation was to be promoted between physics and chemistry, by means of joint projects, and enhanced by the construction of a new research institute in physics and chemistry. He noted that, at MIT, chemistry and physics were becoming indistinguishable, and he had only praise for the work done in physical chemistry under Noyes (Servos 1990). He stressed the importance of the work in the equation of state of gases for physics especially as atomic theory was becoming capable of dealing with these problems theoretically. There was already a combined program under way between the theoretical physicists and the chemists on this subject, and "the Institute should develop into one of the principle [sic] centers for this field."[157] He noted that physics was reaching out into all branches of chemistry, and that several members of the chemical staff, for example in photochemistry and organic chemistry, had begun considering combined research programs with the theoretical physicists and the spectroscopists.

Having accepted Compton's challenge, Slater tried immediately to get Pauling to join his team. Previously, while still at Harvard and undecided as to his future, Slater had hoped Pauling might join the Department of Chemistry, a factor that would increase considerably his willingness to stay at Harvard. Then, in a letter to his parents, Slater commented that Pauling called himself a chemist, but his actual line of thought and research was very similar to his own.[158] Now, he attempted once more to attract Pauling. In the formal invitation sent to Pauling, Slater reiterated his hope that Cambridge would become the "scientific center of the country"[159] and offered Pauling a full professorship in physics, chemistry, or both, depending on Pauling's preferences. Slater once more recalled how the Department of Chemistry had already "a strong leaning toward physics" and added: "I am finding that I work with the chemists a great deal, and they are very co-operative, and well informed on modern physics. The Department of Physics, particularly in the two fields of theoretical physics and crystal structure, would fit in beautifully with your interests."[160] Although Pauling was attracted by the prospects of working with Slater on the structural problems that interested both, and even while confessing that "there is no theoretical physicist whose work interests me more than yours,"[161] he decided to stay at Caltech. However, Pauling visited MIT during April and May 1932, and his stay overlapped with Debye's. This had been part of Slater's plan to make "the general field of structure of solids and molecules rather a feature of the spring semester."[162]

The publication of the textbook *Introduction to Chemical Physics* in 1939 should therefore be seen as the culmination of Slater's project of unification of the two sciences of physics and chemistry. Together with the companion volume *Introduction to Theoretical Physics* (1933) by Slater and Nathaniel Hermann Frank, the two textbooks stood as the pedagogical complement to Slater's academic agenda and administrative activities. *Introduction to Chemical Physics* was an attempt to bridge the gap that had grown in the past between a largely empirical and nonmathematical chemistry and a discipline of physics so far unable to deal with atomic forces. This awkward situation was largely attributed to different scientific traditions and practices, but not to essentially different subject matters. Slater pointed out that things started to change with the development of physical chemistry, statistical mechanics, and kinetic theory, and finally with the appearance of quantum theory. The two sciences were welding together through an exchange in ideas, concepts, and even apparatus with ever more physical instruments finding their ways into chemical laboratories.

Chemists and physicists should start therefore to have a common scientific education in an overlapping field: "For want of a better name, since Physical Chemistry is already preempted, we may call this common field Chemical Physics" (Slater 1939, v). A serious study of chemical physics should start, according to Slater, by a discussion of the fundamental principles of mechanics, electromagnetism, followed by quantum theory and wave mechanics. In this way, the scientist was prepared to attack the structure of atoms and molecules. For the understanding of large collections of molecules, thermodynamics and statistical mechanics were needed, and at last, one could proceed to the discussion of different states of matter and "the explanation of its physical and chemical properties in terms of physical principles." Part of the topics had been already addressed in the companion volume, so that the strategy in this textbook was to offer "the maximum knowledge of chemical physics with the minimum of theory" (Slater 1939, vi).

Quantum chemistry would become the (necessary) intermediary in the metamorphosis of the current scientist into the scientist of the future, who was idealized as neither a physicist nor a chemist, but a sort of hybrid of the two. Such a scientist would transcend the typical physicist or chemist and needed training in empirical chemistry, in physical chemistry, in metallurgy, in crystal structure, as well as in theoretical physics, including mechanics and electromagnetic theory, and in particular in quantum theory, wave mechanics, the structure of atoms and molecules, in thermodynamics, statistical mechanics, and finally in what Slater called chemical physics.

Slater was one of the first scientists to realize that in qualitative discussions of molecule formation, the best procedure to follow was to compare critically the results of the two different methods developed—the valence bond and the molecular orbital viewpoint, and to point out, as already mentioned, that the choice between the two should be made on the basis of convenience rather than correctness. However, in *Introduction to Chemical Physics*, just a few lines are devoted to such a central topic in the context of Slater's contributions to quantum chemistry as well as in the development of the discipline itself. In the chapter on "Interatomic and Intermolecular Forces," in the section about "Exchange Interactions Between Atoms and Molecules," Slater stated that the problems of quantum chemistry are among the most complicated of quantum theory and that the theory itself will not be treated in an analytical manner. He considered the Heitler–London approach and the molecular orbital approach as two different approximate methods of calculation used in wave mechanics. He believed that these two methods do not differ in their "fundamentals, but in the precise nature of the analytical steps used." And he proposed to study "the fundamental physical processes behind the intermolecular actions and we shall find that we can understand them in terms of fundamental principles, without reference to exact methods of calculation" (Slater 1939, 368).

This last sentence makes clear Slater's attitude toward quantum chemistry. Contrary to Pauling, who was definitely in favor of the valence bond method as the mathematical expression of a chemical theory such as resonance, Slater considered both methods as mathematical approximations used in order to gain access to more fundamental physical principles. What really attracted Slater was the search for a unified view of the problem of molecule formation, in which both approximate methods would be treated so that one would manage to get at the essential physical features of the problem situation, forgetting, in the process, the particular method of approximation used.

Having been deeply influenced by Bridgman's pragmatism (as was Pauling), Slater avoided as much as possible any kind of philosophizing. But, nevertheless, he could not avoid having a philosophical agenda. His advocacy of reductionism was not only expressed in the above statement, but also was bluntly stated in a manuscript of a lecture on "Philosophy and Physics" he planned to deliver at the Franklin Institute in Philadelphia in 1937, but which, for unknown reasons, was never delivered.[163] There, his belief in the hierarchy of the sciences and in the supremacy of physics leaves no doubt about the way in which he envisioned the unification of chemistry and physics. With quantum chemistry, chemistry had been given a theory that it formerly lacked. Quantum chemistry was not considered to be a subdiscipline of chemistry but an instance of the application of quantum mechanics to chemical problems. Unification was therefore to be attained through reduction of chemistry to physics, and maybe this would just be a necessary intermediate step toward the creation of a more fundamental science.

It is not hard to understand why Slater grew progressively more disenchanted with the turn quantum chemistry was taking. Many years later, he considered that "if I go back into the field of chemical valence, I fear I won't get much help from the recent writings of the chemists, but will have to start in pretty much from the point of view where I left theory quite a while ago." He further argued that it would be good to get "the chemists to write so as to bring back the interest of the physicists in problems of chemical physics. That is something that I am highly in favour of."[164] It was the acknowledgment of a lifetime project's failure!

Pauling: Reforming Chemistry from the Standpoint of Resonance Theory
By 1935, Pauling believed that he had acquired an "essentially complete understanding of the nature of the chemical bond."[165] Always eager to get his contributions recognized quickly among his peers, Pauling used all communication channels in order to reach as many people as possible. The *Introduction to Quantum Mechanics with*

Applications to Chemistry (1935), written jointly with Wilson, was addressed to chemists, experimental physicists, and beginning students of theoretical physics and did not presuppose much mathematical background on the part of its readers. The book became popular even among those for whom quantum theory was not unknown territory (Pauling and Wilson 1935).[166]

During his tenure as George Fisher Baker Non-resident Professor of Chemistry at Cornell University in the fall semester of 1937, Pauling reorganized for publication in a textbook all his published papers and unpublished notes on the chemical bond. *The Nature of the Chemical Bond* appeared in 1939 and sold so well that another edition came out in the following year (Pauling 1939). It was dedicated to Lewis who was overjoyed by the fact: "I have returned from a short vacation for which the only books I took were a half dozen detective stories and your "Chemical Bond." I found yours the most exciting of the lot. I cannot tell you how much I appreciate having a book dedicated to me, which is such a very important contribution. I think your treatment comes nearer to my own views than that of any other authors I know and there are very few places where I could possibly disagree with you; and those perhaps because I have not thought about the thing sufficiently."[167]

Pauling's classic *The Nature of the Chemical Bond* deserves a special place in any discussion of the early textbooks of quantum chemistry. The reasons for its popularity and persuasiveness are quite complex, and they are not independent of the expressed assertiveness of physical chemistry in the United States, the rather articulate expression of American pragmatism and operationalism in Pauling's book, as well as the deadlock of the program of analytical calculations started by Heitler and London and continued by others.

In *The Nature of the Chemical Bond*, Pauling presented the major questions he discussed in his papers in a language more appropriate for a larger audience of students and fellow scientists. Pauling's greatest achievement was to present a coherent treatment of the chemical bond that was appealing to the chemists because of its frequent reliance on the "chemists' intuition" and the use of a lot of existing experimental data to be able to explain or predict other experimental data. Though it was repeatedly stressed that the understanding of the nature of the chemical bond was possible only because of the developments due to quantum mechanics, his use of detailed mathematical formulations was reduced to a bare minimum. He did not aim at "proving" theorems, but rather at "devising" rules that did not follow in any rigorous way from more general principles, yet they seemed reasonable, were partially justified by quantum mechanics, and, most significantly, they could be used to get results. It was, as Pauling often said, a pragmatic approach to chemistry, a semiempirical treatment of the problems, and an overall attitude that was so dear to the chemists' traditions. In this manner, the book articulated a language for quantum chemistry soon to be enthusiastically adopted by the chemists.

Pauling was aware of the difficulties faced by chemists in understanding such unfamiliar concepts as the quantum mechanical concept of resonance and resonance of molecules among several valence bond structures (Park 1999; Mosini 2000). He noted the existence of an "element of arbitrariness" in the use of the concept of resonance as a result of the choice of canonical structures in discussing the state of the system, but he argued forcefully that "the convenience and usefulness of the concept of resonance in the discussion of chemical problems are so as to make the disadvantage of the element of arbitrariness of little significance" (Pauling 1939, 12). This, as he repeatedly stated, was his constructive criterion for theory building in chemistry. Besides, he reminded his readers that an equivalent element of arbitrariness occurred in essentially the same way in the classical resonance phenomenon.

Finally, he contrasted resonance with traditional chemical concepts such as mesomerism and tautomerism and discussed the reality of canonical structures. Even such a clear and succinct writer as Pauling could not avoid making apparently contradictory statements. As to the relation between resonance and tautomerism, Pauling seemed to be claiming at times that they were the same: "There is no sharp distinction which can be made between tautomerism and resonance." Elsewhere he claimed that they were distinct: "It is convenient in practice to make a distinction between the two which is applicable to all except the border-line cases" (Pauling 1939, 404), differing in the following way: "Whereas a tautomeric substance is a mixture of two types of molecules, differing in configuration, in general the molecules of a substance showing electronic resonance are all alike in configuration and structure" (Pauling 1939, 407). The same ambiguity arose in discussing the reality of different canonical structures. Is it the case that the two Kekulé structures associated with the benzene molecule are real? Pauling claimed that "there is one sense in which this question may be answered in the affirmative," but immediately added that "the answer is definitely negative if the usual chemical significance is attributed to the structures. A substance showing resonance between two or more valence-bond structures does not contain molecules with the configurations and properties usually associated with these structures" (Pauling 1939, 408). Having these linguistic ambiguities in mind, one cannot but wonder about their repercussion in the subsequent arguments over the significance of resonance.

The Nature of the Chemical Bond had a tremendous impact not only on research but also in the teaching of chemistry. Joseph Mayer prepared a review of the book,[168] in which he considered it to be

[U]nfortunate that this treatise will almost certainly tend to fix, even more than has been done by the author's excellent papers, the viewpoint of most chemists on this, and only this one, approach to the problem of the chemical bond. It appears likely that the Heitler-London-Slater-Pauling method will entirely eclipse, in the minds of chemists, the single electron molecular orbital picture, not primarily by virtue of its greater applicability or usefulness, but solely by the brilliance of its presentation.

In effect, throughout his career Pauling published on valence theory with almost total disregard for alternative approaches. However, during 1936–1937, Pauling and Wheland prepared a joint book with the tentative title *Quantum Mechanics of Organic Molecules*. Pauling, uncharacteristically, planned to make an extensive comparison of the valence bond and the molecular orbital methods. The book was to start with an introduction to quantum mechanical principles, a quantum mechanical study of simple molecules from the point of view of valence bond theory, and a detailed analysis of the concept of resonance and its effects on chemical properties.[169] This first introductory part was followed by an application of the "semi-empirical valence-bond method" to organic molecules and a comparison of its results with the results obtained by application of the "semi-empirical molecular orbital method." Wheland prepared the introductory chapters based on the valence bond method, and Pauling was supposed to prepare the chapters on the evaluation of both methods. Wheland's part was soon completed and revised by Pauling, who never managed to finish his share in the project. His attention was drifting toward applications to larger molecules of biological interest.[170] Later on, Pauling tried to revive the project, regretting his "dilatoriness" in pushing the book forward in the first place, but by then too much revision and restructuring would have been needed.[171]

But the expressed "dilatoriness" may not have been the exclusive reason for the failure of the common project. It can, in fact, be argued that another—perhaps major—reason to understand the impasse is to be found in the difference of opinion of the two authors concerning the significance and the character of the concept of resonance. Pauling's work had raised resonance to a chemical category, and the concept was neither a heuristic device nor an algorithm nor a metaphor nor simply a pedagogically expedient method for understanding quantum chemistry. Disagreements over its ontological status were the object of a revealing exchange of letters between Pauling and Wheland, which we shall discuss in the next section. Pauling and Wheland debated questions often addressed by philosophers of science, by examining the extent to which the reality of resonance as a chemical category could be ascertained. Pauling deemed the topic so important that he made his position public in *Perspectives in Organic Chemistry* (1956), and later on in the third edition of *The Nature of the Chemical Bond* (1960). More than the question of the artificiality of the resonance concept, to which he alluded briefly in his 1954 Nobel lecture (Pauling 1954), he wanted, once and for all, to state as clearly a possible his views on theory building. In the preface to the last edition of 1960, Pauling pointed out that the theory of resonance involves "the same amounts of idealization and arbitrariness as the classical valence-bond theory." A whole section was added to discuss this question bearing the revealing name "The Nature of the Theory of Resonance." There, he argued that the objection concerning the artificiality of concepts applied equally to resonance theory as to classical structure theory. To abandon the resonance theory was tantamount to abandoning the classical structure theory of organic chemistry. Were chemists willing to do that? According to Pauling, chemists should keep both theories because they were chemical theories and as such possessed "an essentially empirical (inductive) basis."

I feel that the greatest advantage of the theory of resonance, as compared with other ways (such as the molecular-orbital method) of discussing the structure of molecules for which a single valence-bond structure is not enough, is that it makes use of structural elements with which the chemist is familiar. The theory should not be assessed as inadequate because of its occasional unskillful application. It becomes more and more powerful, just as does classical structure theory, as the chemist develops a better and better chemical intuition about it . . . The theory of resonance in chemistry is an essentially qualitative theory, which, like the classical structure theory, depends for its successful application largely upon a chemical feeling that is developed through practice. (Pauling 1956, 219–220)

The publication of successive editions of *The Nature of the Chemical Bond* should also be assessed in the context of an ambitious strategy by its author. Together with two other textbooks, *Introduction to Quantum Mechanics with Applications to Chemistry* (1935) and later *General Chemistry* (1947) (Nye 2000a), Pauling wished to implement an agenda aimed at nothing less than reforming the whole science of chemistry from the point of view of quantum chemistry. This agenda had also far-reaching implications in what concerned the status of chemistry within the hierarchy of the sciences. Pauling believed in the "integration" of the sciences (Marinacci 1995, 107–111), which he deemed to be achieved through the transfer of tools and methods, the most important kind of transfer being what he called the "technique of thinking." It is in this respect that he came to view chemistry, and specifically resonance theory, as playing a pivotal role within the physical and biological sciences in a manner analogous with his claim of a central place for chemistry, a place formerly held by physics (Nye 2001).

Disagreements on the meaning of resonance were at the center of reactions to Pauling's proposal and were made public even before the publication of *The Nature of the Chemical Bond*. If some were voiced by collaborators, most were put forward by opponents or critics. Determined to address the organic chemists, Hückel wrote a review article in 1937 criticizing Pauling's resonance. The crux of his criticism was that the concept of resonance as articulated by Pauling suffered from an unjustifiable analogy between mechanics and quantum theory. According to Hückel, the Kekulé structures that Pauling started from can be considered as a "formal analogy to two swinging pendula that are uncoupled and have the same frequency." But this was thoroughly misplaced: Because both structures exist simultaneously, neither one has a specific energy, and, thus, no frequency, and this is similar to the case with coupled pendula (Hückel 1937, 1937a).

In fact, Hückel (1937, 767, 764) thought that the very term "resonance" was misleading and had to be dispensed with, preferring instead the term "mesomerism,"[172] which he thought represented in a better way the ensuing molecular state as something in between the initial fictitious states that corresponded with canonical structures. Pauling, of course, in *The Nature of the Chemical Bond*, advanced the argument that the new state was not something in between, but a completely different state that had come to be realized as a result of resonance.[173] But Hückel was still unconvinced 20 years later (Hückel 1957), even after Pauling had, through an ingenious rhetorical strategy, made "his" resonance a household name in the chemical community. Hückel repeated what he considered as crucial arguments against the use of the term "resonance." In his opinion, he wrote, what was basically involved was not a process in which "resonance" arose between different "structures" as in classical physics, but merely "an analogy with a mathematical calculation procedure—a purely mathematical formalism" (Hückel 1957, 872–873) for solving the secular problem. This must not be confused with a real physical phenomenon. He expressed his worry that many chemists attach "inappropriate meanings" to terms like resonance, resonance energy, and resonance stabilization. One of the basic reasons for such a confusion was the attempted formal mathematical analogy with classical pendula, and that resonance itself was a "physical process." He insisted that it would be better to "speak of '*line diagrams*' rather than of *structures*" (Hückel 1957, 873).

But Hückel was going against a culture of the chemical community that was formed for over a century. Visualizability had become one of the defining characteristics of the chemists' culture, and nonvisualizable configuration space was alien territory. In this small incident in the history of quantum theory, what became evident was what is so clear to almost all cultural historians and sociologists. Cultural trends die hard and, perhaps, are the obstacles not for change, truth, and progress, but for discussing and assessing the merit of new ideas and practices. If one is to judge things, Hückel was right. After all, even some advocates of resonance, such as Wheland, thought the same.

Disagreements with Hückel and later with Wheland on the question of the ontological status of resonance were not the only things Pauling had to face. Pauling's theory of resonance was viciously attacked in 1951 by a group of chemists in the Soviet Union in their Report of the Commission of the Institute of Organic Chemistry of the Academy of Sciences (Kursanov et al. 1952; Tatevskii and Shakhparanov 1952; Hunsberger 1954).[174] As they themselves stressed, their main objection was methodological. They could not accept that by starting from conditions and structures that did not correspond with reality, one could be led to meaningful results. Of course, they discussed analytically the work of Aleksandr M. Butlerov, who in 1861 had proposed a materialist conception of chemical structure: this was the distribution of the action of the chemical force, known as affinity, by which atoms are united into molecules. They insisted that any derived formula should express a real substance, a real situation. According to the report, Pauling was moving along different directions. For him a chemical bond between atoms existed if the forces acting between them were such as to lead to the formation of an aggregate with sufficient stability to make it convenient for the chemist to consider it as an independent molecular species.

After this incident, Coulson was designated as a kind of mediator for a proposed discussion on the theory of resonance between Pauling and Soviet chemists suggested by The New York Chapter of the National Council of Arts, Sciences and Professions.[175] It was proposed that a meeting should take place with the form of a debate where N. D. Sokolov from Moscow, Coulson, and Pauling would each contribute a paper and there would follow a discussion of the points raised in the communications. Coulson believed that the best way would be for Sokolov and Pauling to present their viewpoints and that he would make a series of comments. Each party would be asked to provide answers to the following questions: What is the resonance theory? What is the evidence in proof or disproof of the resonance theory? Is the convenience of the theory a proof or a corroboration of the theory? Is the resonance theory essentially a theory with physical meaning or a mathematical technique or both? Has the resonance theory a basis in related sciences, such as physics? Is the resonance theory applicable in all aspects of chemical valence or is it in conflict? The meeting did not take place basically because of the unwillingness of the Soviets, but the points that each party would have had to address were indicative of the uncertainties involved as to the methodological significance and ontological status of resonance in quantum chemistry.

George W. Wheland: Extending the Scope of Resonance Theory

George W. Wheland (1907–1972) obtained his doctoral degree with J. B. Conant at Harvard University and then moved to Caltech as a postdoctoral student (1932–1936) to work with Pauling in the extension of resonance to organic molecules (Mosini 1999; Park 1999). As we have seen, he co-authored the fifth paper of Pauling's series on the "The Nature of the Chemical Bond" and became a staunch advocate of resonance theory. In 1936, Wheland went to the United Kingdom as a Guggenheim Fellow to spend a year at the Department of Chemistry of the University of London headed by Ingold. He also used part of his time to work for some months in Oxford with Cyril Hinshelwood and to visit Lennard-Jones in Cambridge, Sommerfeld in Munich, and Hückel in Stuttgart.

In 1937 he was back in America. He had to decide about several offers made to him: to continue as a postdoctoral student at Caltech; to accept an assistant professorship at the University of California, Los Angeles; or to accept a position at the University of Chicago where a strong program in physical organic chemistry was under way. This last option was his choice. His constant contact with physical organic chemists alerted him to the urgency of clarifying the building blocks of resonance theory and discussing common misunderstandings. Teaching became his preaching platform and writing a textbook a necessary complement to his activities.

Wheland's *The Theory of Resonance and its Application to Organic Chemistry* first appeared in 1944, and it attempted to make as complete a presentation of resonance theory as was then possible. It was a partisan textbook. At a time when there was, really, no outstanding experimental reasons to choose between resonance theory and molecular orbital theory, Wheland's first lines in his preface left no doubts about the "correct" approach: "the theory of resonance is the most important addition to chemical structural theory that has been made since the concept of the shared-electron bond was introduced by G. N. Lewis" (Wheland 1944, iii).

Wheland believed that the general acceptance of resonance theory had been delayed because there was no comprehensive account of the subject, and he intended with the book to provide such an account. Very quickly, however, he expressed the main difficulty of such an undertaking. He considered that although the most interesting applications of the theory are in organic chemistry, "its basis lies in the mathematical depths of quantum mechanics." Its precise presentation can only be achieved by using complicated mathematical language. But being aware that such preconditions could not be expected of organic chemists and that, in general, they were not particularly welcome by the chemists' culture, Wheland suggested that "some sort of working compromise must be reached." Wheland's rhetoric is perhaps the most articulate expression of what—despite the talk of "compromise"—was in store for the chemists.

It is inevitable that the final result should be heavily weighted in favor of the more qualitative and descriptive approach. Experience has shown, however, that often more difficulties are created than are avoided if the attempt is made to ignore entirely the underlying physical basis of the theory. Indeed, many of the present misunderstandings of the theory seem to be directly attributable to the fact that practically all the discussions of it that have been published in the past (including, it must be admitted, some written by myself) have too drastically oversimplified the treatment. Consequently, a rather detailed, but non-technical and actually non-mathematical, discussion of the essential fundamental principles is given [and] no effort has been spared to make this discussion completely rigorous, even at the risk of sometimes going into boring detail in regard to apparently trivial matters. (Wheland 1944, iii)

The message was loud and clear: When it comes to be convincing about the adoption of resonance theory, qualitative treatments alone are by no means sufficient, and organic chemists, especially, will have to get used to the idea that mathematics and quantum mechanics should be on their agenda from now on.

This changing sentiment is clearly displayed when the second and improved edition of the book is compared with the first (Wheland 1955). Though Wheland was still strongly in favor of the resonance point of view, large sections of the book dealt with the molecular orbital method, an inclusion necessitated by the growing interest in and recognition of equivalence of the alternative approach. Despite the fact that he relegated the molecular orbital viewpoint to a secondary role, he declared that it was at least as important as the resonance viewpoint and, in some instances, more useful. The two viewpoints are mathematically equivalent "when carried to their logical extremes," and when applied in their approximate forms, "their relative reliabilities are difficult to judge" (Wheland 1955, viii).

He continued, nevertheless, to be a strong supporter of resonance theory: He insisted that "as an organic chemist, I believe that the resonance approach is clearer and more congenial to the great majority of other organic chemists than the alternative one" (Wheland 1955, viii). He thought that in many cases the theory of resonance was an extension of structure theory, whereas the adoption of the molecular orbital approach leads to the abandonment of structure theory.

Wheland mentioned that the book dealt explicitly with organic chemistry, as approached and interpreted from the resonance viewpoint. His agenda was clearly defined in both editions of the book: The adoption of the resonance viewpoint is the only way for quantum chemistry to become a subdiscipline of chemistry and not remain one of the many instances in the applications of quantum mechanics. Wheland, just like Pauling, was a master in appropriating quantum mechanics for chemistry and using it to further accentuate the autonomous status of quantum chemistry. In fact, though at the beginning he was somewhat apologetic for including a very long and technical chapter on quantum mechanics, he was also quite adamant in telling his fellow chemists that he was convinced that "an understanding of quantum mechanics cannot be acquired by any process of intellectual osmosis, but can be obtained only at the cost of a certain amount of conscious effort. Consequently, I do not apologize for the fact that this new chapter will require much study and the frequent use of pencil and paper" (Wheland 1955, ix).

The argument put forth to convince about the criteria of choice between the two approaches, which are mathematically equivalent and empirically equally satisfying, was exclusively dependent on issues of the shared culture of organic chemists. The resonance viewpoint was preferred over the molecular orbital viewpoint, not because the former was theoretically more appealing, nor because it is mathematically more precise, and not even because it was empirically more satisfying. The bottom line was that the resonance viewpoint was preferred because it was more congenial to the organic chemists and because it bore a close affinity to the organic chemists' culture. This was, of course, a far cry both from related arguments by physicists and from a number of proposals by philosophers of science.

We mentioned in the previous section that differences in the assessment of the methodological and ontological status of resonance were the object of a dispute between Pauling and Wheland. In his book dedicated to Pauling, Wheland argued that the resonance concept was a "man-made-concept" in a more fundamental way than in most other physical theories. This was his way to counter the widespread view that resonance was "a real phenomenon with real physical significance,"[176] which he classified as one example of the nonsense organic chemists in many instances were prone to. For him, resonance was not an intrinsic property of a molecule that is described as a resonance hybrid. It was not something that the hybrid does or that can be "seen" with a sufficiently sensitive apparatus. Instead, it was something deliberately added by the chemist or the physicist who is talking about the molecule. It was simply a description of the way that the physicist or chemist arbitrarily chose for the approximate specification of the true state of affairs. He went on to illustrate his viewpoint by means of the following analogy[177]:

In anthropomorphic terms, I might say that the molecule does not know about resonance in the same sense in which it knows about its weight, energy, size, shape, and other properties that have what I call real physical significance. Similarly . . . a hybrid molecule does not know how its total energy is divided between bond energy and resonance energy. Even the double bond in ethylene seems to me less "man-made" than the resonance in benzene. The statement that the ethylene contains a double bond can be regarded as an indirect and approximate description of such real properties as interatomic distance, force constant, charge distribution, chemical reactivity, and the like; on the other hand, the statement that benzene is a hybrid of the two Kekulé structures does not describe the properties of the molecule so much as the mental processes of the person who makes the statement. Consequently, an ethylene molecule could be said to know about its double bond, whereas a benzene molecule cannot be said, with the same justification, to know about its resonance . . .

Pauling could not disagree more. For him, the double bond in ethylene was as "man-made" as resonance in benzene. Pauling summarized their divergent viewpoints by saying that for Wheland, there was a "quantitative difference" in the man-made character of resonance theory compared with ordinary structure theory—a difference he could not find anywhere. He further asserted that his former student made a disservice to resonance theory by overemphasizing its "man-made character."[178] Wheland conceded that resonance theory and classical structural theory were qualitatively alike, but he still defended, contrary to Pauling, that there was a "quantitative difference" between the two.

Nevertheless, acknowledging or denying the existence of differences between resonance theory and classical structural theory was dependent on the different assessment of the two authors of the role of alternative methods to study molecular structure. Wheland, who considered resonance theory equivalent to the valence bond method, viewed them both as alternatives to the molecular orbital method. Pauling conceded that the valence bond method could be compared with the molecular orbital method, but not with resonance theory, which was largely independent of the valence bond method. For Pauling, the theory of resonance was not merely a computational scheme. It was an extension of the classical structure theory, and as such it shared with its predecessor the same conceptual framework. If one accepted the concepts and ideas of classical structure theory, one had to accept the theory of resonance. And, how could one reject their common conceptual base if they had been largely induced from experiment?[179]

> I think that the theory of resonance is independent of the valence-bond method of approximate solution of the Schrödinger wave equation for molecules. I think that it was an accident in the development of the sciences of physics and chemistry that resonance theory was not completely formulated before quantum mechanics. It was, of course, partially formulated before quantum mechanics was discovered; and the aspects of resonance theory that were introduced after quantum mechanics, and as a result of quantum mechanical argument, might well have been induced from chemical facts a number of years earlier.

Despite their disagreements, the appeal of resonance theory and the line of argument they put forth in its defense was to be found in their differing assessments concerning the affinity resonance theory had to the chemists' traditions and culture. As the years passed, it appeared that Pauling could adopt a less assertive stand concerning resonance. Despite his stubborn lifelong insistence on the unmatchable role of resonance theory, in the abridged version of *The Nature of the Chemical Bond*, published in 1967, and specifically addressed to students, Pauling made a small concession. In the sections "The Hydrogen Molecule and the Electron Bond" and "The Structure of Aromatic Compounds," he introduced students to the molecular orbital approach. Simultaneously, he stripped the textbook of all considerations as to the nature of resonance. But 13 years later, Pauling disparagingly commented that "it was a real tragedy when the writers of elementary textbooks of chemistry were so impressed by the molecular orbital method as to decide to put it into these textbooks" (Pauling 1980, 40).

Two Nobel Prizes Worlds Apart

Mulliken was awarded the Nobel Prize in Chemistry for 1966 "for his fundamental work concerning chemical bonds and the electronic structure of molecules by the molecular orbital method." Barely more than a decade separated this award from the first to reward foundational work in quantum chemistry, which was attributed to Pauling, back in 1954. It had been such a lively decade that by now the two prizes seemed to belong to two worlds many miles apart. As customary, both Nobel award lectures included extensive historical digressions. Both speakers had made these flashbacks part of their steady contributions to the discipline, so that the arguments voiced in the lectures were not new. But they brought again to light what for each were particularly crucial points.

Pauling centered his lecture on past accomplishments and future promises of the modern structural chemistry—*his* resonance theory. He recalled its lineage in the structural theory of organic and inorganic chemistry, and especially its relation to Lewis's electron pair. He recalled the unjustified charge as to the arbitrary character of resonance theory, which was a mild drawback compared with its usefulness, recently reiterated by its extension to metallic and intermetallic compounds, and especially to the complex substances making up living organisms. It was this last promise, to which Pauling had dedicated himself steadfastly for the past decades, which justified his permanent belief in semiempirical explanations based on many principles and rules, just partially justifiable exactly, and in a long tradition of chemical experience. Progressively, he substituted resonance theory for structural chemistry, as he always envisioned his own contributions as a continuous extension of the tradition of structural theory. He anticipated that whenever in the future chemists will get involved with large and complex molecules, they will have to rely on "the new structural chemistry and the rigorous application of the new structural principles" (Pauling 1954, 436–437). He was confident that such an approach will greatly facilitate the resolution of problems in biology and medicine.

No wonder, then, that Pauling participated only in few of the events that shaped quantum chemistry after the end of the Second World War. As we shall see in chapter 4, he was a guest star at the 1948 Paris Conference as he was at the first Valadalen summer school, which took place in 1958; he was only later added to the list of invitees to the 1951 Shelter Island Conference but declined to attend; no Sanibel Island Symposium of the initial series was organized in his honor, although one of the Sanibel meetings on quantum biology was dedicated to him. If no one denied the outstanding role Pauling played in the early decades of quantum chemistry, the founder became increasingly estranged by post–Second World War developments. Despite his staunch belief that theoretical chemistry was the clue to a complete reform of chemistry, his soul laid elsewhere. He could not find any special relevance in computers and exact calculations.

The opposite had been happening with his long-time rival.[180] Mulliken appropriated computers into the practice of his group, entrusting the intruder with high expectations. In his Nobel lecture, he offered a long analysis of the history of molecular orbitals since its very beginning, but he also dedicated considerable attention to the accomplishments of the theoretical subgroup of the Laboratory of Molecular Structure and Spectra, the group he created in Chicago after the Second World War. They gravitated mostly around breaking "bottlenecks" of difficult molecular integrals, performing ab initio calculations, and even writing computer programs. No wonder Mulliken ended his Nobel lecture with a plea for computers in quantum chemistry in consonance with his steadily implemented group agenda.

He was certain that the era of computers, of computing chemists, and of numerical experiments had come to stay. He boasted: "I would like to emphasize strongly my belief that the era of computing chemists, when hundreds if not thousands of chemists will go to the computing machine instead of the laboratory for increasingly more facets of information, is already at hand" (Mulliken 1966, 159). Maybe excitement brought the restrained Mulliken a little too far in predicting that computers would replace experiments. Their increasing relevance in many branches of chemistry came to depend on a growing interplay between computational and experimental chemistry. In any case, chemistry's landscape was being irreversibly changed by recent undertakings in quantum chemistry. If Mulliken repeated arguments, he did not hesitate to go much further confronting such a distinguished and powerful audience. He continued his reasoning by saying: "there is only one obstacle, namely that someone has to pay for the computing time" (Mulliken 1966, 159). He added that the justification for government and other organizations to allocate funds to computing molecular structures should be certainly as high as the justification to support research on nuclear and high-energy particle research and space exploration. It should be, but it was not. Cyclotrons, betatrons, linear accelerators, and rockets took the lead, not computers. Mulliken therefore criticized recent trends in scientific policy. He ended his long speech with a harsh criticism. He believed that progress in chemistry and solid-state physics was stalled by lack of funds for computer time, even though what was required was "trivially small compared with the amounts now being spent on nuclear and high-energy problems and on outer space" (Mulliken 1966, 159).

Mulliken believed that quantum chemistry should be at the center of individual, institutional, and government attempts to reshape chemistry by turning it into a computational science. He realized that quantum chemistry had come all the way from a marginal subdiscipline that defied some of the central characteristics of chemistry (chemistry as a laboratory science, chemistry as a visual science, etc.) to a subdiscipline central to the reshaping of chemistry, both conceptually and socially. He probably cherished the thought that chemistry could even enter the world of Big Science. In any case, later developments to build a national computational center for quantum chemistry took advantage of the momentum created by Mulliken's exhortations. We address these questions in chapter 4.

Some Further Remarks

Though the success of the Heitler–London paper heralded, at the same time, the deadlock of their approach, the very next phase in the development of quantum chemistry was full of new promises: two theoretical schemata established themselves and vied for dominance. The molecular orbital approach initiated by Hund in Germany, but perfected by Mulliken in the United States, was, at the beginning, considered to be antagonistic to the resonance approach developed by Pauling. Despite the interventions of many physicists and especially those by Van Vleck, who underlined the complementarity of the two approaches, in many (chemical) quarters the two approaches continued to be perceived as antagonistic.

One can almost marvel at the ability and way of those who worked on molecular orbitals to make sense of the molecular spectral data. And, yet, the apparently simplistic approach to "duplicate" Bohr's *aufbau* program for molecules turned out to be a great success. Furthermore, it adhered to a basic chemical characteristic: that of being visualizable. The molecule conceived as "united atoms" helped in visualizing the new bonding mechanisms. Notably, one did not need to make use of the Schrödinger equation, even though the Pauli principle was absolutely crucial.

Pauling proceeded to a rather ingenious use of a quantum mechanical notion, that of resonance, formulating another approach and a new "theory of valence"—a most idiosyncratic theory that became close to the heart for the chemists, also as a result of the incessant efforts of its inventor, a most able propagandist. Neither the methodology nor the relatively intricate mathematics were part of the chemists' culture. But if one is allowed to talk in terms of a reformation of a community's culture, it is Pauling's theory that brought about deep changes, by convincing chemists that mathematics will have to be part of their culture. He talked directly to the chemists, and he would not be bothered by any objections by the physicists. He kept on repeating that what he did was in the same spirit as structural theory. He asked chemists to develop a "sense for theory." And, he claimed that what he was doing was, in effect, the theoretical justification of what Lewis, the doyen of American chemists, had already suggested so successfully nearly 20 years earlier: an explanation for the otherwise mysterious electron pair mechanism. Pauling was able to deliver. And he became the hegemonic presence of quantum chemistry, culminating in the publication of his classic *The Nature of the Chemical Bond*.

Here one witnesses the intriguing aspects of contingency at work. Things, it is clear, could have developed differently. The community had distinct choices, both schemata had serious empirical backing, and both schemata shared theoretical virtues. And what counted for the specific developments were certain technical details, the discussions in the community concerning the legitimacy of the semiempirical approaches as well as personalities, decisions of key individuals, and rhetorical strategies of protagonists.

Almost all of the protagonists were aware and worried about the pitfalls concerning the ontological status of the various theoretical entities. Were orbitals real? Was resonance real? What was interesting was that developments did not become dependent on the (particular) answers being given to these questions, but the community was becoming aware that the development of quantum chemistry was not impervious to such considerations of a basically philosophical character. Nor were they indifferent to questions such as those of visualizability, which, in a way, were also related to issues of styles: what could be visualized may, perhaps, be real. It is, again, interesting that such issues were coming up again and again in review articles, popular writings, and public addresses.

3 Quantum Chemistry *qua* Applied Mathematics: Approximation Methods and Crunching Numbers

Starting in the mid to late 1930s when quantum chemistry was already delineated as a distinct subdiscipline, there was in Britain a group of people whose contributions to the further entrenchment of the disciplinary boundaries of quantum chemistry proved rather decisive. If the physicists' approach inaugurated by London, Heitler, Hund, Hückel, and Hellmann emphasized the application of first principles of quantum mechanics to chemistry, and if the chemically oriented approach of Pauling and Mulliken was characterized by a pragmatism combined with a creative disregard toward strict obedience to the first principles of quantum mechanics, then the British John Edward Lennard-Jones, Douglas Rayner Hartree, and Charles Alfred Coulson succeeded to enlarge the domain of applied mathematics so as to include quantum chemistry.

In this chapter we analyze some of the contributions of the British quantum chemists paying particular attention to the influence of the Cambridge tradition of mathematical physics/applied mathematics in shaping their immersion in the new subdiscipline. Ralph Howard Fowler's work expressed the receptivity observed among some Cambridge researchers to the possibilities for chemistry offered by the new quantum mechanics. When the new quantum mechanics was first formulated, Fowler was 37 and immediately became an enthusiastic convert to the new ideas. In 1932, the year he was appointed professor of mathematical physics, two of the students he supervised became professors at the University of Cambridge: Dirac became the Lucasian Professor in Natural Philosophy and Lennard-Jones the first professor of theoretical chemistry.

By 1931, in a report delivered at the Centenary Meeting of the British Association for the Advancement of Science, Fowler expressed his belief that a full quantum mechanical explanation of the valence rules of the quantum chemist was to be reached in the near future. Lennard-Jones in an article in *Nature* in 1931, which also echoed the views he expressed in lectures at the physical and also mathematical societies in London, considered the connection between the pairing of electrons with the "valency rules of the chemist" as a consequence of the same "mathematical and physical principles which have been formulated for other branches of physics" (Lennard-Jones

1931, 462). He was convinced that the general principles behind the different forces were understood and that such insights may come to be regarded as one of the greatest achievements of the then-current formulation of quantum mechanics. What was now required was mathematical techniques to be applied to particular cases. The three Cambridge professors were adopting a rather strong reductionist program for dealing with quantum chemistry.

In 1932, Coulson started his doctorate. He first was a student of Fowler and was later nominally supervised by Lennard-Jones. Coulson's research, though deeply grounded in this Cambridge tradition, showed a characteristic resistance against being lured by the excesses of this program. Coulson, the mathematical physicist, would refuse to become the long hand of physics in chemistry. There is ample evidence that Coulson was progressively displaying an increased sensitivity to the needs of the chemists themselves rather than adopting a patronizing attitude as to what their needs should be from the point of view of a physicist. It was he who legitimized the use of heavy—by the chemists' criteria—mathematics in chemistry and managed to have a rather wide recognition by the chemical community when, eventually, by the early 1950s his textbook *Valence* brought to an end the reign of *The Nature of the Chemical Bond*.

The British quantum chemists perceived the problems of quantum chemistry first and foremost as calculational problems, and by devising novel approximation methods, they tried to bring quantum chemistry within the realm of applied mathematics. Their strategy was one of developing as well as legitimizing formal (mathematical) techniques and methods to be used in chemical problems. For the members of this group and for Coulson in particular, the demand to make a discipline more rigorous meant developing a variety of approximation methods at the beginning and getting involved with computers later—though Coulson, himself, never surrendered to the charms of the new instrument that would change radically the way quantum chemists worked.

The 1923 Faraday Society Meeting and Its Aftermath: Sensing the Road Ahead

The 1923 Faraday Society Meeting

While Lewis was reading the proofs of his textbook *Valence and the Structure of Atoms and Molecules*, he was invited to give the opening address at the general meeting of the Faraday Society held in Cambridge, England, on July 13–14, 1923. Its title was "The Electronic Theory of Valence." The symposium was attended by the physicists J. J. Thomson, William H. Bragg, and Fowler and by several of the most outstanding physical and organic chemists in Britain and the United States, such as Lewis, Robert Robertson, Thomas M. Lowry, Arthur Lapworth, Noyes, and Sidgwick.[1] The opening address by Lewis (1923a) was a forceful summary of his *Valence*. He argued for the reconciliation of the physical and the chemical atom and the formation of the electron pair, which he dubbed "the cardinal phenomenon of all chemistry." Lewis's contributions and the questions debated at the meeting reveal an acute awareness that the fruitful course open to the chemist was indeed the explanation of the chemical facts of valence and molecular structure in terms of the concepts of atomic and molecular physics, so that the mastery of the laws of physics was an essential precondition for being successful in that endeavor.

Fowler and Sidgwick spoke along the same lines as Lewis. Both tried to show how Bohr's theory of atomic structure could be used to clarify the physical nature of valence. Fowler (1923) started by pointing out that there was not as yet a safe guide to molecular structure that would play the role Bohr's theory of the hydrogen atom did in relation to atomic structure. Then, he suggested that the next step in the development of a theory of the electronic structure of molecules would possibly be based on chemical evidence as to the nature of valences. Sidgwick (1923, 469) added to Bohr's theory of the atom the hypothesis that "the orbit of each 'shared' electron includes both of the attached nuclei," and then explained how such a conception of the nonpolar link as shared electrons occupying binuclear orbits enabled one to derive known chemical facts.

The application of the electronic theory of valence was discussed in the section on organic chemistry as well. The opening remarks by Robertson (1923) and the introductory address by Lowry (1923) emphasized the new era initiated by the application of physical ideas to valence. They both voiced the necessity of cooperation between physicists and chemists—"probably a team containing representatives of both groups" (Lowry 1923, 485)—in order to determine the electronic structure of molecules. It was not clear that such a cooperation would bring something new to chemistry, but history tells us that "whenever a clearer conception of molecular structure has arisen, chemists have always found a new way of regarding old facts, and even a new nomenclature for them has provided a powerful stimulus to investigation and has led to a great outbreak of new researches" (Lowry 1923, 485).

Ralph Howard Fowler: Quantum Physics in Cambridge

Among those attending the 1923 Faraday Society meeting, two participants, the physicist Fowler and the chemist Sidgwick, were particularly effective in preparing the ground for quantum chemistry in the United Kingdom.

Ralph Howard Fowler (1889–1944) was the leading and lone figure in mathematical physics/applied mathematics in Cambridge in the interwar period, and his interest in the old quantum theory facilitated a positive, even enthusiastic, reception of quantum mechanics, soon followed by its application to various areas of mathematical physics, quantum chemistry included.

Fowler was educated at Winchester and at Trinity College, Cambridge. He won several prizes in mathematics before completing his B.A. degree in 1911. His publications on the theory of solutions of second-order differential equations, which he was subsequently to apply to the classification of configurations of gaseous and partly gaseous stellar atmospheres, won him a fellowship at Trinity in 1914. As a result of his involvement with war research, he shifted his interests from pure mathematics to mathematical physics. During the First World War, he took part in the Gallipoli campaign as a gunnery officer with the Royal Marine Artillery. In 1916, he was invited to join the group of scientists, headed by A. V. Hill, which was doing research on military problems involving the computation of trajectories of cannon shells and recording the flight of airplanes. The members of the team included E. A. Milne, William Hartree, the father of Douglas Rayner Hartree, who was later to join the team of "brigands" (Milne 1945–1948, 65), and H. W. Richmond. According to Milne (1945–1948, 66), Hill and Fowler "fitted like hand and glove": Hill was the inspirer of most of the research problems investigated and Fowler worked out the solutions. It was in this unusual setting that Fowler was introduced to physical problems. In 1919, he returned to Cambridge as a fellow of Trinity College, in 1920 he was appointed college lecturer in mathematics at Trinity, and in 1921 he married the only daughter of Ernest Rutherford, who, in 1919, had been appointed as the Cavendish Professor. In 1932, Fowler was elected to the Plummer Chair of Mathematical Physics and in 1938 was appointed director of the National Physical Laboratory, succeeding Sir Lawrence Bragg. Fowler was involved in the work of governmental departments and during the Second World War he served as consultant to the Ordnance Board and the Admiralty. He was knighted in 1942.

Upon resuming academic life after the Great War, Fowler's scientific interests broadened considerably. He started working on problems of quantum theory, statistical mechanics, and the kinetic theory of gases, magnetism, nuclear physics, astrophysics, and physical chemistry. His own interest in questions at the interface between physics and chemistry and the work of some of the doctoral students he supervised—most notably, Lennard-Jones, Hartree, and Coulson—contributed to creating the background that facilitated the later developments in quantum chemistry in Britain.

Fowler was particularly keen in learning all there was to know about quantum theory and the theory of relativity. He even attended some of the courses being offered in Cambridge, including E. Cunningham's lectures on the special theory of relativity (Sanchez-Ron 1987; Warwick 1987, 1989, 2003). In Fowler's (1921) first contribution to quantum theory, Fourier's integral theorem was extended to quanta. In 1922, he started collaborating with C. G. Darwin on a series of papers on statistical mechanics. In their joint papers, they developed methods to compute the partition function associated with the distribution of energy in quantum systems and further extended these methods to deal with the equilibrium states of ionized gases at high temperatures (Darwin and Fowler 1922, 1922a, 1922b, 1923). In 1923–1924, Fowler was awarded the Adams Prize for an essay that included most of his contributions to statistical mechanics.² Extending some of the methods he had developed for statistical mechanics, he applied them to assemblies undergoing chemical dissociation and to the high-temperature dissociation of atoms into ions and electrons at low pressures. He was also involved with the interpretation of stellar spectra, developing a new method for the prediction of pressures and temperatures in the interior of stars. Fowler's (1926, 114) suggestion that white dwarfs were made of a "degenerate" gas of free electrons in a strongly ionized environment of very high density, somehow "like a gigantic molecule in its lowest quantum state," was one of the earliest applications of the new Fermi–Dirac quantum statistics. Another area to which Fowler contributed was that of strong electrolytes, a topic at the borderline of physics and chemistry in which he applied his newly developed methods of statistical mechanics.

An enthusiastic supporter of quantum theory, Fowler (1926a) pioneered in the application of quantum statistics to the study of gases in stars and in exploring a general form of statistical mechanics of which the classical, the Bose–Einstein, and the Fermi–Dirac forms were special cases. Among his lectures, those on "Quantum Theory and Spectra" and "Recent Developments of Quantum Theory" included topics that were just being discussed in the scientific literature.³ He addressed the recent developments concerning scattering, dispersion theory, and intensities of spectral lines and discussed the ideas of Heisenberg and Pauli "in their later speculations."⁴

The new quantum mechanics found in Fowler a committed follower. In a letter to Kramers, in which Fowler congratulated him on his recent appointment as professor of theoretical physics at the University of Utrecht, he wrote: "There is a man here Dirac who has got on with the development in a most interesting way though he seems not to have done much in effect different from the Göttingen crowd."⁵ Kramers replied, pointing to a mistake done by Dirac and cheerfully commenting: "Does it not please you to see how the mathematical operations with matrices afford the natural means of expressing Heisenberg's theory?"⁶ Dirac's very first lectures on quantum mechanics, given during the summer of 1926, were instigated by Fowler, who "attended, but kept in the background" (McCrea 1986, 277).⁷ Among others who attended were J. A. Gaunt, Hartree, Neville Mott, Bertha Swirles, J. M. Whittaker, A. H. Wilson, and William McCrea.

In a long letter published in *Nature*, and aiming at wider audiences, Fowler (1927, 241) explained the conceptual and mathematical differences between matrix and wave mechanics, noting that "however abstract the new mechanics may yet seem to us, however incomplete our grasp of its fundamental principles, it is impossible to overestimate its value to theoretical physics." Fowler introduced problems of quantum theory into the discussions of the experimentally oriented physicists who were at the Cavendish Laboratory, the Kapitza Club, and the Del-squared V Club. Mott noted that he was a model of what a mathematical physicist cooperating with the Cavendish Laboratory ought to be—"someone who knows what the experimental work is and where quantum mechanics can help it."[8] Fowler helped translate into English some of the papers that appeared in the *Zeitschrift für Physik* so that students who could not read German could have access to the new ideas. He also, quite often, invited foreign scientists to lecture at Cambridge. Such was the case with Kronig and Heisenberg.[9] He himself was also invited to other countries to deliver lectures. In sum, he played a very active role in disseminating and discussing quantum theory and quantum mechanics in Britain and became one of the well-known British members of the quickly expanding enthusiasts of quantum mechanics,[10] helping to create an environment where quantum mechanics played center stage and where the borderline between physics and chemistry was to be crossed at an ever increasing pace.

First Incursions into Atomic and Molecular Calculations

John Edward Lennard-Jones and His Molecular Fields

Born in 1894, John Edward Lennard-Jones (1894–1954) attended the University of Manchester from 1912 to 1915 (Mott 1955; Brush 1970). In 1915, he joined the Royal Flying Corps as a pilot. After the end of the Great War, he returned to his alma mater as a lecturer in mathematics and took his D.Sc. degree, working with Sydney Chapman on vibrations in gases. Between 1922 and 1925, Lennard-Jones was a Senior 1851 Exhibitioner at Trinity College, Cambridge, and completed his Ph.D. dissertation (1924) under the supervision of Fowler. For a while he considered the offer to go back to Manchester to replace Chapman who had succeeded A. N. Whitehead at Imperial College,[11] but he eventually accepted the readership in mathematical physics offered to him by the University of Bristol in 1925.[12] He got married and took from his wife the French surname Lennard, thereby changing his name from what he thought was a rather banal John Edward Jones to the fancier John Edward Lennard-Jones. (His students called him L-J.) In 1927 he became professor of theoretical physics. He stayed in Bristol until 1932, and in 1929 he visited the University of Göttingen as a Rockefeller Fellow. He was dean of the Faculty of Science at Bristol from 1930 to 1932.

Lennard-Jones's interest in the kinetic aspects of gases stimulated by his relationship with Chapman in Manchester was further enhanced by Fowler's contributions to the theory. His stay with Max Born in Göttingen in 1929 was quite decisive in his becoming thoroughly acquainted with the new mechanics and reinforced his belief that quantum mechanics would help clarify a host of physical problems. He became convinced that quantum mechanics would help him deal with the problem that had vexed him since the beginning of his career—the nature of the forces exerted between the atoms and ions of gases and crystals. In his very first papers of the series "On the Determination of Molecular Fields,"[13] he attempted to devise new methods in order to, indirectly, derive information about these forces because the existing methods made it nearly impossible to proceed to a "direct calculation of the nature of the forces called into play during an encounter between molecules in a gas" ([Lennard-]Jones 1924, 441). Lennard-Jones proposed a "new molecular model" whereby molecules are repelled by an nth inverse power law and are attracted by the inverse third power where the intermolecular distance was the variable. He was very pragmatic about it: No justification was given for choosing the particular form of the attractive part except that it rendered the integrals tractable! The formula he derived was a more general formula than the ones that had hitherto been derived. The force between the two molecules was expressed by

$$f_{12} = (c_n)r^{-n} - (c_m)r^{-m},$$

where the first term represented the repulsive forces and the second term the attractive.

When the theoretical results were tested against the experimental measurements, good agreement was obtained with greatly differing values for the coefficient n. However, the two sets of experimental results, one from viscosity measurements and the other from the virial coefficient, could not be used to build a molecular model that would reproduce both sets of results. Lennard-Jones noted that it might have been the case that the molecular fields determined by the two methods might not have been comparable. After all, in the case of the calculations of viscosity, the forces are those that come into the fore during the actual encounter of molecules. On the contrary, in the case of the virial coefficient, what is calculated is a statistical average of all forces on any one molecule due to all the others surrounding it ([Lennard-]Jones 1924a).

Lennard-Jones thought that X-ray measurements of crystals might shed some light on the actual values of the force constants. Though the general case was still elusive, it became possible to find values for individual substances. Information derived from considerations of kinetic theory was compared with X-ray measurements of interatomic distances in crystals of argon, and, thus, values of the constants were fixed for argon ([Lennard-]Jones 1924b). Similarly, values were fixed in the cases of helium, neon, hydrogen, and nitrogen as well as for krypton-like and xenon-like ions ([Lennard-]Jones 1925; Lennard-Jones 1925a; Lennard-Jones and Cook 1926). Further elaborate calculations were performed to derive the compressibility and elasticity of crystals (Lennard-Jones and Taylor 1925) and in order to explore fully the possibilities of the central field of force, and in particular that of the inverse power law, "probably the most general form in which the force is ever likely to be expressed" (Lennard-Jones and Ingham 1925, 636). Using the inverse power law for the interatomic forces had many more advantages over the treatment of the atoms and ions as rigid spheres with definite diameters, because it permitted the correlation of the physical properties of a gas with those of some crystals. Again, Lennard-Jones stressed that there was no attempt to try to justify his conclusions on any theoretical grounds, and he thought it quite interesting that methods developed in his various papers were shown to give quite satisfactory results. From the calculations and comparisons with the various experimental results, Lennard-Jones concluded that for argon-like ions, the repulsive forces vary according to an inverse 9th power law and for neon-like ions according to the 11th power. The attractive forces were consistently found to be small.

The importance of Lennard-Jones's work on interatomic forces was soon acknowledged by Fowler, who invited him to contribute a chapter on the same topic for the book he was preparing on statistical mechanics. The chapter summarized the state of the art in the field. It started by a sentence that left no doubt as to Lennard-Jones's scientific agenda: "It will no doubt be possible one day, probably soon, to calculate the forces between atoms in terms of their electronic structure, and thus to bridge one of the gaps which still separate molar physics from atomic physics. At present we have to rely entirely on indirect methods for such knowledge as we have of intermolecular fields" (Fowler 1929, 217). Unification and reduction were themata dear to him. Among intermolecular forces, Lennard-Jones was, later, going to concentrate on the homopolar forces in the hope of subsuming chemistry to atomic physics.

Douglas Rayner Hartree: "A Computing and Classifying Physicist"

Hartree (1897–1958) was the great grandson of the famous social reformer and writer Samuel Smiles, the author of the book *Self Help* ([1859] 2006), which became an English classic (Darwin 1958; Lindsay 1970; Fischer 2004). His mother was the first woman mayor of Cambridge; his father, William Hartree, was an engineer and a mathematician with an interest in biology who taught engineering at the university (Hill 1943). He attended the University of Cambridge as an undergraduate from 1915 to 1921, with an interruption during the war years, and became a fellow of St. John's College during the period from 1924 to 1927. He attended lectures by Fowler, Edward Appleton, and Bohr who made a strong impression on him, and he received a First Class in Part I of the Mathematical Tripos. Under the influence of Bohr's lectures, he published one of his first papers in which he studied the propagation of an electromagnetic wave that was not uniform over the wave front. Hartree (1923) wanted to understand the character of a light quantum.[14] Heisenberg visited Cambridge in summer 1925, and during the next summer Dirac gave the first course of lectures on recent developments in quantum mechanics. Together with Dirac and many others, Hartree belonged to the informal Del-squared V Club, formed with the aim of discussing theoretical physics. In summer 1926, Hartree received his Ph.D. His supervisor was Fowler, whom he had known for more than a decade due to his involvement, together with his father, in Hill's experimental artillery research group during the Great War. His main task in the group was to integrate the differential equations for the trajectories of high-angle projectiles used in anti-aircraft gunnery. His innovation was to consider time instead of the angle of elevation as the independent variable. This change facilitated considerably the integration of the equations. The research on ballistics done in this context involved him in much numerical work, the sort of work in which he, eventually, became a leader.

In the first part of his Ph.D. dissertation entitled "Some Quantitative Applications of Bohr's Theory of Spectra," Hartree determined, still in the framework of the old quantum theory, the central field of an atom or ion, which could account for the main features of the X-ray and optical terms of its spectra. Although the hypothesis of the central field was clearly just an approximation to the true atomic field, Fowler nonetheless considered Hartree's calculations "the first successful attempt to establish the *quantitative* as well as the qualitative validity of Bohr's theory" (Jeffreys 1987, 190).

In the report prepared on Hartree's dissertation, Fowler recalled that the application of Bohr's 1922 generalized theory of spectra and atomic constitution had opened the way for more quantitative applications and noted that as such application involved much numerical computation, it was "likely to appear unattractive to anyone not well trained in such work, who must at the same time possess an intimate knowledge of modern physical theory." And Fowler went on to remark: "[Hartree] is to be judged as a computing and classifying physicist—a type of worker for whom a great deal of demand appears to be developing" (Jeffreys 1987, 191).

Hartree and his family spent the winter of 1928–1929 in Copenhagen at Bohr's institute, where Hartree's dislike of heated environments was the reason for his office being known as "Hartree's north pole" (Jeffreys 1987, 192). As soon as Heisenberg's and Schrödinger's papers came out, Hartree decided to have a new look at the description of many-electron atoms. The self-consistent field approximation method was developed in 1928 in the two papers titled "The Wave Mechanics of an Atom with a Non-Coulomb Central Field" (Hartree 1928, 1928a).[15] In the first paper, Hartree addressed the problem already discussed in his dissertation, but now in the framework of quantum mechanics. He continued to use the approximation he had already developed in the context of the old quantum theory, which he called "orbital" mechanics by opposition to "wave" mechanics to stress the abandonment of the concept of orbit. His aim was to analyze the motion of electrons in a many-electron atom by assuming that their effect on the other electrons could be represented by a central non-Coulomb field of force. He adopted Schrödinger's wave mechanics as "the most suitable form of the new quantum theory to use for this purpose." It is, though, interesting to note that he still interpreted $|\psi|^2$ as giving the volume density of charge in the state described by ψ. He commented that it is doubtful whether such an interpretation is always valid, but for the wave functions corresponding with closed orbits of electrons in an atom, with which his paper was exclusively concerned, it had the advantage that it gave something of a model both for the stationary states, for the process of radiation, and it also gave a simple interpretation of the formula of the perturbation theory (Hartree 1928, 89–90). He felt that Schrödinger's interpretation of ψ made it possible to consider the internal field of the atom as resulting from the distribution of charge given by the eigenfunctions for the core electrons, hence, allowing one to find the field of force such that the total distribution of charge, given by the eigenfunctions in this field, reproduced the field. Such was the aim of the quantitative work Hartree set out to accomplish. In the first part of the paper, he went over the theory and methods of integrating Schrödinger's wave equation for the motion of a point electron with total energy E and potential energy V in a static field, in order to find its characteristic eigenfunctions and characteristic eigenvalues. The second part involved the determination of V with the assumption of a central non-Coulomb field, modifications of the equations suitable for numerical work, an outline of methods for integrating the equations numerically, as well as a discussion of results for some atoms.

It was in the second paper that the "self-consistent field" approximation was properly introduced. The question was, of course, to find a reliable method to get an approximate solution of Schrödinger's equation for a many-electron atom, having in mind that the equation could not be solved exactly for structures beyond the simplest cases. The idea of the self-consistent field is to imagine that each electron is moving in a sort of "average" field due to all the others and takes into consideration that in a heavy atom, the core electrons have filled up many of the lower-level groups completely, and, therefore, the influence of their field on the rest is comparatively easy to estimate. The procedure starts with a guess of what Hartree called the "initial" field; proceeds to a correction of the field of the core electrons that assumes that the distributed charge of an electron must be omitted in order to find the field acting on it; derives the radial wave function associated with each of the electrons in the corrected field by means of the numerical solution of ordinary differential equations; then the computed wave functions yield a distribution of electric charge, which is likely to be rather different from its first guessed estimate, and from this distribution new values for the field are calculated. This is called the "final" field. The calculation is repeated for the new field until the two processes of finding the wave functions and their electric fields are mutually consistent. By a process of successive approximations, one obtains a final field that is the same as the initial field. This field, which is characteristic of the atom under consideration or of its state of ionization, is the "self-consistent" field. It is interesting what Hartree (1928a, 114) was anticipating concerning the further possibilities of such an approach: "when time is ripe for practical evaluation of the exact solution of the many-electron problem, the self-consistent fields calculated by the methods given here may be helpful as providing first approximations."

The technique was applied to the atoms of He, Rb^+, Na^+, Cl^-. And it was the justification of Hartree's method that got Slater (1929) to think more about the theory of complex spectra, introducing determinants and the variational method for deriving analytically the self-consistent field equations with the right symmetry properties, as we have already discussed in chapter 2. Furthermore, Vladimir Fock (1930) also improved the method by taking into consideration that exchange forces arise due to the indistinguishability of electrons. Hartree was overjoyed by Slater's paper, and in a letter to Slater he declared: "I am very pleased at your justification of my S.C. field method, and especially that you have convinced the 'pure' theoretical physicists (I mean the ones like Wigner who have the attitude of a pure mathematician rather than a physicist) that there's something in it—more, I admit, that there was intended to be when I started."[16] In fact, physicists and mathematicians such as Wigner, Weyl, Heitler, and London were applying group-theoretical methods to the classification of symmetries in polyelectronic atoms. However, British physicists were not well acquainted with group theory. Hartree writing to London from Denmark noted that, having studied physics, he found group theory rather unfamiliar and "do not feel I understand properly what people are doing when they use it."[17] Hartree had been rather reserved with the possibilities offered by group theory, but he soon changed his mind after Mulliken showed in the early 1930s how indispensable group theory was for simplifying problems of molecular structure. Lennard-Jones and Coulson in Cambridge immediately set themselves to the task of mastering group theory.

Throughout the late 1920s and 1930s, Hartree's expertise on numerical analysis was used to develop ingenious approximation methods for the rapid evaluation of the self-consistent fields of atoms of increasing atomic numbers. In some of them, he was joined by his father, who very much enjoyed the numerical work that he did with the help of a desk calculating machine they familiarly called "Brunsviga," to remind of its place of birth in Brunswick, Germany, or alternatively "the crasher" (Jeffreys 1987, 193). This was the period when Slater moved to the Massachusetts Institute of Technology with the plan of "developing the department along the lines of modern physics." He gathered around him several theoreticians interested in atomic wave functions. In a letter to Hartree, he talked about the differential analyzer that Vannevar Bush, then a member of the Department of Electrical Engineering, was developing (Kevles 1971). The general idea behind such a machine was due to Lord Kelvin, but its practical design was due essentially to Bush. This "astonishing product" was a machine for solving differential or integral equations, and it was able to handle one-dimensional wave equations in a very satisfactory manner. It was enormous and very complicated, and it took "a good while to get familiar with it, and to set it up for a problem, but once that is done it is only a question of ten minutes to carry out the numerical integration of the equations."[18] Slater planned to use self-consistent fields, or some modification of them, in carrying out various numerical integrations on the machine, and so they exchanged many pages of information on wave functions and self-consistent fields calculated by Hartree.[19]

The central problem the machine set out to solve was not new. It consisted in finding a mechanical method to evaluate the solutions of differential equations, which often appear in pure and applied science, and for which no formal solution in terms of quadratures or of tabulated functions can be found. The available graphical methods did not have the scope and accuracy required, and the numerical methods developed so far were cumbersome and became increasingly more laborious as the equations grew more complicated. The differential analyzer provided a mechanical method for the numerical solution of differential equations that was rapid, accurate, and applicable to a wide range of the differential equations that occur in a variety of scientific and technical problems (Hartree 1935).

Upon his return from the United States, where Hartree went to see the new machine, he had a Meccano model of it exhibited at the University of Manchester (figure 3.1). As his daughter recalled:

In the early to mid 1930s my Father introduced us to the wonders of Meccano and stimulated our interest in building more and more elaborate structures as we grew older, and gained more dexterity. However, parts kept disappearing; new boxes were given to us for birthdays and at Christmas; and more parts disappeared! How were we to know they were going to construct the

Figure 3.1
Douglas Rayner Hartree (left) and Arthur Porter (right) viewing the Meccano differential analyzer in 1935.
Source: AIP Emilio Segrè Visual Archives, Hartree Collection.

model for the Differential Analyzer which was later built by Metro Vickers of Manchester in 1935. We were taken to the University to see the machine installed—our reward for contributing the parts for the model.[20]

The 1929 Faraday Society Meeting and the 1931 British Association for the Advancement of Science Meeting

The growing interest shown by the British physicists and physical chemists for atomic and molecular questions was expressed in the organization of nearly all the important meetings. The 1929 meeting of the Faraday Society addressed the topic of molecular spectra and molecular structure: It was held at the University of Bristol where Lennard-Jones was reader in mathematical physics, and it was co-organized by Lennard-Jones and his colleague W. E. Garner. This was also the first meeting in which the British, and specifically Lennard-Jones, began to contribute results of their own to questions of quantum chemistry. It was a meeting that marked a transition from a state of interest in problems of quantum chemistry to a state of active work on specific problems.

The organizers, who were in charge of introducing, summarizing, and assessing the results of the meeting, wrote in a note to *Nature* that it was a "valuable discussion on subjects of interest to physicists and chemists" (Garner and Lennard-Jones 1929, 584).[21] Hund (1929) noted that the concept of chemical valence did not provide an explanation of chemical bonding and that London's quantum theoretical conception of valence showed the same drawback. He then attempted to supplement London's theory, which gave special emphasis to symmetry considerations, with an analysis of the energy changes involved in molecule formation. He proceeded to give a comparative analysis of the conditions that might explain the formation of H_2 and the probable nonoccurrence of HeH with the purpose of finding possible criteria for chemical bonding. It was suggested that the characteristic difference between the two cases might be ascribed to the change in binding of the electrons in the two hypothetical processes, $H + H \to H_2 \to He$ and $He + H \to HeH \to Li$. In the first case, the binding of the electrons increased in the transition from the separated atoms that formed a molecule, whereas in the second case, the binding of one of the three electrons was considerably diminished. As we noted in the previous chapter, Mulliken had already discussed similar considerations.

Garner and Lennard-Jones (1929a, 620–624) referred to the possibilities offered by the new quantum mechanics when recalling the contributions of Heitler and London and their relation to Lewis's ideas, as well as when summarizing Hund's paper. They stressed the need to have simpler methods in order to know when atoms can form molecules because the analytical quantum mechanical calculations were so complicated. The unifying role of quantum mechanics in all branches of spectroscopy was considered to have foreseeable implications for chemistry to the extent that chemical change is associated with electronic transitions. They concluded that spectroscopic knowledge of the visible and ultraviolet regions would contribute to the development of theories of chemical change and of valence. In contrast, structural chemistry and, in particular, knowledge of the structure of organic molecules was likely to be advanced principally by studies in the infrared region (Garner and Lennard-Jones 1929a, 627). During the meeting, it became possible to reach an agreement as to the nomenclature and notation for molecular spectra (Garner and Lennard-Jones 1929b).

In another paper, Lennard-Jones dealt with the electronic structure of some diatomic molecules. He reiterated that a "knowledge of the electronic structure of molecules is of the greatest importance both to the physicist and the chemist" (Lennard-Jones 1929, 668)[22]: to the physicist because molecular spectra could only be interpreted in terms of electronic structure, and to the chemist because a detailed knowledge of electronic structure was a precondition to the knowledge of the exact conditions under which atoms combine to form molecules. Again, he called attention to the role of quantum mechanics in providing the theoretical framework for recent advances. He suggested a possible explanation for the paramagnetism of the oxygen molecule and proposed an *aufbau* principle for certain diatomic molecules. In this description, what happens to electrons in molecules was related to what happens to the molecule when considered as a single atom—the united atom—a condition, of course, which could not be realized. For Lennard-Jones, what mattered was the ability to predict the exact states of excitation of the component parts of a molecule on dissociation, and this was achieved with his *aufbau* principle, in which shared electrons were ascribed to molecular orbitals and unshared electrons to atomic orbitals. This modification of Mulliken's scheme was consistent with the requirements of the group theoretical considerations of Heitler and London. He further proposed a set of four rules for the assignment of quantum numbers to electrons in molecules in the ground state (Lennard-Jones 1929, 680, 680).

By focusing on the relation between the molecule and its products of dissociation, this treatment opened the way for the mathematical representation of molecular orbitals as combinations of atomic orbitals.[23] Lennard-Jones's introduction of atomic orbitals for unshared electrons was an important and clarifying correction to the procedure of using only molecular orbitals. It was something whose necessity became increasingly evident when dealing with heavier or polyatomic molecules. Mulliken considered that Lennard-Jones's proposal provided a good explanation of the bonding and antibonding properties of molecular orbitals, which "I could see empirically but which I tried to explain in terms of promoted and unpromoted orbitals . . . ."[24]

The conference was the first public forum where Lennard-Jones explicitly discussed problems of quantum chemistry. His training in physics was one of the reasons he found the molecular orbital approach more appealing. He had initially started to work on various problems of molecular forces using the techniques of classical physics and particularly those of the kinetic theory of gases, and impressed by the power of quantum mechanics, he expanded his techniques and proceeded to examine systematically the forces responsible for covalent bonds. In a series of lectures and articles published in 1931 and intended for a wider audience than the audience of his strictly professional articles,[25] Lennard-Jones (1931, 1931a) expressed his enthusiasm about the promises of quantum mechanics for understanding the various cohesive forces whose treatment has been "profoundly modified by the advent of wave mechanics": the chemical homopolar forces, the van der Waals forces, ionic cohesion, metallic cohesion. His argumentation followed the "logic" of a physicist's point of view, considering the connection between the pairing of electrons with the "valency rules of the chemist" as a consequence of the same "mathematical and physical principles which have been formulated for other branches of physics" (Lennard-Jones 1931, 462).[26] He was convinced that the general principles behind the different forces were understood and that such insights might come to be regarded as one of the greatest achievements of the then-current formulation of quantum mechanics. Mathematical techniques capable of applying them to particular cases were now required. A unified treatment of all cohesive forces was therefore the aim of his overarching program, a program that bore striking similarities to what Dirac claimed in 1929.

In the 1931 Centenary Meeting of the British Association for the Advancement of Science, one of the subjects discussed in the Chemical Section was the structure of simple molecules. Among the speakers were Lennard-Jones, Fowler, Heisenberg, Born, Debye, W. L. Bragg, and V. Henri.

Lennard-Jones offered a detailed review of molecular spectra based on the classification of the various contributions due to rotation, vibration, and electronic motions as well as their theoretical underpinning. The intensive study of molecular spectra was deemed to open a "new and rich vein of discovery in *chemistry*" (Lennard-Jones 1932, 210).

Fowler talked specifically about the quantum mechanical interpretation of homopolar valence and discussed extensively Pauling's first paper on "The Nature of the Chemical Bond," which had recently been published. He introduced his talk by claiming that the chemical theory of valence was not independent from physical theory but just a beautiful part of nonrelativistic quantum mechanics. He further added that the recent developments had not simply shown that "there is some sense in valencies" (Fowler 1932, 226), but that in the process of interaction of quantum mechanics and chemistry, it was the former that had been glorified by the successes in theoretical chemistry. His careful examination of simple systems led him to suggest that the set of rules Pauling had laid down would be valid for all electron-pair bonds. "With some limitations and qualifications perhaps, these guesses seem reasonable and likely to be replaced eventually by rigorous theorems of similar content" (Fowler 1932, 236). He hoped that a full quantum mechanical explanation of the valence rules was to be reached in the near future. Fowler's optimism was severely damped by Heisenberg's comments (1932, 247), which aimed at making "the present quantum theory of valence still more suspicious to the chemist than it already is." He did not believe that quantum mechanics would have been able to derive the chemical results about valence without prior knowledge of these results.[27]

The 1933 Faraday Society Meeting and the 1934 International Conference in Physics

In 1933 and 1934, another two meetings were held in Great Britain in which a discussion of the advantages and shortcomings of the two approaches to the quantum theory of chemical bonding—the valence bond (VB) method and the molecular orbital (MO) method—took place, together with an assessment of their possible application to the new theory of the solids.

The first of the meetings was, again, a Faraday Society Discussion dedicated to the topic of "Free Radicals."[28] As Lennard-Jones remarked, the theme of this discussion was in a sense "cognate" to that of the discussion on "Molecular Spectra and Molecular Structure" sponsored by the society back in 1929. In the preceding 5 years, considerable progress had been made in the understanding of the electronic structure of molecules and radicals, usually unstable aggregates of atoms, which can play an important role in chemical reactions. In his contribution, Lennard-Jones extended the method of molecular orbitals to a number of simple radicals. A detailed study of their electronic structure and their spectral states enabled classification of dissociation processes in two classes, one of which was accompanied by an "energy of reorganization" (Lennard-Jones 1934). He acknowledged that in the valence bond method, the atoms are considered as the building blocks of molecules taken to be represented by combinations of atomic orbitals. It met with considerable success in the explanation of the covalent bond as a result of pairing of electrons with opposite spins belonging to different atomic orbitals, the directionality of bonds as a result of hybridization of atomic orbitals, and aromatic properties of certain organic molecules as a result of resonance among several valence bond structures. However, his preference lay with the molecular orbital approach. Based on the idea of nonlocalized molecular orbitals extending throughout the entire molecular framework, the method had accommodated in the meantime the fact that molecular orbitals showed different degrees of localization and had been used to explain the properties of the double bond in unsaturated hydrocarbons and the structure of benzene and its derivatives. The exploration of the effect of symmetry properties of orbitals was being applied to the classification of polyatomic molecules. The MO method was, therefore, considered to offer greater promise than the VB method in its extension to polyatomic molecules with increasing number of atoms and in the discussion of excited states.

Erich Hückel (1934) discussed the properties and stability of radicals, that is, aggregates containing atoms with unusual valences such as "trivalent" carbon or "divalent" nitrogen. The necessary condition for stability was taken to be the linkage of the atom with the free valence with aromatic or unsaturated substituents. Hückel also professed his allegiance to the MO method. Although both methods were approximate methods whose adequacy to deal with certain problems depended on the particular case under consideration in organic chemistry, the molecular orbital method proved to be generally more satisfactory than the valence bond method.

Some of the participants in the Faraday Society Discussion met again, 1 year later, during the International Conference in Physics held in London in autumn 1934, which was organized by the International Union of Pure and Applied Physics and the Physical Society. The conference convened to discuss two different topics—nuclear physics and the solid state of matter—and contributed to the consolidation of these recent areas as independent subdisciplines. This was a large meeting gathering more than 200 participants. One of the topics that was extensively discussed was the relation between quantum chemistry and solid-state physics. Slater, for example, had contributed initially to quantum chemistry but withdrew from it to concentrate on solid-state physics. Some of the methods developed in the framework of quantum chemistry increasingly attracted the attention of physicists as they showed similarities to those developed in the context of solid-state theory. Their adaptation to the study of the properties of crystals, lattices, and metals was explored. Slater had been aware of this intimate relation since 1930 when he wrote that a crystal of a metal was nothing but an "enormous molecule" (Slater 1930, 509). He then pointed to the similarities between Heisenberg's approach to ferromagnetism and Heitler and London's approach to the molecule, both starting by separate atoms.

In the opening survey to the session on "The Structure of Molecules and of the Ideal Lattice," William H. Bragg (1935, 6) considered that both Hückel and Hund showed how investigations of a more mathematical character, using the "most recent and most powerful forms" of mathematical analysis, were yielding new results as details of molecular structure became clearer.

Besides reviewing the major results of his publications of the past 4 years, including what he had spoken about at the 1933 Faraday Society Discussion, Hückel pointed to the limitations of the classical theory of valence to the study of ethylene and benzene, and generally of aromatic and unsaturated compounds. Only a quantum mechanical treatment could account for the binding conditions peculiar to these compounds. He proceeded to discuss how the VB and the MO methods could account for such properties. If he conceded that the VB method extended the classical valence theory of the chemist, he thought that this "advantage is naturally not sufficient if the quantitative results cannot be brought into agreement with experience" (Hückel 1935, 15). The MO method did not show such disagreement between predictions and experiment

(Hückel 1935, 22). As Hund did some years earlier, Hückel insisted in the inherent nonvisualizability of the representation of aromatic molecules.

In Hund's presentation of the binding forces in molecules within the framework of quantum mechanics, Hund outlined what he considered to be the three "different degrees" in a "theory of cohering matter" depending on the extent of the application of quantum theory, "whether we only introduce general facts of the quantum theory, and think, in the main on classical lines, or whether we employ complete quantum mechanics for the phenomena" (Hund 1935, 37). To the first stage belonged the classical kinetic theory of matter, which takes atoms as given and only assumes the existence of attractive and repulsive forces between them. This step did not explain the special properties of different substances nor, and above all, the rules of chemistry. At the next stage, the individual properties of atoms are deduced from quantum theory, but the explanation of the aggregation of matter is still classical and pictorial. The chief points left unexplained were the properties of homopolar molecules, metals, and certain insulators like diamond. In the third stage, a quantum theory of the molecule and the crystal lattice is developed in nonpictorial terms. This stage accounted at last for the facts of homopolar chemistry. Therefore, the chemistry of covalent compounds (and that of solid atomic lattices) was explained only in the framework of a theory (that of quantum chemistry), which is inherently nonclassical and nonpictorial.[29]

Further Developments in Molecular and Atomic Calculations

Lennard-Jones and the First Chair of Theoretical Chemistry

Lennard-Jones moved to Cambridge in 1932 and became the first Plummer Professor of Theoretical Chemistry. On November 7, 1931, at the bequest of J. H. Plummer, the John Humphrey Plummer Professorship of Inorganic Chemistry was established. In a meeting held on July 19, 1932, the search committee decided to offer the chair to Lennard-Jones, who had not even applied for the position, bypassing in this way nine candidates.[30] Lennard-Jones accepted, and on December 19, 1932, the title of the chair was changed: *Inorganic Chemistry* was replaced by *Theoretical Chemistry*.[31] It became the first such chair in Great Britain and in the world (Hall 1991, 5).

Coulson, who was one of Lennard-Jones's former students, wrote in 1972 when offering to the University of Cambridge Library his lecture notes, which included those on the course *Quantum Theory and Molecular Structure* taught by Lennard-Jones for the first time in the 1933 Michaelmas term: "I believe that these lectures were the first ever to be given in Britain (and perhaps anywhere in the world for an undergraduate course) dealing with quantum chemistry."[32] Although there were already lectures in the mathematics faculty dealing with quantum mechanics, practically nothing had been taught previously about molecules and valence. The course *Quantum Theory and Molecular Structure* began with a review of the principles of wave mechanics, including topics such as orbital angular momentum, interaction with light, atomic units, and wave functions for any spherically symmetrical potential field. The second major theme dealt with the symmetry properties of many-electron atoms, methods of calculation of orbital wave functions, including the variational method and perturbation theory, as well as applications to simple cases. The last major topic was, of course, the formation of molecules and included a survey of the valence bond and the molecular orbital approach and ended with references to group theoretical methods.

Among the extant manuscripts of lecture notes, it is particularly instructive to look at the notes for the course on *Theoretical Chemistry* because they provide us some interesting insights on what Lennard-Jones considered to be the main issues to be discussed in such a course.[33] In the introductory passage titled "The object of theoretical chemistry—the explanation of chemical phenomena," he proceeded to a historical overview of the more important theoretical ideas. These were the early theory of molecular structure by Dalton and Avogadro, the early ideas of valence by Kekulé, van't Hoff, and Le Bel, the structural formula of Palerno [sic], the early developments in organic compounds, the tetrahedral valences of carbon, the stereochemical ideas in organic chemistry, Werner's proposals for metallic compounds, thermodynamic ideas and their applications to dilute solutions, Arrhenius' concept of electrolytic dissociation, and the dynamical theories of chemical reactions. He, then, went on to a presentation of developments in the twentieth century, where almost everything was around wave mechanics, and a discussion was added about directed valences and the concept of resonance. The last section of this introductory passage is titled "The unifying principles of chemistry," and it discussed the concept of stationary states, the exclusion principle, the principle of the identity of electrons, the principle of minimum energy, exploring at the same time the way these principles contributed to the resolution of a number of problems in theoretical chemistry, organic and inorganic chemistry, as well as metallurgy. The rest of the lectures were a standard course in introductory quantum mechanics. They end by examining a number of special problems including the electronic structure of diatomic molecules, elements from the valence bond method and the molecular orbital method, directed atomic orbitals and angles between chemical bonds, the nature of the double bond, and resonance energy and its importance in conjugated molecules.

As professor of theoretical chemistry at Cambridge, Lennard-Jones found himself in a unique position in the traditional academic city. His research and teaching was on what everybody considered as a mathematical subject but it took place within a department of chemistry. In fact, he had a room in the chemical laboratories, and he headed a subdepartment, including a few staff members and premises for staff and students. Before the war, Lennard-Jones's group was working on topics whose solution depended on mechanical computations. This situation led Lennard-Jones to construct a small differential analyzer for his group (Mott 1955, 177) and convinced him of the need to create in Cambridge a mathematical laboratory. Therefore, it is no wonder that he took an active role in the events that brought about such a new structure. By the end of 1936, the Faculty Board of Mathematics issued a 6-page report that recommended that a computing laboratory be established and be associated with the Faculty of Mathematics, that the faculty allocate a given sum toward the initial cost of the laboratory, that the first stage in its equipment should be the installation of a differential analyzer, an 8-unit Bush integrating machine, and, finally, that a standing committee be set up to promote the installation of this computing laboratory and to consider the method of financing it as well as its further development.[34]

A few months later, Lennard-Jones wrote a memo in which he discussed how the new laboratory should be named. He disagreed with naming it Calculating Laboratory and preferred the designation Computing Laboratory. But he argued instead for the designation Mathematical Laboratory. For him, to calculate was then usually associated with the possibility of arriving at an exact result by mathematical reasoning, whereas to compute referred to situations in which it was just possible to arrive at an approximate, not an exact, answer. However, he considered that the processes like those of the differential analyzer were not conveniently described by either, and he predicted that machines succeeding to the differential analyzer in the future would perform functions accounted for by neither the words *compute* nor *calculate*. He suggested that it should be called the Mathematical Laboratory.[35]

In 1949, Lennard-Jones started the publication of a series of papers on "The Molecular Orbital Theory of Chemical Valency."[36] In the first (1949), he emphasized that the theories of chemical valence had been successful to the extent that the principles of wave mechanics could be satisfactorily applied to problems of molecular structure. But the difficulties and limitations of these theories were not so much the result of any "intrinsic deficiency of the principles involved," but rather stemmed "from the intractable nature of the calculations to which they lead"—echoing Dirac's 1929 note. Accordingly, the simplification of the methods of calculation and the equations involved became the central problem of theoretical chemistry. Lennard-Jones compared the merits of the two methods of calculation, as he referred to the two different modes for dealing with valence. As we will see, his line of attack had much in common with Coulson's approach. The generalization of the electron-pair method had resulted in a close understanding of the structure of a wide range of molecules and was also amenable for quantitative, albeit approximate, treatment of the interactions of atoms in molecules. One of its advantages was that the method was based on various states of a molecule such as can be "represented by conventional structural formulae." The end result was the representation of a molecular state as a superposition of "more standard" structural formulas and the interpretation of results in terms of these more basic structures. He considered the prediction of directed valences as an additional advantage of this approach. Though both methods could be used interchangeably supplementing each other, the molecular orbital method was considered as being the more fundamental of the two, basically because the electrons were treated as belonging to the whole molecule rather than to atoms or localized parts of the molecule. There are two comments to be made. The first is that nowhere was the term "resonance" mentioned. The second is that Lennard-Jones repeatedly extolled the electron-pair method as providing the chemists with a satisfying picture of the processes involved. Fifteen years earlier, Lennard-Jones had interesting things to say about visualization, in the opening paragraph of a talk he gave at the Chemical Society. The new theories, he believed, had swept away Bohr's orbits, which had the merit of "being picturesque and tangible." They were substituted by "something more elusive *which leaves unsatisfied the natural craving of the human mind for pictorial representations*" (Lennard-Jones 1934a, 223, emphasis ours). But now he reconsidered: the new theories did lend themselves under certain conditions to "visual presentation in a way which is actually more useful and satisfying—at any rate as far as chemistry is concerned—than the older pictures of solar systems." And when he advanced his new treatment, he was aware that a theory of molecular structure based on nonlocalized electrons "seems remote from the chemist's picture" of localized bonds and their geometrical relation to each other (Hall and Lennard-Jones 1951, 357).

Lennard-Jones appeared to consider the electron-pair method as a physicist's method to the extent that he considered atomic physics the prerogative of the physicists, and the molecular orbital method as a chemical method. In his assessment he reversed Mulliken's former evaluation of the valence bond theory as following the "ideology of chemistry" and the molecular orbital theory as departing from it. More than a decade had passed since Mulliken's assessment, and by now one was witnessing a reversal in the potential of both methods for further extension and adaptation to computation. It is not coincidental that he talked of orbitals: in the case of the electron-pair method, these orbitals were assumed to be localized between an atom and its interacting neighbor, whereas in the case of the molecular orbital method, they were distributed throughout the molecular field.

Lennard-Jones proceeded to the development of a theory of transformation from molecular orbitals to sets of equivalent orbitals for the case of molecules whose constituent atoms possessed inner shells and lone pairs of electrons.[37] In this particular treatment, it was concluded that the interaction of electrons in molecules might be interpreted as due mainly to the repulsion of charges distributed in equivalent orbitals. It was further surmised that the change of the angles between bonds might change the disposition of lone pairs of electrons relative to each other and, hence, lead to a change in the ways they repelled each other. In this respect, lone pairs may exercise a stabilizing influence on bonds and help determine angles between them: "lone pairs, may, in fact, be more important in determining molecular structure than has yet been recognized" (Lennard-Jones and Pople 1950, 167). All this, of course, was the expression in chemical language of the mathematical result that equivalent orbitals have a more localized distribution than the molecular orbitals from which they are derived, and their interpretation was that they corresponded with localized chemical bonds or with lone-pair electron distributions or inner shells of electrons.

Though there had been various references to particular molecules in the previous papers, it was only in the seventh paper of the series that Lennard-Jones appeared to have satisfactorily completed the necessary mathematical formalism to start applying his results to particular molecules such as methane, ammonia, water, hydrogen fluoride, ethylene, butadiene, benzene, the boron hydrides, and acetylene. Realizing the significant role of the electron pairs, Lennard-Jones noted the proposals by Lewis in 1916 and their subsequent development by Langmuir in 1919, but there was no mention either of the Heitler–London paper or of Pauling's work. At the end, Lennard-Jones felt confident to state those points that he considered to be incontrovertible evidence for the understanding of molecular structure. First, it became possible to give a rigorous description of the molecule in terms of equivalent orbitals that could be considered as approximately localized, and they could be transformed into molecular orbitals and vice versa without changing the value of the wave function. Second, a chemical bond corresponded with an equivalent orbital in a saturated molecule concentrated around two nuclei. In fact, Lennard-Jones's work showed that equivalent orbitals could sometimes be identified "with the chemists's conception of a chemical bond" (Lennard-Jones and Pople 1951, 191), but because they were not always strictly localized between the two nuclei, they could be considered as being more general than those bonds. Third, the equivalent orbital bonds included in themselves effects of delocalization. Finally, Lennard-Jones stated the next part of his program: Because the equivalent orbitals could be transformed into molecular orbitals without loss of accuracy, it was possible to use the equivalent orbital analysis as a starting point for a molecular orbital treatment.

Together with two of his students, John Pople—who would receive the Nobel Prize in Chemistry for 1998—and George G. Hall, they surveyed the principles used to determine the structure and properties of molecules. They expressed their worry that many of the theories already formulated explaining molecular structure appeared to have inconsistencies and tended to obscure the nature of the forces involved. Their main assertion was that it was indeed possible to provide a qualitative picture of the structure of molecules by considering only the electrostatic repulsions operating in conjunction with the antisymmetry principle (Lennard-Jones and Hall 1951; Lennard-Jones and Pople 1951a). The awareness that limited progress had been accomplished in making reliably quantitative calculations made it a pressing task to have a deeper knowledge of the fundamental principles of the subject to facilitate the discovery of those improvements needed in order to overcome such an incapacitating state of affairs (Lennard-Jones 1953).

On September 2, 1954, Lennard-Jones spoke before the Chemistry Section of the British Association for the Advancement of Science in Oxford. It would be his last public talk on these topics, as he died later that year. He recalled some of the new ideas "drawn from physics" that had been introduced in the theory of molecular structure in recent years. He asserted his belief that "the scope of the theory has been widened so that the physical and chemical properties of both atoms and molecules can be explained by the same unifying hypotheses." He further asserted that progress in science was mainly the result of two processes at work. One was the "steady accumulation of factual knowledge," the other the interpretation of "the facts in terms of unifying principles." Advances came about through the "interplay of theory with experiment," which often led to new predictions. The most impressive advances, though, were "produced when general hypotheses are put forward which bring different branches of science within the same comprehensive scheme." Such had been the case in the emergence of quantum chemistry. Among those guiding hypotheses, the exclusion principle stood as "the cornerstone of chemistry" (Lennard-Jones 1955, 175). Resonance, molecular orbitals, equivalent orbitals, π-electrons, and lone pairs had all contributed to the clarification of many of the properties of organic molecules.

Lennard-Jones expressed the thought that even if one claimed that the principles governing the behavior of molecules were known, the same could not be said of rigorous methods of calculations. The main difficulty was deemed to be the calculation of the effect of the repulsive fields of the electrons on each other, a topic that Lennard-Jones had addressed in his later research papers. In the process of increased mathematization of the discipline, one should never forget that "the task remains clear and insistent, now as [in the time of Faraday] . . . that the attempt be made to express as simply and clearly as possible the physical meaning of the mathematical theories that have swept through the whole of chemistry" (Lennard-Jones 1955, 184). Subsequent developments in quantum chemistry were to show how the increased possibilities of calculation provided by ever more powerful digital computers could easily lead many chemists to forget this article of faith.

The Extension of Hartree's Self-Consistent Field Method
In 1932, the same year when Lennard-Jones was appointed to the chair of theoretical physics at Cambridge, Hartree was elected a fellow of the Royal Society of London. Since 1929, he had been at the University of Manchester where he held the chair of applied mathematics until 1937. He was then appointed professor of theoretical physics, a post he held until 1946.

Life at the Department of Mathematics did not leave Hartree much spare time. The department was small, with two professors, four lecturers, and two assistant lecturers (Jeffreys 1987, 195). Nevertheless, he did interact professionally with people from the Department of Physics—namely with W. L. Bragg, Mott (who was visiting in 1929–1930), and Hans Bethe and Rudolf Peierls (who were appointed in 1933). This interaction helped to orient him to applications of quantum mechanics. He also was on close and friendly terms with P. M. S. Blackett and had a great admiration for C. T. R. Wilson and Coulson.[38] His scientific life was devoted to applied mathematical work, and he showed an impressive lack of interest in "pure mathematics, Hardy and others."[39]

Besides Slater, the other close American friend of Hartree was Robert Bruce Lindsay. Lindsay was a physicist working at the time on the same topics as Hartree and shared with him an interest in the "foundations of dynamics and physics."[40] Once Hartree revealed to him that "there is a colloquium which meets in London about six times a year to try to keep physicists in this country in touch with one another and with what is going on; except at Cambridge, there are not enough in one place to get a representative colloquium together, and it is valuable to meet people from other universities now and then."[41] In another letter Hartree confided that "our physicists here are mainly experimental; there are only about 4 of us who have any acquaintance with Dirac's equations, for example."[42]

During Lindsay's visit to Manchester in 1935, they made plans to work together on a common research project. Hartree considered three different possibilities.[43] One was the extension of the self-consistent field method to include exchange, relativistic effects, or both. Another possibility was to include configuration interaction (a name he disliked and preferred instead "interference of configurations" to stress the analogy with interference in the optical sense) (Hartree and Hartree 1935; Hartree 1937). Despite Hartree's attachment to the self-consistent field method, he suggested in a paper in *Nature* (1936) that the variational techniques were "more analytical" than the self-consistent field method. The third line of research included applications of the differential analyzer to technical problems such as "retardation" or "time-lag" problems.[44] In fact, Bertha Swirles from his group was dealing with the general theory of the self-consistent field including relativity effects. She was also trying to include retardation and spin–spin interactions, but no quantitative work was then possible. This was an instance in which work developed concomitantly on the differential analyzer gave hope for the future applications of the self-consistent field (SCF) method. For tactical reasons, Hartree was concentrating on strictly technical applications in the hope of getting help from industry for further improvements of the machine.

The development of suitable instrumentation was crucial to the extension of Hartree's research program, but in turn the availability of ever more sophisticated machines led Hartree to consider problems in mathematical physics that otherwise would probably not have called his attention. He confided to Lindsay, "After all, I am interested in other things besides atomic structure computations, and actually, since I have had the help of my father in that work . . . I have been doing less and less of the actual computing work in that field myself; and the differential analyser has been bringing me into contact with various branches of pure and applied physics with which I was barely acquainted with previously, and I am finding the experience both stimulating and interesting."[45]

After their meeting in 1935, those who gathered around Hartree and his friend Lindsay exchanged information and articulated their future lines of research. By the end of the 1930s, Hartree had no collaborators and little time to devote to his own research. Conditions were becoming increasingly gloomier, and Hartree worried that another world war was imminent. In a letter written the day Germany invaded Poland, he confided to Slater his concern about saving the computations of atomic wave structures, which he and, mostly, his father had been doing all these years (figure 3.2). He suggested that all the calculations "which have taken a good deal of time and work to obtain" be sent to Slater, including results not yet published. He would leave the

Figure 3.2
William Hartree, father of Douglas Rayner Hartree, with a spiral and a linear slide rule, late 1930s.
Source: AIP Emilio Segré Visual Archives.

decision of what to make of them to Slater. "All I am concerned for is that the results should not be lost to science."[46] And he had no qualms about what appeared to be coming soon. "For all your country's criticism of our policy in the last 20 years, with much of which I would agree, I hope we will have at least your moral support in the struggle against what seems to us to be the use of armed force as the sole instrument of international politics."[47]

This letter—where he also confided that it was becoming increasingly difficult to do any serious work in pure physics and that he was not at all sure of what he will be doing—sounded too much like a last will, and Slater did not feel comfortable about it. He suggested that the proper thing to do as soon as he got hold of the calculations was to put them in proper order, have the Na$^+$ carefully checked, and publish them in the *Physical Review*, under the signatures of the Hartrees, followed by an editorial note by himself. Slater offered his own views on the war. He talked about a generalized sentiment against the Americans entering the war and believed that in that respect Roosevelt did not represent the country.[48] He apologized for writing a pessimistic and unfriendly letter, but Hartree was a scientist and he owed him total frankness. Then he talked about the work at the Massachusetts Institute of Technology. And he finished: "I am curious to know how completely science in England will be disrupted by the war."[49]

Hartree could not help worrying that the Censor's Office might regard the letters as some code message, and not just what they professed to be. "If so, someone must have had a tough and unremunerative job trying to decode them!" As to Slater's opinion on the war, "I do not agree with you that we should take such an opportunity as arises to get out of it, if this means making peace with the present German government . . . I think the alternative now to continuing the war is not peace, but another 'twilight between peace and war' for 6 months, with a 98% chance at the end of it of another war with ourselves starting in a weaker position than we were in at present . . . ." Concerning the research activities in physics, Hartree noted that even though there was still some experimental work, it appeared that it was the activity most directly affected by war compared with theoretical physics.[50] A month later the parcel with the data had not yet arrived, and Slater feared that Hartree's suspicions about the delays in the Censor's Office were justified. He asked Hartree to trace them and to try sending them by diplomatic mail, either British or American. He commented, again, on the war, this time referring to Hermann Rauschning's *The Revolution of Nihilism*, a book he had just finished reading. "Rauschning is a former Nazi, from Danzig, who left when he became disillusioned about the movement, and his thesis is that the Nazi revolution is purely destructive, without any constructive aims whatever, and that they are trying, with great ingenuity and cynicism, to break down European civilization . . . ."[51]

In a few weeks, the notes with the calculations arrived safely. There had been simply a mixup, and Hartree's father did not send the calculations when he was supposed to. Hartree thought that his father had completed almost all the cases with exchange that were manageable. He was pessimistic about the possibility of calculations with exchange for atoms much heavier than Cu, even though there was now a proper framework in order to make the appropriate extrapolations. "Someday I would like to see a really heavy atom (Hg for example) worked out with exchange and Dirac wave functions, but this would probably be a two-year job!"[52] Electronic computers were going to turn Hartree's expectations into reality faster than he could imagine.

The electronic digital computer appeared not very long after the differential analyzer. Hartree played an important role in this new stage and was called for advice when the first of the new generation of the computers ENIAC, the acronym for Electronic Numerical Integrator and Calculator, was set up in Maryland. This was right after the end of the war. Contrary to the differential analyzer, which was an analog machine, ENIAC belonged to the class of digital instruments and offered the prospects of much greater speed and accuracy. Hartree offered his help to all those who were fighting for the installation of new machines as he did for the installation of Electronic Delay Storage Automatic Calculator (EDSAC) at Cambridge, first while still in Manchester and then after having moved to Cambridge in 1946 to become the Plummer Professor of Mathematical Physics, succeeding Fowler—who was his doctoral supervisor. Hartree consistently supported the activities of the team working under Maurice V. Wilkes's leadership in the Mathematical Laboratory. The first machine program was run in May 1949 (Smith and Sutcliffe 1997, 278).

Hartree's inaugural lecture at Cambridge was about what he strongly believed in: the role of high-speed computing in the future development of science (Hartree 1947, 1948). One of his biographers recalled that he compared the impact on civilization of high-speed digital computers to the advent of nuclear power (Darwin 1958, 111), but such a statement is nowhere to be found in the printed version of the lecture. For the short term, Hartree wondered whether mathematical physicists were ready to seize and explore the opportunity to handle the new range of problems whose study these machines made practicable. "This needs the right kind of insight and imagination" (Hartree 1947, 30). He had already called attention to the fact that machines were no substitute for the "thought of organizing the computations, only for the labour of carrying them out. This point seems to me of great importance, and to be missed entirely by those who speak of a machine of this kind as an 'electronic brain'" (Hartree 1947, 21).

In a report for the Physical Society published in 1948—the further elaboration of which led to a book in 1957—Hartree underlined the significance of the newly developing calculating machines through a dramatic numerical demonstration.

It has been said that the tabulation of a function of one variable requires a page, of two variables a volume, and of three variables a library; but the full specification of a single wave function of neutral Fe is a function of seventy-eight variables. It would be rather crude to restrict to ten the number of values of each variable at which to tabulate this function, but even so, full tabulation of it would require 10^{78} entries, and even if this number could be reduced somewhat from considerations of symmetry, there would still not be enough atoms in the whole solar system to provide the material for printing such a table. (Hartree 1948a, 113)

Hartree's self-consistent field method became particularly important when extended to molecular problems, and his pioneering work in machine building heralded the era of the application of computers to quantum chemistry. Quantum chemistry which for about 20 years was a witness to the impossibility of exact solutions was to provide roughly in the decade after Hartree's report the very ground that would falsify such state of affairs. The strong belief in such impossibility was experiencing the start of a strong jolt that would soon displace it from being the dominant constitutive aspect of quantum chemists' culture. A new instrument and new laboratory spaces accompanied such dramatic changes.

Charles Alfred Coulson: A New Research Agenda

Coulson in Cambridge: First Steps into Molecular Orbital Theory
Charles Alfred Coulson (1910–1974) is credited among his many other virtues with instantiating a program for quantum chemistry by developing its mathematical quantum mechanical framework and achieving the seemingly incompatible possibility of its pictorial representation.

Coulson attended Clifton College in Bristol where his "wise and cunning mathematics master" H. C. Beaven, a Balliol graduate, invited him to a lecture at the University of Bristol by Selig Brodetsky, a rather well-known applied mathematician.[53] Years later, Coulson remembered how flattered he was to be asked to a lecture at the university and also how impressed he was by what he heard. The speaker explained how a specific root-squaring process could be used to find the roots of a polynomial equation. The event marked him for life.

Professor Brodetsky was showing all this, and reminding us that in the design (at that time) of a new aeroplane it was sometimes necessary to find the roots of such an equation of up to the sixteenth degree, when suddenly I saw the order, the power, the skill of the process; and by the time I left that University lecture room I said: "I am going to become a mathematician." Never mind now that the root squaring process has dropped out of our mathematical tool-kit. It did its job for one young lanky schoolboy, and the spell it cast remains to this day. (Coulson 1969, 228–229)

Notably, developing methods for finding numerical solutions was later to become the hallmark of his contributions in quantum chemistry. Beaven's influence extended to other matters as well. After Coulson handed in the assignment sheets, Beaven would compare the solutions with the best solutions that he had gathered over the years, and more often than not he would say "'Coulson can't you find a nicer way of doing this?' And so one more youthful mathematician was led to consider not just the substance, but the form, of what he had done" (Coulson 1969, 233). Ever since his school days, he confessed to be under the spell of the "power and beauty of properly organised numerical methods" (Coulson 1969, 235).

Not long after this first experience, Coulson was again reminded of the importance of beauty in mathematics, and generally of beauty as a constitutive aspect of science: at the University of Cambridge he was taught electricity and magnetism from J. H. Jeans's book. He did not like it because he realized that it was not in vector form, and this hid its beauty. During his summer vacation, he rewrote the whole book in vector notation. By the time he completed it, he had learned something about mathematics that, up to that time, was known to him only at second hand: "that mathematics which is not beautiful has no right to be allowed to continue to exist."[54]

In 1928, Coulson was elected to an Entrance Scholarship in Mathematics at Trinity College, Cambridge. At the Tripos he excelled, taking First Class in Mathematics and in Physics. By the end of his undergraduate days, Coulson had a very sound knowledge of the main areas of mathematics and physics. He took courses by Mott, Dirac, and Lennard-Jones.[55] At Cambridge, Coulson experienced what he later called his "conversion." He had been a member of the Methodist church since 1928 becoming a lay preacher in 1929. After this "conversion," religion acquired a constant and fundamental dimension in his life and work. The three scientists who mostly influenced Coulson in this process were the physicist Alex Wood, the naturalist and theologian Charles Raven, and the astrophysicist Arthur Eddington from whom, as Coulson confessed later, he learned to improve his ability to communicate with lay audiences (Coulson 1958, 2).[56] His friend Douglas Edward Lea, who became a mathematician, physicist, and biologist, also exerted a profound influence on him. But it was his father, as Coulson acknowledged in the dedication of *Science and Christian Belief*, who was the first "to show me the unity of science and faith" (Coulson 1955).

From 1932 to 1936, Coulson worked toward his Ph.D. at Cambridge. At first he was supervised by the physicist Fowler. Coulson later remembered Fowler "who managed to communicate to me something of his own most evident excitement about it [quantum mechanics]" (Coulson 1955a, 2069). When Coulson was starting graduate work, he asked Fowler what topic to choose for his own research. He had expressed an interest in Mulliken's work on band spectra but was advised to keep away from them: "Don't take that up," Fowler said, "unless you are prepared to spend all your life at it. It's a full time job, and if you want to be free to dabble in other topics as they arise, keep away from band spectra" (Coulson 1964, 3). Upon Fowler's sabbatical leave, he became the first student of Lennard-Jones. Relations between professor and student left a lot to be desired. Lennard-Jones did not approve of Coulson's informal way of dressing[57] and demanded that Coulson visit him more often.[58] In 1934, Coulson was elected to a Prize Fellowship at Trinity (1934–1938) together with his friend Lea. The dissertation he submitted dealt with the application of Bessel functions to some methods of integration. The mathematician G. H. Hardy was one his examiners.

Coulson's initial interest in molecular orbital theory appears to have started in 1933. His training as a physicist and mathematician made it much easier for him to understand the group theoretical approach to problems of valence. Coulson's early wanderings among the problems of molecular structure are found in a letter written to Fowler, in which he reported on work done from June 1933 to March 1934.[59] In this letter, Coulson purposefully included what he had already done together with what was half done and what was, yet, an idea he planned to develop. All the work mentioned had "to do with the method of molecular orbitals for the structure of molecules." As to the other methods, they "come in merely for comparison." Coulson even added "that does not mean that I may not find some attractive little sideline in the other theories and spend a while trying to develop it, as occasion prompts."

The report had three sections, one on "Methane," another one on "Other molecules than CH_4," and finally a more theoretically oriented section on "General Theory of Molecular Orbitals." The clarification of the electronic structure of methane was the chief problem that occupied him during this period. As Coulson pointed out, the problem had been treated along two lines, in each case assuming tentatively a tetrahedral structure. For small nuclear separations, the molecule acts like a perturbed neon atom. Van Vleck was among the Americans whose work exerted a profound influence in shaping Coulson's agenda. As we discussed in chapter 2, in the paper on methane published in the first issue of the *Journal of Chemical Physics*, Van Vleck already addressed the question in the context of molecular orbitals showing a tendency for a directional effect, but Coulson criticized his sweeping approximations and the preliminary character of his conclusions. The papers on methane and on ammonia, together with the review paper written in 1935 jointly with Sherman played a crucial role in this process (Van Vleck 1933, 1933a, 1934; Van Vleck and Sherman 1935). Coulson's first incursions into quantum chemistry included some of the topics of Van Vleck's papers. Coulson was looking at the same molecule, providing a critical analysis of Van Vleck's assumptions with the aim of replacing some of them by more general ones. Lennard-Jones worried about the possibility that Van Vleck and his group might anticipate Coulson's own research results: "I am very anxious that you should publish soon. I hear that Van Vleck's people are busy working out molecular orbitals and you may be anticipated if you are not careful."[60]

These preliminary thoughts were further extended in the thesis submitted to the Fellowship Examination at Trinity College in 1934.[61] Its title was "The Electronic Structure of Molecules from the Standpoint of the Theory of Molecular Orbitals." The thesis was concerned with the examination and application of molecular orbital theory, although in many instances comparison with the valence bond method was found instructive. The work of Heitler and London was discussed in its connection with Born's extension and its development into the theory of electron pairs by Pauling and Slater. Curiously, the use of group theory for problems of valence by Heitler as well as London was never outlined. Mulliken's contributions were highlighted, and, notably, the application of the methods of group theory by Mulliken was presented as "in some ways the most interesting discovery" deserving, therefore, special attention. As Coulson eagerly noted, scientists were being confronted with an instance in which "existing work on pure mathematics has been called to their aid by the theoretical chemists."

Mulliken was another of the Americans who played a significant role in shaping Coulson's career. In 1933, Coulson attended a lecture by Mulliken at Cambridge (Mulliken 1989, 86–87). Later he recalled "how neat, and in a sense how obviously satisfying" (Coulson 1970, 258) Mulliken's explanation of the formation of the double bond in ethylene was, by bringing two CH_2 groups together with σ- and π-orbitals. Mulliken's use of group theory had impressed both Coulson and Lennard-Jones. They shared with one another a real sense of excitement when the powerful methods of group theory were shown to provide an explanation of the symmetries of molecular states and of the allowed and forbidden transitions between them. This happened in the same period when Coulson discovered the existence of an ultraviolet-absorption spectrum for benzene. His first attempt was to explain this spectrum by group theory. Though he succeeded in doing so, this work was never published (Coulson 1970, 258). Group theory was particularly appealing for large molecules, and in fact the next step in the framework of molecular orbital theory was its extension to organic molecules, all of which are polyatomic molecules.

By the end of 1936, Coulson was already delineating an extensive program of postdoctoral work on organic molecules, which included the development of new calculational techniques; the conceptual clarification of criteria used for comparing the molecular orbital method with the valence bond method; the comparison of resonance energies of molecules by the molecular orbital and valence bond methods[62]; and the extension of these considerations to polyatomic molecules, that is, the extension of the molecular orbital method and the self-consistent field calculations to polyatomic molecules.

In his first publications, Coulson was able to deal analytically with a number of integrals occurring in certain molecular problems (Coulson 1935) and prove that the criterion of maximum overlapping introduced by Pauling and Slater in 1931, which stated that the energy of a system that is made up of two component wave functions is least when the two wave functions are chosen to overlap as much as possible, was not universally valid (Coulson 1937, 1937a).

He then dealt with the electronic structure of methane (Coulson 1937b). It was an ambitious paper. A methodological clarification was offered in its very first lines. The importance of the methane molecule in quantum chemistry was twofold. It stemmed, first, from its simplicity—it is the simplest molecule to exhibit the characteristic tetravalence of carbon. Second, methane was important as a privileged test molecule that could be used as a probe into organic chemistry. In fact, Van Vleck had shown that both VB and MO methods agreed in predicting a tetrahedral model for CH_4. It was therefore clear that a "fuller treatment, which would be intermediate between the two, would also predict such a model" (Coulson 1937b, 388). Nevertheless, when one calculated the energy values, the two methods did not give satisfactory results. Coulson's new approach to the problem was taken from the point of view of molecular orbital theory, and he considered the results to be tentative. He emphasized that before a complete solution is attempted, it is "desirable that the problem should be discussed from as many different angles as possible, so that the effects of different simplifying assumptions may be clearly understood" (Coulson 1937b, 389).

The recourse to the molecular orbital approach had the further advantage of enabling other calculations such as the effect of electrical and magnetic fields on methane and the scattering of electrons. Together with the extension of the molecular orbital theory to polyatomic molecules, another important characteristic of Coulson's approach was the numerical investigation of the electronic structure of molecules. In the paper on methane, a novel technique was introduced to compute three-center integrals with recourse to Legendre polynomials and Bessel functions. The extension of molecular orbital theory to organic molecules became the main topic of Coulson's publications from 1938 onwards. Following the research program of Lennard-Jones (Lennard-Jones 1937; Lennard-Jones and Turkevich 1937; Penney 1937), he addressed the question of the nature of the links of certain free radicals. He aimed at extending the molecular orbital approach to the case of chain molecules and radicals that have an odd number of carbon atoms, concluding that "there is one electron which does not form a bond, in the usual picture of the chemist" (Coulson 1938, 383). For those molecules, which can be represented by the formula $C_{2n+1}H_{2n+3}$, he was able to write down general formulas for the lengths of their links and to show that the effect of resonance was to remove some of the characteristic properties of alternate single and double bonds.

Coulson (1939) contributed still another paper in which the notion of fractional bond orders was introduced for the first time in the context of molecular orbital theory. Fractional bond orders formalized the notion entertained by many chemists for a long time that in many substances, of which benzene is the paradigmatic example, bonds appear to have an intermediate character between single and double bonds. Hückel was the first to introduce the distinction between σ- and π-electrons in benzene, and it was in the framework of valence bond theory that the first definition of π-fractional bond order was given (Pauling, Brockway, and Beach 1935). Though some objections with this definition were expressed quite soon (Van Vleck and Sherman 1935; Penney 1937), the concept continued to be discussed in the literature for a long while. Coulson's original definition, in which the order of the bond was dependent on the coefficients of the wave function in the occupied molecular orbitals, could be grasped without a deep knowledge of quantum mechanics.

The idea was the following. If one represents by $\phi_i (i = 1, 2, ..., 6)$ the orbitals of the six π-electrons in benzene, then the total wave function is

$$\Psi = \sum_{i=1}^{6} c_i \phi_i.$$

The six energy eigenvalues $E^{(m)}$ and the six sets of coefficients $c_i^{(m)}$ each corresponding with one molecular orbital $\Psi^{(m)}$ of the above form could be calculated with the variational method. The six mobile electrons would distribute themselves between the six molecular orbitals. Coulson argued that the probability of finding one electron in the ij bond of the m molecular orbital was given by $\operatorname{Re} c_i^{(m)*} c_j^{(m)}$, and therefore he suggested for the definition of bond order the expression

$$p_{ij} = \operatorname{Re} \sum_{m=1}^{6} a^{(m)} c_i^{(m)*} c_j^{(m)},$$

where $a^{(m)}$ is the occupation number of the m molecular orbital: It is 1 if the molecular orbital is occupied by one electron, it is 2 if the molecular orbital is full, and it is 0 if the molecular orbital is empty.

Coulson then proposed that the total energy of the mobile electrons could be written in the form

$$E = \sum_i q_i \alpha_i + 2 \sum_{i<j} p_{ij} \beta_{ij},$$

where $q_i = p_{ii}$, α_i is the Coulomb integral, and β_{ij} is the exchange integral

$$\alpha_i = \int \phi_i H \phi_i d\tau$$

$$\beta_{ij} = \int \phi_i H \phi_j d\tau,$$

and H is the Hamiltonian of the system.

The expression obtained for the energy makes it clear that an increase in the total bonding energy implies a concomitant increase in the bond order. It has the further advantage of enabling one to relate bond orders to bond lengths. This proposal about fractional bond orders contributed to the further success and acceptance of the molecular orbital theory during the 1940s.

Coulson continued to compare the different methods of calculation as applied to more complicated molecules. He attempted to extend the theory of the screening effect from atoms to molecules and, therefore, to analyze the extent to which it was possible to establish rules for writing down approximate wave functions in molecules (Coulson 1937c). In his last paper from his Cambridge days, Coulson extended for the first time the self-consistent field method to molecules. As we saw, this method was originally suggested by Hartree and improved by Fock and enabled one to calculate approximate wave functions of atoms containing many electrons. In the atomic case, there is spherical symmetry, and therefore the mathematical problem reduces to the calculation of the radial wave function, whereas in the molecular case such a simplification is no longer valid. Coulson (1938a) looked at the case of molecular hydrogen in the ground state and applied the self-consistent field method to a discussion of its energy and approximate wave functions.

These first papers hinted at what was to become Coulson's characteristic approach to quantum chemistry. He attempted to provide answers to issues not yet completely settled in the context of the newly emerging discipline by extending the range of validity of calculational procedures that had already proved effective, by developing new techniques of calculation, and by clarifying the conditions of validity of certain criteria used for the critical evaluation of the results obtained with the different approaches. His contributions helped to accentuate further the quantitative possibilities of the theory and extend its domain to organic molecules. Though the emerging community of quantum chemists continued to discuss the relative merits and disadvantages of the valence bond and molecular orbital theory, mostly with the aim of choosing one of the two, Coulson argued for their *complementarity* and their mathematical equivalence when each method is adequately extended. By the 1940s, he was adamantly putting forth the view that it no longer made any sense to be partisan in accepting one and rejecting the other method.

Coulson Pushes Molecular Orbital Theory Forward

In 1938, Coulson ended his stay at Cambridge. The war years found Coulson away from the center, in the periphery at Dundee, Scotland, where he received a senior lectureship in mathematics at University College. Coulson's refusal to participate in any war activities because of his religious beliefs was an exception: Many quantum chemists participated in the war effort often in activities that had no relation to their expertise. Coulson, in fact, was pressed by Lennard-Jones to join him and work for the Ministry of Supply. Soon after he moved to Dundee at the end of 1938, Coulson decided to become a conscientious objector. He served on the Committee of Fellowship of Reconciliation and the Methodist Peace Fellowship. Simultaneously, air raids were a source of constant stress and fear. In the correspondence with Mulliken, they discussed how the war will affect and slow down research work not directly related to the war effort. Coulson hoped that some fundamental research tradition could still be preserved at the universities, even though he was rather pessimistic about it as the pressure of work was such that they may have to stop all research.[63] Mulliken was in complete agreement.[64]

Coulson felt miserable in his almost nonexisting "scientific environment." He complained to Lennard-Jones and asked advice on a job offer at Belfast to which he applied unsuccessfully. He was anxious to leave Dundee for a place where there is "more doing in the research line (they've never had a research student in Dundee since the beginning of time!)."[65] One of his few satisfactions in this scientifically barren environment was to travel to Edinburgh and discuss physics with Born—another pacifist.[66] Graduate students were rare, but still, he was able to collaborate with W. E. Duncanson, then at University College, London, and S. Rushbrooke, who came to Dundee for a third year as a graduate student.[67]

While at Dundee, Coulson (1939a) completed his work on the lengths of the links of unsaturated hydrocarbons.[68] Lennard-Jones and Coulson had already shown how the lengths of these links could be calculated in the framework of molecular orbitals (Lennard-Jones 1937; Lennard-Jones and Turkevich 1937; Coulson 1938). They had also been computed using the valence bond theory. In both cases, the calculations were "extremely cumbersome" (Coulson 1939a, 1069). Therefore, Coulson chose to outline a method for the calculation of those lengths by assuming an empirical expression for the interaction energy between each pair of contiguous carbon–carbon bonds. The lengths obtained were in good agreement with those calculated with the valence bond method. He went on to assess the relative merits of the two methods by agreeing that his own empirical interaction term, much like resonance, is one of the ways of expressing the mutual influence of trigonal bonds, even though resonance is in a sense artificial because it results from a forced separation of variables. Much like resonance where various unknown parameters have to be obtained by correlation with certain observed quantities, his method was also empirical, and again unknown parameters had to be obtained in the same way. He believed that both treatments were essentially simplifications of the full quantum mechanical analysis, which was much too complicated to permit accurate calculations. He then suggested that his own calculations should not displace the ones done by using resonance because the latter are able to take account of other properties, such as excitation, fine structure, electrical and magnetic polarizabilities, and so forth, which his own model was unable to predict.

He sent the paper to the *Journal of Chemical Physics* in 1939. The correspondence with Harold C. Urey, the first editor of the journal, is revealing. Urey sent to Coulson some comments by the referee in which he acknowledged that Coulson found a "simple completely empirical method" of predicting values of bond lengths that agreed approximately with those predicted by quantum mechanical calculations. However, the referee believed there was not a clear outline of the method, as he stated that the lengths of the rings can be determined by purely classical considerations, in which resonance plays no part. The referee believed that the empirical interaction terms used by Coulson represented essentially the "phenomenon which is conventionally called resonance." The main objection of the referee, though, was that the discussion in the paper was unsatisfactory due to the "incorrect point of view of the author." He suggested omitting the "well understood argument regarding the artificiality of the concept of resonance." In conclusion, the referee suggested the paper should be shortened and that the author should concentrate on a description of the empirical scheme for making rough calculations of bond lengths.[69]

Less than a month later, Coulson returned the manuscript for reevaluation. He tried to incorporate all suggestions, although he considered that "most of them arise from a misunderstanding of the purpose of the paper." He included a letter addressed to the referee where it was considered that the referee's criticisms were mainly due to "questions of interpretation and meaning." His reply attempted to elucidate three points: the use of the word *classical*; the relation of the empirical interaction terms with resonance; and the argument about the artificiality of the concept of resonance. He noted that in naming his method classical, he never meant to imply that quantum theory was not required, but that his own calculations did not rely on the resonance technique. He disagreed with the referee's suggestion that the empirical interaction terms represented essentially the resonance phenomenon. He thought that it was just a matter of terms, "since, as you say, resonance is an artificial concept, it hardly matters whether we say that these empirical terms represent resonance or not." He reminded the referee that both approaches were approximations. But if his approach had any validity and if the theory of resonance was equally valid, "then the two must inescapably represent what is essentially the same phenomenon." He was careful in his paper not to criticize the theory of resonance and not to give the impression that his method was intended to displace any other approach. "After all everyone knows that resonance is an excellent and useful approximation, which enables quite a large number of deductions and predictions to be made; but it would be a pity if anyone ever came to believe that resonance was real, and that therefore without appealing to its phraseology and technique we could not interpret many of the facts of theoretical chemistry."[70] He continued to have the same views when, some years later, he went on to explore the notion of the chemical bond in the framework of molecular orbital theory (Coulson 1947b) at the same time that, in an article written for a semi-popular magazine, he became increasingly critical of the notion of resonance (Coulson 1947a) in the sense that, for him, resonance was not a well-defined molecular property but just a heuristic device.

Is resonance a *real* phenomenon? The answer is quite definitely no. We cannot say that the molecule has either one or the other structure or even that it oscillates between them . . . Putting it in mathematical terms, there is just one full, complete and proper solution of the Schrödinger wave equation which describes the motion of the electrons. Resonance is merely a way of dissecting this solution: or, indeed, since the full solution is too complicated to work out in detail, resonance is one way—and then not the only way—of describing the approximate solution. It is a "calculus," if by calculus we mean a method of calculation; but it has no physical reality. It has grown up because chemists have become used to the idea of localized electron pair bonds that they are loath to abandon it, and prefer to speak of a superposition of definite structures, each of which contains familiar single or double bonds and can be easily visualizable. (Coulson 1947a, 47)

Even if Pauling was not the referee, there is much in these exchanges to herald the future discussions between Coulson and Pauling on the ontological status of concepts and, especially, of resonance. In the final version, Coulson (1939a) opted to omit the argument about the artificiality of resonance although he was still of the opinion that the matter was of such relevance as to warrant another discussion. A few years later, Coulson formulated a response to Pauling's program outlined in *The Nature of the Chemical Bond*. In a paper deliberately called "Quantum Theory of the Chemical Bond," Coulson (1941b) appropriated Pauling's concepts of hybridization and of maximum overlapping and translated them into the language of molecular orbitals.[71] While Mulliken was adamant that there was no such thing as a chemical bond, for Coulson molecular orbital theory did not have to abandon a pictorial interpretation of chemical bonds. On the contrary! He filled his paper with diagrams, which later became familiar to any college chemistry student, depicting the formation of the molecular orbital in water, the formation of "double-streamers" in ethylene, and the formation of the chemical bonds in benzene depicted by what came to be known as the "doughnut" model.

Coulson in Oxford and London: Conceptual Clarifications

Sidgwick at Oxford was quite decisive in creating a milieu where the possibilities opened by quantum mechanics for chemistry were actively sought. Although 53 years old when the new quantum mechanics was first formulated, he was immediately converted to it. As we saw in chapter 2, through his book *Some Physical Properties of the Covalent Link in Chemistry* (1933), his annual reports, and his presidential addresses, he became one of the most effective advocates of the immense usefulness of resonance for chemistry. But Sidgwick's contributions toward the establishment of quantum chemistry in Britain were not limited either to his persistent propaganda in favor of resonance theory or to his strategies to convince organic chemists. He also contributed to the "modernization" of the chemical tradition at Oxford. In this endeavor, he was helped by his student Leslie E. Sutton. Especially after the late 1930s, when Sidgwick was already well into his sixties, Sutton took upon himself the task of turning Sidgwick's former "dipole group" into a "molecular research group." He was also the behind the creation of the Physical Chemistry Laboratory.[72]

Coulson profited from this new scientific environment when, at the end of 1944, he followed the advice of R. P. Bell from Balliol College and the Physical Chemistry Laboratory and applied for the newly established I.C.I. Fellowship to come to the

University of Oxford as a theoretical physicist to work with the people from the Department of Chemistry. The idea of a theoretical physicist in a chemistry department was, of course, quite unusual. Coulson received the fellowship and was appointed lecturer in mathematics at University College, starting at the same time a small research group at the Physical Chemistry Laboratory with graduate students (mainly chemists, but also some physicists) who would also be keen in computational techniques. Among his first graduate students were H. C. Longuet-Higgins, Bill Moffitt, and Roy McWeeny.

While at Oxford, Coulson's belief that organic chemistry was a particularly promising area for the application of quantum mechanics was reinforced—the organic molecules are sufficiently stable and form regular crystals, facilitating a theoretical analysis of their structure. Coulson (1946, 1947, 1949) started thinking of ways to improve the mathematical representation of the old chemical concept of free valence in the context of molecular orbital theory. He was probably unaware that this topic had already been addressed by the French Raymond Daudel and Alberte Pullman, who were among the latecomers to the discipline (see chapter 4). The concept of free valence had proved to be very useful in many chemical applications. Coulson's new idea was to compute the total bond order number N_i of each carbon atom in a hydrocarbon molecule by adding up the orders of all bonds to it. Experiments with carbon atoms in different positions in different hydrocarbons showed that there was a value N_{max} for the total bond order. Defining $F_i = N_{max} - N_i$, Coulson concluded that a large F_i meant a carbon atom with high reactivity, whereas a small value of F_i meant the opposite because most of the electrons of the carbon atoms are bound and thus unavailable for reactions.

Together with his student Bill Moffitt, Coulson also looked for an interpretation of the old chemical concept of bent bond (Coulson and Moffitt 1949). In the molecule of cyclopropane (C_3H_6), each carbon atom is located at the vertices of an equilateral triangle and is surrounded by four ligands (two carbon and two hydrogen atoms). By analogy with what happens with methane, one should expect tetrahedral bond angles of about 109°. However, the geometric configuration of the carbon frame implies that the angles between two adjacent C–C bonds are instead 60°. If in normal molecules one expects bond hybrids to be oriented in the direction of the bonds to ensure maximum overlapping, Coulson and Moffitt argued that in molecules such as cyclopropane, this tendency had to be balanced against the need to form a wider angle of the hybrid. Minimizing the energy of the molecule, they found that the bond hybrids pointed toward a different direction than the bond direction.

With another student of his, Longuet-Higgins, Coulson was hard at work developing a general theory of conjugated systems (Coulson and Longuet-Higgins 1947, 1947a, 1948, 1948a, 1948b). What they meant by "general theory" was the theory of molecular orbitals. The electron densities and bond orders were computed as first order derivatives of the energy of the mobile electrons with respect to the energy integrals appearing in the secular equations. Other relations were also established, with the purpose of corroborating the validity of the definitions made. The concepts of atom–atom and bond–bond polarizability were here introduced for the first time. They discussed the mutual polarizability of two atoms or two bonds or an atom and a bond and their importance in determining the effect of structural changes on chemical reactivity; that is, how a perturbation of an atom or bond affects another atom or bond. They then related the results derived to the interpretation of the chemical behavior of the simpler types of conjugated systems; to predict how the reactivity of different positions in more complex molecules is altered when an atom is changed; to the change of bond orders as a result of the alteration of resonance integrals or Coulomb terms; to the variation of the force constant of a bond and the interaction constant between a pair of bonds; and finally to the discussion of the interaction of two conjugated systems across a conjugated single bond. One mathematical novelty was the extensive recourse to Green's functions to obtain the solution of the polynomial equation to which one can reduce the determinantal equation for the energy. Those solutions were not possible with the limited computing facilities of the time, and so without recourse to the contour integration method, no results would have been obtained except for very small molecules.

Not long after being settled in Oxford, Coulson was asked by J. P. Randall, who had just been appointed to the Wheatstone Chair of Physics at King's College, University of London, whether he would consider moving to the chair of theoretical physics at King's College. Coulson's dislike of London did not turn this into an appealing prospect. He had moved to Oxford because he felt that it had the best chemistry department in the country. Though he had learned a lot, much remained to be learned from his colleagues at Oxford (Altmann and Bowen 1974, 82). But a little push by his wife soon made him change his mind. They stayed in London from 1947 to 1951—a period during which the scientific community was readjusting to the post-war realities.

By 1948, when all their joint papers were published, Longuet-Higgins was still at Oxford. They wrote to each other quite often. Longuet-Higgins discussed the possibility of going to Princeton or Cornell—deciding "to take the bull by the horns" and see whether they would be prepared to let him study theoretical physics.[73] Coulson considered that an alternative might be a job as lecturer in theoretical physics with special emphasis on statistical mechanics at King's College where he had himself moved. As far as Longuet-Higgins was concerned, this seemed to him a good alternative to Cambridge or Leeds, as it offered the possibility of building a research group. Coulson added "as you know, theoretical physics with me is taken to include chemistry; and will continue to do so."[74]

In the same year, Lennard-Jones asked his former student Coulson, now a reputed quantum chemist, advice on who to consider for a senior position at the Theoretical

Chemistry Laboratory at the University of Cambridge. Coulson suggested three rather young men with a chemistry background: Longuet-Higgins, who was then lecturing on theoretical chemistry at Oxford; the Australian Allan MacColl of University College, London, who was mainly interested in theory; and the Australian D. P. Craig, who was working on theoretical chemistry and was considering with some success "the characteristic divergencies between resonance and molecular orbital approximations." Coulson expressed his dismay that there were "so few candidates for quite nice jobs."[75] Longuet-Higgins finally opted for a post-doc position in Chicago followed by another in Manchester.[76] In 1952, still in his late twenties, he succeeded Coulson as professor of theoretical physics at King's College, London, and in 1954 he succeeded Lennard-Jones as professor of theoretical chemistry at the University of Cambridge.

Soon after the papers appeared, Erich Hückel wrote to Coulson announcing that he was reviewing the series for the *Mathematische Zentralblatt* and commenting that "the way you treated the problem generally and the method developed are very ingenious."[77] A lot of people asked for reprints of the series, and in November 1950 Coulson suggested to Longuet-Higgins that they publish a small volume with their joint papers together thinking that the techniques they developed had "come to stay for a few years."[78] Nothing came of the project.

While in London, Coulson noted that they were not "a learned company and in fact I am the only full-time theoretical physicist on the staff, but there are nuclear physicists under Champion, radio physicists, and biological physicists under Professor Randall, all of them with seven or eight research students."[79] But a large and active research group was soon to gather around Coulson.[80] He started some activities such as the coffee party seminars in the country of tea breaks, and the "centenaries"—celebrations commemorating each set of one hundred publications by the group.

This was also the period in which Coulson became much in demand as a speaker, especially after his extensive involvement with activities related to religion and pacifism. In 1950, Coulson published "The Christian Religion and Contemporary Science," his first major article on science and religion, thereby initiating his amazing activity at broadcasting, giving speeches and sermons. In an intended pun or unintended slip, the media identified him as the holder of the chair of "Theological Physics."[81]

This was also the year in which Coulson was elected a member of the Royal Society of London. Before his election in 1950, an enthusiastic Lennard-Jones—who was the person informing him when he was not elected to become FRS 2 years earlier[82]—was writing to Coulson to tell him about his nomination and to confide how gratifying it was to him that he continued to be so strikingly successful in his work. The discussion at the Royal Society had brought out "how greatly you have contributed to chemistry and how much you have helped the experimentalists. May you go from strength to strength!"[83] Four days before Lennard-Jones's letter, Hinshelwood had also written to Coulson and confided: "the matter is still in that 'semi-confidential' state which remains a little unsatisfactory for everybody concerned; but I feel I must write and congratulate you most heartily on your nomination for election to the Royal Society."[84]

Coulson continued to be preoccupied with the conceptual foundations of the molecular orbital theory, of the valence bond theory, and about the criteria for their comparison. He strongly believed that despite the extensive use of quantum mechanics, it was still not possible to either choose one of the theories as superior to the other or to consider them as competing approaches. This had been in fact one of the main controversies raging over the first decade of quantum chemistry: "we all learned about the molecular-orbital and valence-bond theories, and we became as partisan about them as, in Britain, we are partisan about the Oxford and Cambridge Universities' boat race on the Thames!" (Coulson 1970, 259).

In 1949, Coulson discussed the possibilities and limitations of the molecular orbital theory with the Swedish student Inga Fischer, as he was still uneasy about the fact that the extensive—and successful—use of the method did not contribute much to understanding the reasons for its validity "except on semi-empirical grounds" (Coulson and Fischer 1949).[85] The way they probed the possibilities of the molecular orbital approach was by including in the wave function of the ground state additional terms multiplied by parameters whose values could be determined by using the standard variational method to minimize the energy function. Hence, a value for these parameters was found for each value of the interatomic distance R, and, hence, it was possible to plot the change in these parameters as a function of the internuclear distance. One could proceed to a number of assessments by looking at these graphs. At those internuclear distances for which these parameters are equal to one, it appeared that the molecular orbital approach is quite valid. There are internuclear distances at which one would expect some effect of the additional terms, and, yet, these parameters fall to zero or are quite large—in which case it appeared that the molecular orbital approach cannot give us much useful information. They extended their method to include the excited molecules finding that the description of the excited states as the "sum" of pure configurations is rather naïve and that by including terms that expressed configuration interaction, they improved on the predictions of the molecular orbital approach. Configuration interaction accounted for the situations in which the description of a molecular state in terms of one single configuration wave function broke down, and it occurred when the interactions between configurations are comparable with the energy intervals of spectroscopic interest. The work continued with Craig—who had worked with the idea of configuration interaction before Coulson—and Juliane Jacobs, and they proceeded to analyze further the possibilities provided by configuration interaction in their search for more complex methods of approximation in dealing with π-electron systems than the simple semiempirical methods that had been used with some success (Coulson and Jacobs 1951; Coulson, Craig, and Jacobs 1951). Coulson's probe into the molecular orbital theory had been a combination of the possibilities implied by the pictorial representation of a two-center molecule *and* the limitations posed by the mathematical treatment.

In the introductory paper at the 1950 Faraday Society Discussion on electronic spectra, Coulson expressed his own views on the history of the subject.[86] He laid out what had been already achieved, what were some of the main results to be presented at the meeting, and what were the open problems that had yet to be tackled. During the past 20–25 years, people had concentrated on the relatively simple problem of electronic transitions in diatomic molecules. At the very beginning there was little theoretical guidance except by Lennard-Jones, Hund, Herzberg, and Hückel. Their wave mechanical studies had confirmed almost all the details of the picture largely developed semiempirically by Mulliken, which was based on the united atom viewpoint; the classification of individual electron orbitals into σ- and π-orbitals according to the symmetry they possessed; and the use of the isoelectronic principle (pairs of related molecules such as CO and N_2, which share the same number of valence electrons and have related spectral transitions, are studied together and compared).

Though it was assumed that molecules could be constructed through analogies with the atoms (orbitals, energies, transitions, etc.), the "intractable character of the numerical work was such that no really detailed ab initio calculations could be made except perhaps for molecular hydrogen" (Coulson 1950a, 2)[87] In attempting to study large molecules, the chemists started to realize that many of the valence electrons cannot even be treated as belonging to one bond, with a local behavior similar to that obtained in diatomic molecules, but they must be treated as if they belonged to the whole molecule. This was particularly true of the π-electrons in organic chemistry. The three main ideas he had mentioned for the diatomic molecules—united atom viewpoint, classification according to symmetry of σ- and π-orbitals, isoelectronic principle—could now be applied to these large molecules also, but with an important difference. These ideas had been changed "in the process of adaptation, but I do not believe we shall properly appreciate the papers in this Discussion unless we recognize them in their new guise" (Coulson 1950a, 2).

In concluding his overview of the papers to be presented at the discussion, Coulson stressed that there was no longer a sign of a conflict between the valence bond and molecular orbital theories as it happened just a few years ago.

This is a healthy sign. It does not mean that all methods are in a nice happy state of mutual agreement. But it does mean that we now recognize the very limitations that are inescapable in any attempt at a thoroughgoing calculation of molecular spectra. It is an interesting stage at which we have now arrived, that we regard our calculations as being indicative and corroborative, rather than as settling the issues in dispute. (Coulson 1950a, 2)

He was always very apt at understanding the state of his discipline and finding ways to express the characteristics of the framework with respect to which there could be a consensus within the community.

In the same year as the Faraday Society Discussion, Coulson once again stepped back to offer, along the lines of what he had already done for electronic spectra, a critical review of the merits and shortcomings of the method of ionic–homopolar resonance. He believed that a sound theoretical basis of the method of ionic–homopolar resonance was very far from being established, so that its then-current status should be considered to be more semiempirical than was formerly supposed, stressing at the same time that such a conclusion "must not be allowed to detract us from or to obscure, the astonishing success which the theory has had in correlating a vast field of chemical knowledge and experience" (Coulson 1951, 64). He underlined the significance of hybridized orbitals for the understanding of molecular structure, especially since "this concept . . . above all others, *reveals the basic limitations* of the usual resonance scheme" (Coulson 1951, 67). He concluded his intervention by playfully suggesting that there is a "kind of uncertainty relation about our knowledge of molecular structure: the more closely we try to describe the molecule, the less clear-cut becomes our description of its constituent bonds!" (Coulson 1951, 73).

Back in Oxford: Quantum Chemistry as Applied Mathematics
In 1951, Coulson moved back to Oxford where he stayed until his death in 1974. He succeeded the famous astrophysicist E. A. Milne as the second holder of the Rouse Ball Chair of Applied Mathematics at Oxford. Chapman, E. C. Titchmarsh, and J. H. C. Whitehead were the other members of the Department of Mathematics. Coulson created the Mathematical Institute in 1952, collaborated in the growth of the School of Mathematics, and never missed an opportunity to underline the importance of pure and applied mathematicians talking to each other. He exerted a strong influence on the Chemistry School, starting to collaborate with the more theoretically inclined chemists, and created the University Computing Laboratory. As before, he built a strong research group, started again the Tuesday coffee parties, and was the inspiration for the annual summer picnics. In 1955, on a less social note, he started the Summer School in Theoretical Physics, which gathered initially 35 students, but it quickly grew to 60 people, and began the publication of an annual progress report of the contributions by his group, which was circulated worldwide. In 1971, he became president of the Institute of Mathematics and Its Applications. In 1973, he became the first holder of the chair of theoretical chemistry at the recently created Department of Theoretical Chemistry (figure 3.3).

His appointment to the Rouse Ball Professorship of Applied Mathematics appears to have been an appropriate moment to reflect on what he considered to be the main characteristics of the discipline he so much helped to consolidate in Great Britain. Coulson was in fact the most effective propagandist of the view that quantum chemistry, though an independent subdiscipline, should be very much in contact with the culture of applied mathematics. In his inaugural address delivered on

Figure 3.3
Charles Alfred Coulson in the study. Old Road, Headington, 1959.
Source: Courtesy of the Coulson family.

October 28, 1952, Coulson expressed some of his most articulate views on the subject. Milne in his inaugural address talked about "The Aims of Mathematical Physics." Coulson chose to talk about "The Spirit of Applied Mathematics," introducing in his title a word not so often used in scientific talks, but which hinted at Coulson's serious commitment to a worldview in which religion had also an important role to play.

Coulson started his talk by stating that the nature and the spirit of applied mathematics are very different things. The nature of applied mathematics is to discover the "inner structure and form" of the physical world. Its spirit is that of the "joy and passion of creative activity." He claimed that theoretical physics, as well as theoretical chemistry, are included in the field of applied mathematics. He even speculated that a new discipline such as theoretical biology will soon become a branch of applied mathematics. He positioned applied mathematics between pure mathematics on the one hand and experimental physics and chemistry on the other. Pure mathematics was built upon "unchanged and unchallengeable" theorems, and the facts of physics and chemistry were "equally stubborn." Applied mathematics was built around concepts, and its progress consisted in imagining a "new set of concepts to transcend the old. . . . It is the formulation of the concept that is important, because it is the real function of applied mathematics. After this, we can use pure mathematics to work the implications of our concept" (Coulson 1953a, 7). The introduction of new and more encompassing concepts was the main task of the applied mathematician, but one should not think of it as wholly directed by experiments. Often, concepts are not permanent but ephemeral, a situation that does not diminish in any way the important role they play. Applied mathematics should never become "an appendage of experiment" just as in the same way it should never "degenerate into a bastard form of pure mathematics" (Coulson 1953a, 11). Therefore, an applied mathematician should always rest his feet on solid ground, but his head must be in the clouds (Coulson 1953a, 12).

To illustrate the methodology of applied mathematics, Coulson adopted J. L. Synge's view on how the use of applied mathematics is related to a physical problem: a dive from the world of reality into the world of mathematics; a swim in the world of mathematics; a climb from the world of mathematics back into the world of reality, carrying the prediction in our teeth (Coulson 1953a, 12–13). Coulson noted that the real world is inhabited by physicists who in case they dive into "the ocean of mathematics" must be careful to fill their lungs with air. In contrast, the world of pure mathematics is inhabited by fish with gills, who, in case they stayed too long away from their native environment, would soon die. "Only the applied mathematician is truly amphibious, at home on sea or land. And unless he is at home in both elements, he is no true applied mathematician" (Coulson 1953a, 13).

Coulson went back to this theme in the following years. He chaired a radio discussion with S. L. Altmann, Craig, and M. J. Dewar aired on August 14, 1957. Called "Chemistry by Computation," they introduced quantum chemistry to a general audience, discussing its mathematical aspects, their role, and implications for the discipline's makeup.[88] In his contribution for the Jubilee volume of the Calcutta Mathematical Society, Coulson talked of what he thought to be the three most important features of mathematics: that it communicates a sense of beauty, that it conveys a sense of structure, and that the division into pure and applied mathematics is "both unrealistic and dangerous" (Coulson 1958–1959, 262). He added: "For my own part let me make this one remark; that I have no use for the pure mathematician who claims to like his subject because he believes it is of no use, nor the applied mathematician who likes his subject merely because it is useful" (Coulson 1958–1959, 265). In a program prepared for the Open University, Coulson lectured on "Making Models," once again distinguishing between a pure and an applied mathematician.[89]

If there were disembodied spirits inhabiting some non-material universe, and capable of strict logical thought, then I suppose they could become Pure Mathematicians. But what they could not become would be Applied Mathematicians. This is because Applied Mathematicians are concerned with the real world. They want to describe it, to talk about it, to feel they understand it, and—not least—be able to make predictions about it.

These were themes he had already addressed in previous occasions. In the first inaugural lecture delivered as professor of theoretical physics at King's College, University of London, he reflected on the role of wave mechanics in chemistry (Coulson 1948).[90] There, as many times afterwards, he chose to call attention to the fact that the real contribution of wave mechanics was to be found at the conceptual level and not at the computational level. In the Tilden Lecture delivered at Burlington House on October 18, 1951, Coulson talked again about the role of quantum mechanics, repeating and rephrasing many of his former arguments, to point that quantum mechanics should not be considered the answer to all chemical problems. Contrary to what Dirac asserted in the Bakerian Lecture (1929), the true contribution of quantum mechanics to chemistry was not that it provided its mathematical theory, "as I have to remind myself when I am being urged to start immense schemes for the numerical evaluation of molecular integrals" (Coulson 1955a, 2069). Instead, what quantum mechanics had achieved was to show what was going on at the deepest possible level. First and foremost, quantum mechanics provided insight and understanding. Wave mechanics had given "flesh and blood" to what chemists inherited from their colleagues who worked in the classical tradition. It added "the quality of a deeper understanding. . . . We do see more deeply now into the meaning of our subject—what is really happening in chemistry. That, and not the calculation of a binding energy or a dipole moment, is the contribution of wave mechanics" (Coulson 1955a, 2084).

The Textbook *Valence*

It is, therefore, no wonder that Coulson's insights on the importance of quantum mechanics for chemistry were also expressed in the textbook *Valence* published in 1952 (Simões 2004, 2008a). It was neither his first incursion in textbook writing nor his last, but it was certainly his most famous. Coulson had written a small book called *Waves* (1941) for advanced undergraduate courses, followed by *Electricity*, published in 1948. In both, the unifying role of mathematics was explored. And in many papers for more general audiences, he addressed the relations of mathematics and the mathematical apparatus of quantum mechanics in relation to chemistry. V*alence*, which Coulson began writing during his tenure as a professor of theoretical physics at King's College, London, and which came out in 1952 when he was already back in Oxford, would eventually become a classic, a must for every student of chemistry!

Coulson would emphasize time and again that the enormous progress of valence theory during the past 25 years was due to wave mechanics. It was imperative that every chemist should be sufficiently at ease with the concepts and techniques that lie behind modern valence theory—even if it was not realistic to expect that every chemist would be in a position to make his own theoretical calculations. He argued that the elucidation of a large part of chemistry by quantum mechanics forbade chemists to be happy with an electronic theory of valence couched in pre–quantum mechanical terms. But the book should be understandable by a chemist with no mathematical training, and he was going to provide some introductory aspects of quantum mechanics. In many instances, mathematical results were illustrated or complemented by the extensive use of visual representations, an implicit acknowledgment that visualizability remained one of the more effective ways to communicate.

The message was loud and clear—quantum mechanics had become a hegemonic worldview and provided the fundamental theoretical framework for chemistry and was here to stay. But Coulson wanted to find ways to appropriate quantum mechanics to the chemists' culture. His main purpose was to communicate the fundamental reasons why molecules are what they are and how the theoretician looks at his problems. "Practically no mathematics is needed for this purpose, since almost everything necessary can be put in pictorial terms" (Coulson 1952, v).[91] But this situation did not pose a threat to the relative autonomy of chemistry in relation to physics because "contrary to what is sometimes supposed, the theoretical chemist is not a mathematician, thinking mathematically, but a chemist, thinking chemically" (Coulson 1952, v). Quantum chemistry was not another instance in the application of quantum mechanics but a new subdiscipline of chemistry. Partly echoing Pauling, Coulson proceeded to a rather strong statement: that the particular form of mathematics in theoretical chemistry is suggested "almost invariably" by experimental results. And because it is impossible to have any exact solution of the wave equation for a molecule, the approximations to an exact solution "ought to reflect the ideas, intuitions, and conclusions of the experimental chemist" (Coulson 1952, 108–109). This was a rather intriguing proposition. The approximations to the exact solution were not exclusively the result of mathematics per se, but of the mathematics that had been suggested by the chemical intuition, ideas, and experimental results of the quantum chemist. Hence, the form of mathematics and the chemists' (collective?) experience were in a dialectical relationship: one influenced the other and vice versa.

The presentation of molecular orbital and valence bond theory did not follow their historical order of appearance. Molecular orbital theory was selected as the first topic because it is "conceptually the simplest" (Coulson 1952, 68). Though *Valence* was undoubtedly a book sympathetic to the molecular orbital viewpoint, the valence bond method received a fair treatment. Coulson considered both schemata as approximations whose range of validity had been sufficiently understood "for us to recognize the folly of trusting to either alone" (Coulson 1952, vi). In the book, Coulson acknowledged that resonance represented one of the most powerful ways in which chemical intuition guided one into finding suitable wave functions, and that the appeal of the VB method was in its selection of component wave functions that carry pictorial connotation. No analogous comments can be found for the MO approach. But the book included the new diagrams introduced in the papers already referred to (Coulson 1941b, 1947a, 1947b), thereby offering a visual representation of the formation of bonds in water, ethylene, and benzene (figure 3.4). *Valence* was a hit. It sold very well and at a very reasonable price for students. In the 3 years following its publication, 8000 copies were sold, and the American market contributed to absorb a considerable fraction of the copies printed because many colleges had put the textbook on the reading lists of courses in quantum or theoretical chemistry. The same was to happen with the second edition (1961) (Simões 2004, 315).

As soon as his "little book" *Valence* appeared, Coulson made sure to send a copy to Pauling.[92] In the first days of September 1952, a review of the book by Pauling appeared in *Nature*. The review was definitely hostile. Pauling believed that both the treatment of the mathematics in quantum mechanics and the facts of structural chemistry were brief and sketchy. "This treatment which might be called descriptive quantum mechanics as opposed to theoretical quantum mechanics—may be suited to some readers, but I fear that many would be confused" (Pauling 1952, 384–385). He did not believe that the chemistry students could use the book in order to acquire a solid quantum mechanical framework. He further criticized Coulson for his overenthusiasm for the molecular orbital method, which pushed him to make various unsupported claims. The marginalia in Pauling's copy of *Valence* reveal thoughts he had and that did not find their way into the review. "This is not a book about the broad subject of valence as the chemist understands and uses it but is a book about the quantum mechanical theory of covalence. There is no mention of oxidation number, the ordinary valence used by the chemist in the consideration of oxidation-reduction reactions, nor is there any general discussion of the valencies of the elements in relation to atomic numbers."[93]

In a short while a very positive review written by Wheland (1952) appeared in the *Journal of the American Chemical Society*.[94] Wheland appreciated the fact that Coulson presented the two approaches in a complementary manner and, as if he wanted to answer Pauling's criticism about the unsatisfactory way of introducing quantum mechanics and the relevant mathematics, praised the specific treatment as being "the best rounded and most nearly complete one that is now available." Might it be the case that Wheland used the review to try to settle the unsettled issues between him and Pauling? How else can one understand the following passage: "The book is more convincing as well as more up to date than is Pauling's *The Nature of the Chemical Bond* which of all the older treatments of valence is doubtless the one more closely analogous to it" (Wheland 1952, 5810).

Coulson immediately wrote to Pauling. He confided to Pauling that he had already been criticized on the grounds that the book had either "too much mathematics" or, on the contrary, did not have enough. In Pauling's review, there were two remarks in particular that worried Coulson. Not unexpectedly, one concerned what Pauling believed to be Coulson's "over-enthusiasm for the m[olecular] o[rbital] method," and

Quantum Chemistry *qua* Applied Mathematics

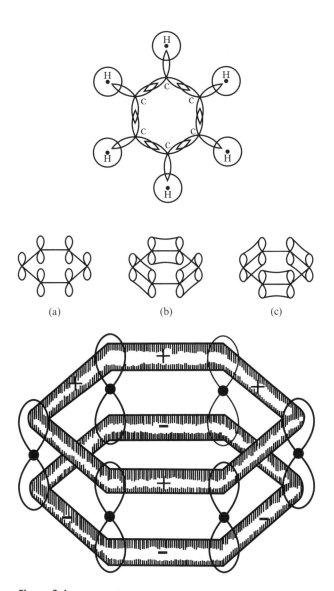

Figure 3.4
(Top) The σ hybrids of the carbon atoms of benzene. The π atomic orbitals in benzene (a), and the Kekulé pairing schemes (b, c). (Bottom) The π molecular orbitals in benzene (double streamers).
Source: Reprinted with permission from Charles Alfred Coulson, *Valence*, Oxford at the Clarendon Press, 1952 (on pp. 224 and 226). Copyright © 1952, Oxford University Press.

the other was his treatment of hybridization, to which a whole chapter was dedicated. Coulson believed that a reason for Pauling's criticism may be the fact that the latter attributed a different meaning to hybridization than most people. Coulson concluded his letter asking for Pauling's help to clarify the whole issue about hybridization.[95] He had mentioned in the book that resonance represents one of the most powerful ways in which chemical intuition may be introduced into the finding of a suitable wave function and that the ingenuity of the valence bond method is that it selects as component structures wave functions that do carry pictorial connotation. No analogous comments are found for the molecular orbital approach. One wonders why Pauling could not have been a little more generous with his comments about Coulson's book.

In his reply, Pauling wrote that he was sorry for not liking the book more. In fact, it was the chapter on "hybridization" that particularly disturbed him. Hybridization was a concept he "discovered (or invented),"[96] the cornerstone of his resonance theory, and he believed that Coulson did not give him proper credit. The first reference to his work appeared already well into mid-chapter, and then only in reference to *spd* hybridization. However, most of the material contained in the chapter was published first in his paper of 1928, in which he hinted at hybridization as a possible explanation for the tetravalency of carbon (Pauling 1928a), and then in the first paper of the series on "The Nature of the Chemical Bond," published in 1931, in which he discussed analytically hybridization (not yet named as such), overlap integrals and strength of bond orbitals, tetrahedral hybrid orbitals, trigonal, digonal hybrids, and so on. Pauling acknowledged that there is an element of arbitrariness in hybridization as there is in resonance theory, but he believed that these notions had constituted a "significant contribution to chemical theory." And finished the letter writing that he had "greater confidence in some of my own simple calculations than in some of the very complicated calculations, involving a larger number of arbitrary decisions, that have been published."[97] Might it just be the case that this statement has the seeds of Pauling's realization that things were drastically changing in quantum chemistry and his hegemony was being questioned?

Pauling also reacted against the discussion of CO for giving the impression that its properties are better understood on the molecular orbital basis than on the valence bond basis. The discussion of electronegativities also displeased him: He insisted that his contribution was based on theoretical (quantum mechanical) arguments and not, as Coulson claimed, on purely empirical grounds, and he also disagreed with Coulson's assessment of Mulliken's scale of electronegativities as being better than Pauling's. Nevertheless, Pauling was willing to confess—in the private correspondence and not in any public way—that his comments should be taken as personal opinions about questions to which there did not appear to be any final answers. And in a subsequent letter, Pauling argued that bond strength is a quantity offering greater reliability for discussing hybridization of orbitals than the approximate detailed calculations involving minimization of energy.[98]

In a letter to Pauling, Coulson apologized for his unintentional unfairness toward some viewpoints associated with Pauling, confessed that "I have learnt so much from your work myself as to make me always grateful," and agreed with Pauling on the importance of the 1931 paper: "I agree with you that the 1931 paper was one of the best things you have ever done; and this is still true even if we discover, as time goes on, that a good many of the details require a certain amount of alteration."[99] In Pauling's reply,[100] the angry tone of his former letters had subsided.

Despite Pauling's insistence on the central role of his structural approach to quantum chemistry, he most probably believed that in Coulson he had found, at long last, a non-antagonistic and understanding interlocutor. In fact, the second edition of *Valence* (1961) incorporated most of Pauling's comments and gave more weight to his two earlier papers. Resonance was dealt with in a more systematic manner. Coulson was also uncharacteristically assertive about the status of resonance. He objected to Pauling's choice to raise resonance into a chemical category, taking resonance as just a heuristic device, an algorithm or a metaphor or simply one pedagogically expedient method (out of various) for understanding quantum chemistry. But despite Coulson's view that nothing good came out of a strong commitment either to the valence bond method or the molecular orbital method, the fact remains that the latter viewpoint received an impressive boost with Coulson's two editions of *Valence*. And, as late as 1970, Coulson would assert that "resonance is a dirty word" (Coulson 1970, 258).

The Unseen Chemical Bond
Quantum chemistry revolved around the problem of the chemical bond. The most famous books in its history refer to it. The theoretical, and perhaps philosophical, status of the bond has been continuously negotiated in order to achieve some kind of consensus concerning this elusive entity or notion, so decisive for the foundation of quantum chemistry. Coulson was among those who were rather assertive in the negotiations. He called the chemical bond a "concept of the imagination." According to Coulson, all chemistry rests on the idea of a chemical bond, and every generation of chemists has tried, in its own way, to describe what is a bond. The different descriptions that have been given show how greatly our understanding of the "real essence of chemistry" has developed in the past since Edward Frankland or Kekulé (Coulson 1952a, 11). For nearly 100 years, chemists noticed the characteristic affinities of one substance for another. Lewis had suggested that this affinity is related to the disposition of two electrons, but "remember, no one has ever seen an electron!" (Coulson 1953a, 20–21). The quantum mechanical underpinning of Lewis's description showed that the shared electrons have their spins pointing in opposite, or antiparallel, directions, but "remember, no one can ever measure the spin of a particular electron!" (Coulson 1953a, 20–21). However, everyone was captivated by "the simplicity of the idea" (Coulson 1970, 287). Then the distribution in space of these electrons is described analytically with closer and closer degrees of precision, but "remember, there is no

way of distinguishing experimentally the density distribution of one electron from another!" (Coulson 1953a, 20–21).

In the meantime, concepts like hybridization, covalent and ionic structures, resonance, and fractional bond orders had been introduced. Coulson was rather uneasy that none of these concepts could be linked to a directly measurable quantity. Nevertheless, "chemical knowledge and, perhaps even more, chemical intuition, find their full expression and their proper setting within the mathematical framework that has now been devised" (Coulson 1952a, 13). The importance of conceptual insightfulness together with the usefulness and truthfulness of concepts is stressed again and again in Coulson's writings. He believed that a bond is "no more real than the square root of −1!" (Coulson 1953a, 20–21). And he continued to hold this view a couple of years later when he noted that:

> Sometimes it seems to me that a bond between two atoms has become so real, so tangible, so friendly that I can almost see it. And then I awake with a little shock: for a chemical bond is not a real thing: it does not exist: no-one has ever seen it, no-one ever can. It is a figment of our own imagination. (Coulson 1955a, 2084)

What was going to happen in the future to the idea of a bond? Coulson gave two possible answers to this interesting question. The work of the next years will have to be more concerned with refining, and perhaps simplifying, the sort of description already worked out (Coulson 1952a, 12). As already hinted at in the introduction, at a symposium commemorating the 50 years of quantum chemistry that took place in 1969, Coulson went much further:

> So to the question: has the chemical bond now done its job? Have we grown to that degree of knowledge and that power of calculation that we do not need it? . . . This a tantalizing question. And only a little can be said by way of comment. Chemistry is concerned to explain, to give us insight, and a sense of understanding. Its concepts operate at an appropriate depth, and are designed for the kind of explanation required and given. If the level of enquiry deepens, as a result of our better understanding, then some of the older concepts no longer keep their relevance. . . . From its very nature a bond is a statement about two electrons, so that if the behaviour of these two electrons is significantly dependent upon, or correlated with, other electrons, our idea of a bond separate from, and independent of, other bonds must be modified. In the beautiful density diagrams of today the simple bond has got lost. It is as if we had outgrown the early clothes in which, as children, we could be dressed, and now needed something bigger. But whether that "something bigger" that should replace the chemical bond, will come to us or not is a subject, not for this Symposium, but for another one to be held in another 50 years time, and bearing for its title: the changing role of chemical theory. (Coulson 1970, 287)

Swan's Song: Coulson and the Chair of Theoretical Chemistry

In 1973, Coulson was appointed as the first professor of theoretical chemistry of the newly created Department of Theoretical Chemistry of the University of Oxford. The inaugural address was titled "Theoretical Chemistry Past and Future," and, as expected, he further elaborated on the points he had raised in his other two inaugural lectures of more than 20 years ago. The occasion, however, called for a clarification of what is meant by theoretical chemistry, and specifically why it deserved an independent department.

Theoretical chemistry is not a recent discipline, he said, but an old subject. Theoretical chemistry is of value only if it rests on concepts that are not only true but also useful. "It is the interplay between theory to give us insight and understanding, and experiment, related to and often suggested by, our theory, that is the role of theoretical chemistry" (Coulson 1974, 6).[101] The relation between facts—which, Coulson insisted, are not independent from theory—and theories is one of interdependence, out of facts there comes theory (although there is "no logical way" to pass from the one to the other), and the theory is the basis of new experiments. Granted the importance of concepts in the development of a scientific discipline, he enquired, "how are we to classify the concepts of chemistry?" Chemistry often has been considered to be a "less fundamental science" than physics basically because the concepts of chemistry have been "less deep" than the ones of physics (Coulson 1974, 8).[102] He emphasized that every science operates at its own characteristic depth, and the fact that chemistry has been considered a less fundamental science than physics should not be considered in any sense derogatory. The increasing interaction of chemistry with mathematics led to the further refinement of the concepts of chemistry. Though Coulson stressed the privileged relation of theoretical chemistry to applied mathematics, he also noted that it was not algebra, but quantum mechanics instead, that had altered the chemist's understanding of his own discipline. This "revolution" had affected all former areas of chemistry and had been accomplished in two different ways, through the exact and approximate solutions of Schrödinger's wave equation: The possibilities offered by the newly developed computers should not discourage the theoretical investigations of chemical problems. These approaches were not in mutual conflict, and both were needed and complemented each other, so that "the particular approach which a person makes to the use of a computer almost determines his judgment on the relative merits of the two types of study" (Coulson 1974, 20).[103] As we will see in the next chapter, Coulson appeared to be less alarmed than he was at the Boulder Conference concerning the divisions of the community of theoretical chemists. Time, no doubt, and his disposition brought out a conciliatory rather than a partisan view.

Some Further Remarks

More than any other group, it was the British theoretical chemists who undermined the uneasy relationship between chemistry and mathematics. The work of Lennard-Jones, Hartree, and, above all, Coulson contributed decisively to transforming what was a problematic relationship into a rather promising symbiosis. Their research papers, talks delivered at meetings, new academic courses, inaugural lectures, Coulson's textbook on *Valence*, as well as their popular presentations, were all different means used tirelessly to explore all kinds of mathematical techniques in order to bring chemical problems into the domain of applied mathematics. They, thus, legitimized a new space for the practice of theoretical chemistry. It was a space that was delineated not only by the successful appropriation of quantum mechanics but also—and this appeared to be of comparable importance—by a host of mathematical and numerical methods.

In the development of quantum chemistry, one can identify at the epistemic level four procedures—conceptual, mathematical, experimental, and pictorial—that have always been complementary to each other, each having at the same time a relative autonomy, and each coming to the rescue of the whole enterprise whenever some of the others were reaching their limits. For members of this group, the development of numerical techniques became important because they contributed toward conceptual clarifications. Simultaneously, for Coulson the exploration of the mathematical potentialities of quantum mechanics went hand in hand, and against all odds, with a plea for its pictorialization. Coulson's consistent practice proved that the mathematization of quantum chemistry and its visual expressions were not incompatible defining characteristics, to such an extent that he was able to usurp one of the long-lasting appeals of Pauling's rhetoric concerning the identity of valence bond theory and the old chemical structural theory. With the appropriation of visual representations by molecular orbital theory and the awareness of its special suitability to further mathematical developments, the work of the British, and especially Coulson, acquired a progressively increasing appeal among chemists.

During the development of quantum chemistry, many mathematicians, physicists, biologists, and, of course, chemists raised conceptual issues not confined to the discussion of the aims and methods of quantum chemistry. Mostly in popular lectures, and often when discussing the interface of science and religion, Coulson in particular addressed questions such as the role of beauty, simplicity, and other such themata associated with the context of discovery as well as the role of predictions, the evaluation of "falsifications" in the assessment of the validity of theories, and the characteristic "incommensurability" of traditions (Simões 2004). Though Coulson's work accelerated the establishment of the ab initionist culture of theoretical chemistry, Coulson himself appears to have been entrenched in the more traditional culture. He was deeply committed to the view that theoretical chemistry was first and foremost an enterprise whereby mathematical notions, numerical methods, experimental measurements, pictorial representations, and, above all, chemical concepts constituted an undivided whole. There was a fine balance among all these aspects, a balance that could not be articulated in any distinct way, and yet, many (quantum) chemists believed, it was the distinctive feature of theoretical chemistry itself.

With the extensive development of numerical techniques, one gets the feeling that something started to change. One wonders whether through the work of Coulson, Lennard-Jones, and Hartree, and certainly against Coulson's expectations, the extensive and systematic development of calculational techniques led to a progressive "deconceptualization" of quantum chemistry. This is not to imply that the work of the British quantum chemists was devoid of any new and novel concepts. It is to stress that understanding the particularities of the practice of the British quantum chemists and of their overall outlook to reformulate the problems of quantum chemistry as problems of applied mathematics, one may appreciate more fully the multiplicity of theoretical cultures that contributed to the development of quantum chemistry. It seems that, in the long run, the work of this group also contributed indirectly but decisively to turn theoretical quantum chemistry into what is now called computational quantum chemistry. Dirac's 1929 dictum might have found, perhaps, a particularly strong resonance with the agenda of his Cambridge colleagues.

4 Quantum Chemistry *qua* Programming: Computers and the Cultures of Quantum Chemistry

Our "story" ends with two conferences: the conference of 1959 held at Boulder, Colorado, and the conference of 1970 held at Bethesda, Maryland. The former dealt with molecular quantum mechanics, and speakers talked about their subject within a totally new rationale compared with that of earlier conferences. It was the rationale formed by the realization that powerful computing machines were making their presence felt in no uncertain terms and that they were becoming an indispensable aspect of the future of quantum chemistry. If the 1959 Conference on Molecular Quantum Mechanics was heralding a new period of quantum chemistry, then the 1970 Conference on Computational Support for Theoretical Chemistry mapped the future of quantum chemistry in terms of the possibilities provided by computers, not simply as machines that would facilitate the calculational work of chemists but as instruments that would act as probes of amazing exactness, at times, even, substituting for the need for experiments. If what was reflected in the deliberations of the conference of 1959 was that computers were to become an indispensable tool for quantum chemists, then the discussions of the 1970 conference reflected a totally new social vista: the amazing development of hardware and software and the pivotal role of quantum chemistry in the development of computer technology as well as its mounting importance within chemistry. Post-war developments, however dramatic in their repercussions, were not limited to the impact calculating machines would have in the practice of quantum chemists. The emergence of two active European groups, one in Paris and one in Uppsala, widened appreciably the scope of quantum chemistry.

The Newcomers: Quantum Chemistry's Forays into New Realms

Raymond Daudel and the 1948 Paris Conference

It is certainly surprising to realize that quantum chemistry started in occupied France in 1943. This was the year in which Raymond Daudel (1920–2006) created the Centre de Chimie Théorique de France, initially a rather informal association that had the patronage of such highly respected figures of the French scientific establishment as the quantum physicist and Nobel laureate Louis de Broglie, the Nobel laureate couple Irène and Frédéric Joliot-Curie, and the physician Antoine Lacassagne. This was also the year in which Daudel co-authored the textbook *La chimie théorique et ses rapports avec la théorie corpusculaire moderne* with the aim of informing French chemists about what quantum mechanics could do for chemistry (Poitier and Daudel 1943).

The delayed emergence of quantum chemistry, as some authors refer to the emergence of the subdiscipline in France, has usually been accounted for by the devastating effect of the First World War and the subsequent isolation of French science, together with the dominant role of experimental organic chemistry in France, and finally the opposition of the physical chemist and Nobel laureate Jean Perrin to the quantum mechanical explanation of the chemical bond (Guéron and Magat 1971; Charpentier-Morize 1997; Blondel-Mégrelis 2001).[1] Why, then, did quantum chemistry appear during such uncongenial times? A clue to this apparent paradox might be the correlation of extreme and adverse conditions and the episodic vulnerability of the otherwise immutable structure of the hierarchical and closed French academic system (Pestre 1992). Extreme conditions call for drastic responses: The pioneers of quantum chemistry in France chose the difficult study of very big molecules.

Raymond Daudel, a "half-physicist, half chemist,"[2] was a student of the École Supérieure de Physique et de Chimie Industrielles de la Ville de Paris. He then became an assistant at the Faculté des Sciences de Paris, Sorbonne, and since the early 1940s was working at the Institut du Radium with Irène Joliot-Curie, who at the time was professor of chemistry at the Sorbonne. Next to them, in a building across the courtyard stood the Pavillon Pasteur of the same institute, created by Marie Curie in the aftermath of the First World War in order to explore the medical applications of radioactivity. Its director, the head of the "cancer people,"[3] was Antoine Lacassagne, professor of medicine at the Collège de France. In fact, Irène Joliot-Curie and Lacassagne supervised Daudel's doctorate on the chemical separation of radioelements formed by neutron bombardment, which was completed in 1944. Both Daudel and Lacassagne met often to discuss scientific topics not necessarily included in Daudel's doctorate. When Lacassagne stumbled upon a paper by the German Otto Schmidt in which Schmidt hypothesized about a relation between the electronic makeup of compounds and their carcinogenic effect (Schmidt 1941),[4] Lacassagne's inability to follow the details of the paper forced him to seek Daudel's advice (Lacassagne 1955).[5] Daudel foresaw the potential of an alternative and more sophisticated approach based on quantum mechanics but could not find the time to tackle the problem, unless Lacassagne could procure a grant to get someone to work on it.

The awardee was Alberte Bucher (born 1920), later Alberte Pullman after her marriage in 1945 to the quantum chemist Bernard Pullman. She was a Sorbonne science undergraduate knowledgeable in mathematics, chemistry, and physics. During her undergraduate years, to make ends meets she got a job at the Institut Poincaré doing all sorts of calculations "by hand, with logarithms, slide rule and mechanical calculating machines,"[6] ranging from computations of trajectories of projectiles to various statistical calculations. The grant offer at the Institut du Radium got her involved in quantum mechanics and, subsequently, in quantum chemistry.

From 1941 and until the end of the war, living in Paris meant surviving in German-occupied territory. The papers of Heitler, London, and Hückel were available but mostly unread because few commanded German. So was Pauling's *The Nature of the Chemical Bond*, through a clandestine microfilm copy.[7] It is no wonder that the first French contributions to quantum chemistry of Daudel's group, including those of Alberte Pullman, followed the methodology of resonance theory, which they named as "mésomerism."

Alberte Pullman developed a pictorial method within the general framework of resonance theory that came to be called "méthode des diagrammes moléculaires" (Daudel and Pullman 1946). Together with Daudel, this method was used to represent quantitatively the distribution of the electronic cloud using one single diagram (Daudel and Pullman 1946a). Besides determining the weight of each component of the wave function associated with each formula, and building on the notion of bond order, they also arrived at a definition of the free valence index (*indice de valence libre*) in the context of resonance theory. This was done independently of Coulson, who, unaware of their work, was to formulate the same concept in the context of molecular orbital theory (Daudel, Bucher, and Moreu 1944; Daudel and Pullman 1945; Coulson 1946, 1947, 1949).

In 1946, Alberte Pullman defended her Ph.D. thesis (A. Pullman 1946). She offered molecular diagrams for several complex organic molecules showing that they enabled prediction of the chemical and biological behavior of the molecule in a more precise and complete way than with any other formula. She established a relationship between electronic structure, and more specifically the charge density in certain molecular regions and the carcinogenic activity of aromatic hydrocarbons. In what came to be known as the K-region theory, she proved that certain molecular regions (K-regions) denoted a positive correlation between their electronic makeup and their carcinogenic potency (A. Pullman 1947).

In 1947, upon visiting and lecturing in the United Kingdom, she met Coulson, who was enthusiastic about her ideas on carcinogenicity and was eager to learn more about the topic. With the help of graduate students, they decided to study the naphthacene molecule ($C_{18}H_{12}$) by the method of molecular orbitals, which turned out to be a much simpler calculation than by using mesomeric formulas (resonance structures) (Berthier et al. 1948). A small aside: They published these notes in *Comptes Rendus* "in French, in France! It was quite an achievement that Coulson consented doing that. Coulson was British, very British, incredibly British!"[8]

While Alberte Pullman completed her Ph.D. work, Bernard Pullman (1919–1996) received his undergraduate degree (license-ès-sciences). They met during their first academic year at the Sorbonne (1938–1939), and were engaged in 1939. When France was overrun by Germany, Bernard joined General de Gaulle's Free French Forces (1940). The 5 years that followed "parallel closely the epic of Free France, through the jungles of Cameroun and Central Africa, the mountains of Erythrea, Syria, and Lebanon, and for nearly three years the deserts of Egypt, Lybia and North Africa." Because of his "scientific background," he became an officer in military engineering, "mining and demining, building bridges and blowing them up." He returned to Paris in February 1945, they were married 3 weeks later, and on May 8 they "were among the millions who completely jammed the Champs-Elysées" (B. Pullman 1979, 35).

After graduating, Bernard took the decision to join Daudel and Alberte in their project of consolidating theoretical chemistry in France by promoting quantum chemistry, but "if Alberte became a quantum chemist by choice, I became one essentially by marriage" (B. Pullman 1979, 35). Nevertheless, he realized he was enrolling in a different sort of fight, this time against "the rather unfriendly attitude of the professorial establishment toward theory in chemistry, and above all quantum theory." And when Alberte started working at the Centre National de la Recherche Scientifique (CNRS), he became the recipient of the grant Lacassagne had secured for her in the past.[9] Two years after completing his license-ès-sciences, Bernard completed his doctorate in 1948 with a dissertation on the effect of substituents upon the electronic structure of conjugated molecules (B. Pullman 1948). His work included both an experimental and a theoretical part. The experimental part dealt with the study of isotopic exchange reactions in relation to substituent effects and was justified partly for its relevance in the research work taking place at the Institut du Radium where the group was located.

The year 1948 was an important year for French quantum chemists. At the institutional level, the Centre de Chimie Théorique founded by Daudel was financially supported by the CNRS; Jean Barriol, whose work dealt with molecular quantum mechanics by the use of group theory, became the holder of the first chair of theoretical chemistry in France, in the University of Nancy; after Bernard Pullman received his doctorate, Alberte Pullman stopped her collaboration with Daudel's group to begin collaborating with her husband. Going their own different ways, Daudel and the Pullmans created two active research groups on quantum chemistry in Paris. The Pullmans' group, which came to include young researchers such as Jeanne Baudet, Gaston Berthier,[10] Hélène Berthod, André Julg, Marcel Mayot, and Paul Rumpf, shifted from valence bond to molecular orbital theory, and Daudel's group, which initially included among its most active researchers Sylvette Besnainou, Hélène Brion, Henri Moureu, Monique Roux, and Simone Odiot, tackled foundational issues involving the clarification of the compatibility of chemical concepts and quantum mechanics. In

Quantum Chemistry *qua* Programming

the same year, the first international meeting on quantum chemistry after the war took place in Paris.[11] The meeting played a triple function. It boosted the reorganization of the quantum chemical community; it secured a position for quantum chemistry within theoretical chemistry at the national level; and it acknowledged the standing of the French quantum chemists in the international community.

The Paris Colloque de la Liaison Chimique (figure 4.1) was sponsored by the CNRS, whose director was Frédéric Joliot-Curie, together with the Rockefeller Foundation, and was organized by Edmond Bauer, a physicist interested in exploring the application of mathematical theories, such as group theory, to quantum mechanics and who had written a small booklet on the quantum theory of measurement with Fritz London (Bauer 1933; London and Bauer 1939). The meeting drew almost all those who were actively involved in quantum chemistry: Coulson, Longuet-Higgins, Sutton, Lennard-Jones, Michael Polanyi, Mulliken, and Pauling. The theoretical chemists from France included Raymond and Pascaline Daudel and Bernard and Alberte Pullman.

Figure 4.1
The 1948 Colloque, most probably the Colloque de la Liaison Chimique.
Source: From Ava Helen and Linus Pauling Papers, Special Collections, Oregon State University.

Bauer delivered a paper on the history of the chemical bond, using the occasion to explore a number of attractive forces other than those forming the chemical bonds in order to revise and clarify the assumptions behind the old notion of the chemical bond. Both Mulliken and Pauling were the stars of the conference. Mulliken's major activity in the preceding year was the preparation of a long review report on molecular orbital theory with a view to future developments and intended for presentation at the Paris Conference. In fact, this was not the first occasion after the war's end in which he publicly took stock, but it was certainly the one with an enduring impact. The talk lasted all morning. Each slide shown by Mulliken (1949), and there were many, was followed by a comment immediately translated into French.[12] In the afternoon it was Pauling's turn. He spoke about the application of valence bond theory to metallic crystals, and he also talked for a long time. But he could not help to quip that Mulliken's talk was so long and boring that no one followed it, except Mulliken himself and the poor translator. In its turn, Mulliken later confirmed their ongoing antagonism by recalling that "I held forth and he held forth."[13] But he also recalled relaxing at a "gorgeous party at the Pullman apartment at which we were provided with endless bottles of champagne" (A. Pullman 1971, 10). In any case, the conference strengthened the relation of the emerging French groups in quantum chemistry with Coulson's and Mulliken's groups and gave them the much needed exposure and legitimization within the dynamic network of post–Second World War quantum chemists.

On a national level, the meeting marked the consolidation of quantum chemistry as part of theoretical chemistry and at the same time the emancipation of theoretical chemistry from physical chemistry, its old predecessor. Granting Bauer's engagement in work on the theoretical aspects of physical chemistry (Guéron and Magat 1971), the work in quantum mechanics and the quantum theory of the chemical bond gave rise to the delineation of a program for quantum chemistry with its own agenda. By entertaining close ties with experiment, by exploring quantitative methods to deal with large molecules but still looking for precision and rigor, by attempting to find exact quantum mechanical explanations of concepts used in quantum chemistry and often inherited from classical chemical theories, and, above all, by their preoccupation with molecules of interest to biologists and biochemists, the French quantum chemists initiated a tradition in quantum chemistry that moved away from what was stipulated by Perrin's program in physical chemistry, which hindered many of the initiatives in quantum chemistry (Pestre 1992; Charpentier-Morize 1997).

The Group of Raymond Daudel
After having been "officially recognized as a branch of science in France" (Rivail and Maigret 1998, 368) in 1948, theoretical and quantum chemistry gained institutional visibility throughout the 1950s. Major events were associated with the move of the

groups of Daudel and the Pullmans out of the Institut du Radium. In 1957, Daudel's Centre de Chimie Théorique was transformed into a CNRS research center called Centre de Mécanique Ondulatoire Appliquée. In 1962, the Centre de Mécanique Ondulatoire Appliquée moved to a building north of Paris in which the CNRS installed a CDC 7600 multipurpose computer, especially dedicated to atomic and molecular calculations. In 1954, the Pullmans moved to an old crumbling building belonging to the Fondation Curie, which was interested in their work on carcinogenesis. The building housed Pasteur in the past and recently served as an apartment house for nurses until it was considered unfit for them, but not for quantum chemists (B. Pullman 1979, 39)![14] While Alberte maintained her permanent position as a CNRS researcher all her life, in 1954 Bernard was offered a professorship in quantum chemistry at the Sorbonne. In 1958 he was invited to establish a laboratory of quantum biochemistry in the old Institut de Biologie Physico-Chimique, one of the best-known institutes of fundamental research in France. It was a private institution funded in 1927 by the Jewish patron Baron Edmond de Rothschild, and its first director was the physical chemist and Nobel laureate Jean Perrin. According to Bernard Pullman's recollections, "it was probably the first institute in the world devoted to molecular biology, a quarter of a century before molecular biology was born" (B. Pullman 1979, 40).

Events were not just taking place in the center. Starting with the chair of theoretical chemistry offered to Barriol in Nancy in 1948 (Blondel-Mégrelis 2001), by the end of the 1950s, several universities including Marseille, Bordeaux, Pau, and Rennes offered positions in theoretical chemistry (Rivail and Maigret 1998, 369).[15]

Daudel's immersion into quantum chemistry had not only the full support of de Broglie but was also strongly influenced by his legacy (Lochak 1992). His initial interaction with Coulson and their discussions about the chemical bond had a "catalytic effect" on his subsequent work (Daudel 1992, 113). Daudel became increasingly convinced that quantum mechanics, which he often referred to as de Broglean mechanics (Poitier and Daudel 1943, 4), was the clue to understanding the structure and dynamics of large molecules and, therefore, of the concomitant necessity of articulating a language for quantum chemistry compatible with the foundations of quantum mechanics. This had become the central aspect of Daudel's agenda for quantum chemistry in its early years.

He emphasized that quantum chemistry was built upon many chemical concepts, often inherited from classical chemical theories. K-electrons, L-electrons, valence electrons, π- or σ-electrons, localized or delocalized electrons, bonds, and so forth, were certainly incompatible with the quantum mechanical ideas of indistinguishability and nonlocality but had proved to be very useful tools for chemists and had been appropriated by quantum chemists (Daudel 1952). Hence, he argued, they should be given alternative formulations compatible with the new physics. Such was the case of the concept of "loge" introduced by Daudel and his students to explain quantum-mechanically the concept of the chemical bond. For the helium atom, they showed that it was possible to associate to each shell a certain domain of space, and that these domains could be decomposed in small partitions (loges) where there is a high probability to find one and only one electron with a specific spin (Daudel, Odiot, and Brion 1954). They then extended their discussion to excited states of the helium atom (Odiot and Daudel 1954), and proved that it was possible to associate with each loge a value of the energy in some ways equivalent to the energy of orbitals in the self-consistent field and giving in simple cases an approximate value for the ionization energies (Brion, Daudel, and Odiot 1954).

Daudel then extended the notion of loge to include not only atoms but also molecules. The loge became a part of space associated with an atom or a molecule in which there is a high probability of finding a certain number n of electrons, and just this number, with certain spins. By analogy with atoms, they distinguished loges in molecules with different properties that could be classed into core loges, bond loges, lone-pair loges, as well as localized and delocalized bond loges. And to obtain the best division of the molecular space into loges, one should look for the division that gave the maximum amount of information about the localizability of electrons. By the early 1970s, Daudel and his collaborators realized that this was tantamount to finding the distribution of loges corresponding with the minimum-information function, as defined by Claude E. Shannon and Léon Brillouin, the only other French scientist who together with de Broglie was sympathetic to the field of the electronic structure of atoms in the 1930s. In the case of the molecule of lithium, Li_2, a partition of the molecular space into loges was obtained by dividing space into three loges: two spheres of equal radius R centered on each nucleus (core loges), and the rest of space (Daudel 1973; Daudel 1992, 113; Daudel 1992a, 632–635). In this new theoretical context, the chemical notion of a bond became clarified and found a proper definition in the framework of quantum mechanics as a region of molecular space extending between some core loges in which there is a high probability of finding a given number n of electrons with specified spins. Therefore, the concept of loge established a bridge linking two incompatible ways of thinking and, at the same time, "saving" chemical intuition along more traditional paths.

Together with the concept of loge, Daudel and his group introduced in the literature the notion of "densité électronique differentielle" and applied it initially to the lithium molecule (Roux and Daudel 1955). The differential electronic density was defined as the difference between the electronic density computed at a point of a molecule and the density that existed at the same point if the atoms were side by side without interacting. This notion revealed the effect of the chemical bond on the electronic distribution density (Daudel, Brion, and Odiot 1955; Roux, Besnainou, and Daudel 1956). A positive difference meant that in the formation of the molecule, there was a region where there was a higher electronic interaction than when the atoms did not interact, an explanation that gave, together with the notion of loge, an additional theoretical support to the chemical notion of a bond. The results of precise calculations were compared with experimental measurements and were found to agree. As computers became more powerful, this notion also lent itself to pictorial representations, recovering in the framework of quantum mechanics one of the traditional components of chemical modes of thought (Daudel 1992a, 635–636).

The Pullmans' Group
The period of the Colloque de la Liaison Chimique marked the growth and consolidation of the Pullmans' group. It also marked the shift into the molecular orbital method with the exploration of its extension into large molecules, specifically large conjugated systems without geometrical restrictions, including aromatic non-benzenoid compounds such as fulvene (C_6H_6) and benzofulvene ($C_{10}H_8$). Notably, this line of research had already been addressed in parallel by Bernard Pullman's work leading to his Ph.D. dissertation, and at this point it was motivated by a criticism to Wheland's textbook *The Theory of Resonance and Its Applications to Organic Molecules* (1944), which prompted Bernard Pullman's interest for aromatic non-benzenoid hydrocarbons. Considering it "one of the best books ever written on the problem" (B. Pullman 1979, 36), he doubted the strictness and uniformity of all rules enumerated to account for the role of resonance in organic chemistry and went on to dispel some of them. In particular, he doubted that increasing the dimensions of conjugated systems always produced a bathochromic shift (shift toward longer wavelengths) in their respective molecular spectra. The Israeli experimental organic chemist Ernst Bergmann, a close friend of Chaim Weizmann, the first president of Israel and himself a chemist, confirmed Pullman's predictions, becoming "a devoted believer" in quantum chemistry (B. Pullman 1979, 36). In 1950, the Pullmans spent a few months with Bergmann in Rehovoth, Israel, starting a collaboration that lasted for many years, produced various publications, gave rise to the creation of the Jerusalem Symposia in Quantum Chemistry and Biochemistry, and accompanied their move to quantum biochemistry.

Although initially the group was influenced by Pauling's agenda, the shift toward the molecular orbital approach involved the constant recourse to a comparative methodology in the context of which priority was given to the comparison of results of both the VB and the MO methods in order to assess their relative advantages and shortcomings.

The pairs fulvene and benzene on the one hand, and benzofulvene and naphthalene on the other, are two groups of isomeric compounds differing only in the relative position of the double bond but revealing very different physical and chemical properties. The study of fulvene using both the valence bond and the molecular orbital method showed agreement between both in what related to the distribution of bond orders and free valences, but a marked difference when looking for the distribution of charge densities (Pullman, Pullman, and Rumpf 1948, 1948a). Comparison of results obtained with the molecular orbital method as well as with the valence bond method underlined their support of Coulson's comparative methodology and their implicit advocacy of methodological pluralism, so much at odds with Pauling's general approach to quantum chemistry.

The same methodological standpoint guided them in exploring the quantitative potential of the molecular orbital method. André Julg and Alberte Pullman (1953) used configuration interaction, investigated initially in Coulson's group for much smaller molecules (Craig 1950; Coulson, Craig, and Jacobs 1951), and consistently at the forefront of concern by many groups especially after the 1953 Nikko Symposium (see corresponding section in this chapter). Berthier (1953) applied the self-consistent field method following the lead of C. C. J. Roothaan in Mulliken's group.[16] Results were compared for fulvene using a limited number of configurations. The two methods agreed when calculating intensities and transitions but differed substantially when computing electric charge distributions and dipolar moments. In this case, the self-consistent field method was considered better as it gave rise to values of the dipolar moment closer to the experimental ones.

The group used experimental values to decide for or against alternative quantitative methods, and in other cases experimental difficulties in the preparation of certain compounds were circumvented by recourse to theoretical studies of their properties. In one instance, Alberte Pullman turned to butadiene, a relatively simple molecule studied previously by Mulliken's and Coulson's groups, to test the effects of introducing configuration interaction in the results of the self-consistent field method (A. Pullman 1954; Pullman and Baudet 1954).

Computers with respect to quantum chemistry were referred to for the first time in publications in France in 1956 by the Pullmans' group (Mayot et al. 1956).[17] Initially, access time was granted to academic groups by hardware companies such as IBM or Bull, but afterwards several academic institutions realized the importance of purchasing their own computers. For example, the University of Nancy bought an IBM 604 in 1957 soon to be replaced by an IBM 650 (Rivail and Maigret 1998, 374), Berthier at the École Normale Supérieure bought a small IBM 1620 computer around 1960, and the Pullmans bought a "really modern computer" for their laboratory around 1961.[18] By the late 1960s, these facilities were considered increasingly insufficient, and chemists strove to create a center devoted to theoretical chemistry or, if things came to the worse, to scientific computing. In fact, the director of the CNRS opted for the second suggestion. A national center for scientific computing was created in 1969. Called the Centre Inter-Regional de Calcul Électronique (CIRCE), it also housed another institution named the Centre Européen de Calculs Atomiques et Moléculaires (CECAM) (Rivail and Maigret 1998, 374).

The 1950s brought many changes in the Pullmans' group related to their rising impact in the national and international realm. Circulation, networking, and training became central to their agenda. The 1955 meeting organized in Sweden by Per-Olov Löwdin and Fischer was the first meeting the couple attended abroad (B. Pullman 1979, 38). It was followed by the participation in the 1956 Texas Symposium, the 1959 Boulder Conference, and the Sanibel Symposia. From then onward, they were present in most key gatherings of quantum chemists. Among many others, they further contributed to the establishment of the Jerusalem Symposia in Quantum Chemistry and Biochemistry and the establishment of the Edmond de Rothschild Schools in Molecular Biophysics at the Weizmann Institute of Science in Rehovoth, Israel (B. Pullman 1979, 40).

Scientifically, the group moved from quantum chemistry to quantum biochemistry (A. Pullman and B. Pullman 1962, 1973). Conjugated systems became the natural link between the initial contributions of the Pullmans' group and their shift into biochemistry. By extending the molecular orbital approach to biochemistry, their work pointed to a correlation between the processes of life and electronic delocalization (B. Pullman and A. Pullman 1962). We will not go into this fascinating topic, which is already beyond the scope of this book. Let us just note that their incursions into the new subdiscipline were informed understandably by the same commitments, centered on methodological pluralism and the emphasis on comparative methodologies, which had guided their work in quantum chemistry. By 1969, reacting to what they believed was an undue stress on ab initio calculations, they persevered on their lifelong commitments. They acknowledged the important role of "non-empirical calculations . . . in lending precision to our fundamental concepts and in deciding between approximate methods." They were convinced, however, that it was premature to conclude that "the results are revolutionary. . . . We hope to have made clear that any method should be used with caution and that hasty critical statements should be avoided" (A. Pullman 1970, 30).

The Role of Textbooks

In the 1940s and 1950s, the leaders of the two Parisian groups put forward different agendas for quantum chemistry, one attempting to clarify various concepts used in quantum chemistry, the other attempting to assess the relative advantages of different approaches to study large conjugated systems. But despite their differences, one common strategy united them. Both were aware of the not too sympathetic (national) context in which quantum chemistry emerged and of the necessity to consolidate theoretical chemistry as a subdiscipline of chemistry and quantum chemistry as one of its various manifestations. Both were eager to strengthen and diversify the research activities of their group members, and both were eager to contribute to the training of younger scientists.

Textbook writing became one of the goals of nearly every group working in quantum chemistry. Raymond Daudel, Bernard and Alberte Pullman, Barriol,[19] and Julg all wrote at least one textbook, and some were also involved in the writing of popularization books. Like quantum chemistry itself, many of the textbooks had the endorsement of de Broglie, and many prefaces were written by him. Heirs to a strong national tradition in textbook writing, initially textbooks were written in French and addressed to a national audience. Soon some foreign publishers expressed their interest in some of them, and English translations were prepared. But, even more surprisingly, from the late 1950s onward, textbooks by these French authors were written originally in English and aimed, from the start, at an international audience. They offered state of the art comprehensive accounts, starting with quantum mechanics and going into detailed discussions of various aspects of quantum chemistry. They included an up-to-date overview of past and recent contributions and revealed a full command of the literature in the field. This situation contrasted strikingly with that of R. Poitier and Daudel's first textbook *La Chimie Théorique* (1943), published during the war in 1943 and which did not include a single reference to foreign contributions, except for a reference to the Heitler–London 1927 paper.

La Chimie Théorique is particularly telling for giving ample evidence of the characteristics of the content in which the new subdiscipline was striving to impose itself. The textbook's introduction included elements typical of a foundational document: The authors made sure to emphasize the French lineage for the corpuscular theory of matter, the French origin for quantum mechanics in the work of de Broglie (to the point of using often the wording "de Broglien mechanics" instead of wave or quantum mechanics), and the patronage of Frédéric Joliot-Curie in supporting a recent series of lectures forming the textbook's origin. All these elements underlined the view that chemistry had been undergoing changes associated with the rise of a new form of theoretical chemistry, and that these changes became possible due to the new corpuscular theory of matter heavily dependent on French contributions. A parallel underlying message was also there: The French, with few exceptions, had unjustifiably ignored the possibilities of these developments for chemistry, despite de fact that "de Brogliean mechanics" offers a "powerful weapon" to "deduce chemical phenomena from atomic conceptions" (Poitier and Daudel 1943, 4). Daudel believed that he should actively seek to change this paralyzing attitude.

In 1959, the year in which the Boulder Conference was convened, after years of intense networking among members of the quantum chemical community, Daudel, together with two colleagues from the Centre de Mécanique Ondulatoire Appliquée, published a textbook titled *Quantum Chemistry*, instead of *Theoretical Chemistry* for which he opted when writing his first textbook co-authored with Poitiers (Daudel, Lefebvre, and Moser 1959).[20] Locally, things had matured, but the textbook was addressed to an international audience rather insensitive to the constraints of the local context and gradually heading toward using quantum chemistry when referring to their subdiscipline. Quantum chemistry was also adopting the systematic use of computers, and the book indirectly adjusted to the new situation. The preface to the textbook was written a few days after Daudel's attendance of the Boulder Conference, and its organization included a brief discussion of Coulson's worries voiced at Boulder, revealing the extent to which the discussions at the conference helped to (re)form the practices of quantum chemists. Furthermore, Daudel wanted to stress that as semiempirical and nonempirical methods were being developed, and as the mathematical background became more sophisticated and calculations became more and more involved, one had to balance the liability of tedious calculations against the usefulness of results—as when he commented on an approximation by Rudolph Pariser and Robert Parr considering it a "middle ground between the very rapid, sometimes incoherent but often successful Hückel theory and the tedious, coherent and unfortunately disappointing nonempirical calculations" (Daudel, Lefebvre, and Moser 1959, 517). The message he wanted to convey was expressed in no uncertain terms: "the study of quantum chemistry will make chemical thinking more supple and will often make possible a more intelligent attempt to resolve complicated chemical problems" (Daudel, Lefebvre, and Moser 1959, 9).

Organic chemistry and conjugated systems were the starting point of the Pullmans and the connecting link between the electronic understanding of carcinogenesis, their move into quantum chemistry, and their next shift into quantum biochemistry, a subdiscipline that they essentially created. Their textbooks reflected their scientific wanderings. Three textbooks signaled the three major areas of their research: *Les théories électroniques de la chimie organique* (B. Pullman and A. Pullman 1952), *Cancérisation par les substances chimiques et structure moléculaire* (A. Pullman and B. Pullman 1955), and *Quantum Biochemistry* (B. Pullman and A. Pullman 1963). Bernard Pullman called them their three scientific babies (B. Pullman 1979, 36, 37, 41).

In the same year in which Coulson published his *Valence*, the French couple published a voluminous 500-page-long textbook *Les theories électroniques de la chimie organique*. It had a success they did not expect. Written in French, the book was widely read not only in France but also abroad (B. Pullman 1979, 36). Both textbooks adopted a comparative methodology, assessing different topics, such as the chemical bond and conjugated molecules, from the standpoint of the valence bond and the molecular orbital approach. Their methodological proximity to Coulson (whose help they acknowledged in their introductory note) might explain the prominence given to the comparison of the two approaches. Bernard recalled in his "Reminiscences" that they sought the active support of de Broglie, whose preface was meant "to shake the thick walls of indifference, if not hostility, toward theoretical chemistry among the still largely old-fashioned masters of organic chemistry in our country" (B. Pullman 1979, 36–37). In fact, de Broglie called attention to the emergence of a new version of theoretical chemistry founded on quantum mechanics, to the innumerable difficulties it faced due to the complexities of the world of molecules it purported to explain, and to the manifold expertise it demanded in mathematical finesse, which had to be guided by vast chemical knowledge. He also emphasized the impact quantum mechanics would have on organic chemistry and praised the status the authors had already secured in the national and international communities. The patronage of de Broglie was certainly not a secondary factor in the difficult process of legitimating quantum chemistry in France.

In the textbook published in 1955 titled *Cancérisation par les substances chimiques et structure moléculaire*, prefaced by Lacassagne and addressed to a variety of scientists, such as organic chemists, biochemists, physical chemists, physicians, and pharmacists, the Pullmans opted for a presentation with relatively simple mathematics yet without sacrificing rigor. In their previous textbook, they had already called attention to the sophisticated mathematical underpinning of quantum chemistry (which they still called theoretical chemistry) and to the possibility of providing explanations with relatively little mathematics involved. But they decided to provide a detailed description of the mathematical methods of quantum chemistry to which readers of the new textbook could refer to (A. Pullman and B. Pullman 1955, 11). They presented a coherent and more complete theory able to account for a substantial portion of the available data on chemical carcinogenesis by polycyclic hydrocarbons. The impact of the book on experimentalists was immediate, and in early 1956 the Pullmans received the Essec Prize of the French League Against Cancer. The award attracted the attention of journalists who did not miss to note the "Frenchness" of successful scientific couples—the Curies, the Joliot-Curies, and, now, the Pullmans (B. Pullman 1979, 38).

The period from 1955 to 1958 marked the Pullmans' shift into quantum biochemistry, culminating with their move to the Institut de Biologie Physico-Chimique. Their exploration of the relationship between carcinogenic activity and chemical reactivity in organic molecules led them to work on problems of cancer chemotherapy and to extend their methodology from the field of chemical carcinogens to the field of compounds active in cancer chemotherapy. In the attempt to correlate structure to ability of certain compounds to behave efficiently as drugs, they became interested in purines and pyrimidines and assembled a considerable amount of data on their electronic structure and properties. They were, thus, well prepared to take advantage of James Watson and Francis Crick's groundbreaking discovery of the double-helical structure of DNA and the central role played by the base-pairing scheme between purines and pyrimidines (A. Pullman and B. Pullman 1962).

In 1958, the Pullmans were invited by the biochemist Albert Szent-Györgyi to spend two months at the Marine Biological Laboratory in Woods Hole, Massachusetts, together with other scientists. For Szent-Györgyi, the mysteries of life were trapped in the structures of biomolecules, and specifically in electrons, and the Pullmans' work immediately caught his attention (Kasha 1962; Wurmser 1962).[21] The electronic aspects of biology became the topic of intense discussions. Their stay at the Marine Biological Laboratory was repeated for four consecutive summers, and there the main themes and ideas included in their textbook *Quantum Biochemistry* started taking shape. The textbook was completed in winter 1961–1962 in Tallahassee, Florida, at Michael Kasha's Institute of Molecular Biophysics (B. Pullman 1979, 41). Written originally in English and addressed to an international audience, the textbook presented results of research conducted at the borderlines of chemistry, physics, biology, and pharmacology. Its plan was twofold. It purported to show how biochemists could profit from quantum mechanics in answering questions related to the structure and action of components of living matter, and it meant to show quantum chemists those aspects of biochemistry to which they could contribute (B. Pullman and A. Pullman 1963, v). Addressed specifically to two different classes of specialists—quantum chemists and biochemists—the textbook strove to show practitioners how their emerging research fields could reinforce each other, thereby consolidating links between two "in-between" subdisciplines.

Despite the authors' insistence on the importance of using both the VB and MO theories in their research and their textbooks, they now made a plea for the exclusive use of the molecular orbital theory in biochemistry. They pointed to the conceptual congeniality of the valence bond method for chemists and biochemists, but they called attention to the complexity of its basic principles and the cumbersomeness of its mathematical techniques when going beyond strict qualitative explanations. Therefore, when dealing with organic molecules, the molecular orbital method was to be preferred on account of its relative simplicity (B. Pullman and A. Pullman 1963a). In the textbook, they discussed various approximations, but they used in most cases the simplest Hückel LCAO approximation, which had proved to be "an extremely powerful tool of investigation" in the field of organic molecules (B. Pullman and A. Pullman 1963, 178). And they compared the situation in which quantum biochemistry found itself to that undergone by quantum chemistry 30 years earlier. They recalled how Van Vleck and Sherman depicted it in their 1935 review paper, contrasting the conflicting mental attitudes of the pessimist and the optimist. The Pullmans were certainly the modern-day optimists as far as the establishment of the (sub)discipline of quantum biochemistry was concerned. And, they appeared to be ready to acknowledge that they had adopted Coulson's outlook, by citing part of his lecture at the 1959 Boulder Conference where he appealed to those quantum chemists who worked on problems of biology not to be too fussy in trying to establish a theoretical understanding of the problems involved. "*A rough track through the jungle precedes the construction of metalled high-way.* . . . In this field the prizes are immense—no less than the understanding and control of life itself" (B. Pullman and A. Pullman 1963, 181, emphasis ours).[22]

The Pullmans and Daudel's group increasingly focused their research interests toward applications to biochemistry, biology, and medicine. Surely, they were not alone in this enterprise. Together with other groups in Europe and the United States, the potential of quantum chemistry as a "donor discipline" became clear to many. And times were ripe in many new domains for such profitable incursions.

The 1951 Shelter Island Conference

In April 1950 at the Detroit meeting of the American Chemical Society, a group of chemists and physicists discussed the inadequacies of valence theory and what might be done to improve it. The most pressing need was that of a table of the difficult integrals in valence calculations. By the end of the year, Mulliken was asking D. A. MacInnes of the Rockefeller Institute, who was chairman of the National Academy of Sciences Committee on Scientific Conferences, whether it would be possible to prepare a conference on "Quantum Mechanical Methods in Valence Theory and Molecular Structure."[23]

MacInnes's answer was very enthusiastic, and despite the fact that there was no complete agreement in the Council of the National Academy of Sciences on whether the Shelter Island Conferences should continue because "they are limited in size and some feelings get hurt," he encouraged Mulliken to go ahead.[24] Mulliken suggested that the Office of Naval Research (ONR), which was already sponsoring the activities of his Laboratory of Molecular Structure and Spectra, might be willing to contribute toward the expenses of the conference.[25] MacInnes would support such a move hoping that the conference would not grow in size because there was a tendency "for everything good to be ruined by getting too big."[26] Mulliken suggested to confer with Slater from the Massachusetts Institute of Technology (MIT) and Parr from the Carnegie Institute of Technology about whom to invite. He also thought that the title of the conference could be changed to "Atomic and Molecular Wave Functions and the Quantum Mechanical Computation of Interatomic and Intermolecular Interaction Energies."[27] MacInnes was not particularly supportive about bringing people from abroad, which increased the expenses, and he doubted whether the contributions of someone coming from abroad for such a short time would justify the expenses.[28] In Mulliken's formal proposal to the ONR, it is noted that the conference would discuss questions similar to those of the ongoing research at the University of Chicago already being financed by the ONR "in view of the Navy's interest and possible military utilization of the basic knowledge accruing from this investigation."[29] It was planned to emphasize computational methods. The invitees were informed that the discussions would be directed toward the better understanding of the forces between atoms and between molecules with emphasis on quantitative and semiquantitative computations. Of special interest would be the discussion of methods for rapid computation of integrals appearing in atomic and molecular electronic structure problems.[30] The participants were encouraged to "think over, beforehand, what results, ideas or questions he can present that are either novel, or controversial, or not widely or clearly enough understood" and present them at the conference.[31]

Problems such as the ones to be raised at the conference were already preoccupying Mulliken and Coulson. A year after the Paris Conference, Mulliken returned to some points he had raised at the meeting. He wanted to pursue his suggestion of a generalization of Coulson's definition of bond order. But, above all, he wanted to plan an "extensive computation program to obtain expressions for and tables of, all the major two-electron two-center integrals (at least for the homopolar case) and overlap integrals started by Roothaan."[32] Worried that the work of the two research groups might overlap, Coulson decided to send copies to Mulliken of two manuscripts to explore the possibilities for improved calculations.[33] They were prepared jointly with Barnett who was able to evaluate numerically five such integrals, starting from scratch, in just one day. Soon afterwards, Coulson and Mulliken were discussing the preparation of tables of overlap integrals.[34]

Coulson had been addressing the problem of the calculation of molecular integrals since his Cambridge days. In 1942, he returned to the topic and presented a table of two-center integrals together with the explicit forms of many of these (Coulson 1942), and later on, in a two-part paper written together with Barnett (Coulson and Barnett, 1951), a new method, the ζ-function method, based on the expansion of exponentials and modified Bessel functions, was introduced to tackle some of the problems still unsolved at the time. They first listed the properties of the functions needed for the two-center integrals and then attempted to apply the results in chemistry. Formulas for more than 180 distinct integrals were listed. The calculations were rather complicated when only two centers of force were involved and the integration was to be carried out over the space of a single electron. An even more complicated situation arose when double space integration was necessary. Multiple integrals involving three or four centers of force had been considered to be intractable except in a few isolated instances. Coulson and Barnett believed that they were attempting to introduce new ways of calculating these integrals, which had been previously regarded as inaccessible. The new method was further improved, and its generalization enabled the calculation of many integrals occurring in the context of molecular structure problems.[35] It included procedures to solve two-, three-, or four-center integrals and culminated in Coulson's contributions to the 1951 Shelter Island Conference.[36]

The conference was to be held between September 8 and 10, 1951, at Ram's Head Inn in a beautiful setting on Shelter Island, Long Island, New York. Twenty-five people attended the meeting, of which 18 were Americans, and 5 were British. From the United States came T. H. Berlin, B. L. Crawford, H. Eyring, J. O. Hirschfelder, G. E. Kimball, D. A. MacInnes, H. Margenau, J. E. Mayer, R. S. Mulliken, R. G. Parr, K. S.

Pitzer, C. C. J. Roothaan, K. Rüdenberg, H. Shull, J. C. Slater, C. W. Ufford, J. H. Van Vleck, and G. W. Wheland. From Great Britain came M. P. Barnett, C. A. Coulson, J. E. Lennard-Jones, W. Moffitt, and L. E. Sutton. Pauling did not seem to be particularly keen on attending the meeting and did not come. MacInnes's recommendations as to size were taken into consideration, but the importance of having participants from abroad, either senior scholars or newcomers to the field, all experts in calculations, was expressing the eagerness to push forward a particular agenda for the discipline in which calculations were to become "a way of life to be adopted by us" (Parr 1990, 327). Two new participants in such international ventures deserve our attention. They were the Japanese Masao Kotani (1906–1993) and the Swedish Per-Olov Löwdin (1916–2000).

Löwdin was then 34 years old, a young man full of energy and ideas, already with very good international connections, and determined to put Sweden, and specifically Uppsala, on the map of quantum chemistry. A newcomer into the discipline, Löwdin was among those who played a decisive role in shaping the agenda for quantum chemistry in the post-war world. He would become an accomplished group leader, founder of journals, conference organizer, and eloquent teacher who, like Coulson, had a rather pronounced sensitivity to the methodological issues concerning quantum chemistry, offering philosophical reflections on the foundations of quantum chemistry, its nature and methods.

He obtained his Swedish degree of filosofie licenciat in 1942 and immediately became a lecturer in Mechanics and Mathematical Physics at the University of Uppsala. He visited Pauli's group in Zürich in 1946 and studied problems in quantum electrodynamics, a field appealing to his mathematical mind. His doctorate was on solid-state physics and was awarded in 1948. It was a very original piece of work that dealt with a first principle calculation of the cohesive energy and elastic constants of ionic crystals, and specifically of the alkali halides. He developed a number of mathematical techniques to deal with the overlap and non-orthogonality problem that involved the calculation of molecular integrals carried out on FACIT desk calculators by a group of students, known as the "student-computers." These were the first successful ab initio calculations on crystals. Immediately after obtaining his Ph.D., he became docent at the University of Uppsala, visited Mott's group in Bristol, and spent the academic year 1950–1951 in the United States, with Sponer at Duke University, Mulliken's group in Chicago, and Slater's group at MIT (Ohno 1976, 2–3). This visit and attending the Shelter Island Conference led him to the decision to concentrate his research on problems of quantum chemistry.

Ten years older than Löwdin, Kotani by 1951 had already a successful career as a theoretical and mathematical physicist behind him. He obtained his B.Sc. degree in physics in 1929 at the Tokyo Imperial University, became a lecturer at the Faculty of Engineering in the same university, and 3 years later was appointed associate professor in the Department of Physics. In 1943 he received his doctoral degree and was promoted to a full professorship in the same department. During his undergraduate years, he took mostly courses in classical physics and mathematics, and together with a friend, T. Inui, using journals reaching the university library via Siberia, they taught themselves quantum mechanics. Already as a lecturer, the duo became a trio by the inclusion of another assistant professor, Yamanouchi, and they continued their autodidactic studies. It was probably in this period that Kotani became aware of the 1927 Heitler–London paper on the hydrogen molecule, which, according to his assessment, solved the riddle of the chemical bond and promised to unite chemistry and physics on a common foundation (Kotani 1984, 13). Although still working mainly on problems of mathematical physics, specifically on hydrodynamics and aerodynamics, in the late 1930s Kotani used group theory to study the electronic properties of polyatomic molecules with special emphasis on the methane molecule (Kotani 1984, 14; Ohno 1995).

During the 1930s, after the invitation to join the Committee for the Investigation of Catalytic Action, newly established under the Japan Society for Promotion of Science, he opted for molecular science and decided to tackle the calculation of molecular integrals. This decision was also taken after consultation with Katayama, the committee's chairman, and a senior professor of physical chemistry at the Department of Chemistry at the University of Tokyo. In this way, Kotani became a participant in the tradition inaugurated by Yoshikatsu Sugiura in 1927 when he provided the first numerical calculation of the exchange integral appearing in Heitler and London's paper (Park 2009). Although technical constraints inhibited scientists from extending Sugiura's work and proceeding with molecular integral calculations, in the pre-war years mechanical desk calculators were already available, and the computation of molecular integrals became a time-consuming yet feasible undertaking. Kotani used Tiger calculators operated exclusively by male technicians. His results were included in a book titled *Tables of Molecular Integrals* (Kotani et al. 1955), in which he discussed eloquently his method of treating many-electron systems together with an explanation of how to build representation matrices of a symmetric group of permutations.

Because of the war, Kotani traveled abroad for the first time in 1950. He visited Paris and was relieved to hear the confirmation of the president of the International Union of Pure and Applied Physics that Japan had continued to be a member even during war years. It was suggested by the president that an international conference on theoretical physics should be organized in Japan. Kotani took this advice seriously, and in 3 years time the conference took place. In 1951, Kotani made a longer journey not only to Europe but also to the United States. He met first Coulson in London and then traveled to the United States to attend the Shelter Island Conference, where he finally got acquainted with many molecular scientists he knew only through their publications. He stayed in the United States for a couple of months afterwards, visiting several universities, including the University of Chicago and Mulliken's group there.

The conference was a great success. MacInnes was informed "how the last evening session on Sunday night ran well into Monday morning, with a resulting clarification of the several calculational programs which alone justified the conference."[37] In his autobiography, Mulliken (1989, 136) designated the Shelter Island Conference "a watershed." In a letter to Coulson, he considered that it had an "interestingly activating effect."[38] The promise of success of the new era ahead was sensed by everybody involved.

Fifty-two papers were presented at the conference. Parr and Crawford, formerly Ph.D. student and supervisor, acted as conference secretaries and prepared a summary in which they classified the papers presented into six major categories. The papers of the first group dealt essentially with straightforward applications of theory, including calculations for some atoms, but concentrating on molecular problems ranging from the molecule of hydrogen to pyrene "in order of increasing complexity of molecule and increasing empiricism of approach" (Parr and Crawford 1952, 548). Notably, there was little discussion at the conference on the subject of aromatic molecules, which most probably reflected a consensual satisfaction with the π-electron theory developed thus far. The papers of the second group dealt with the quantum mechanical interpretation of chemical concepts such as electron pairs, bond energies and bond orders, hybridization and chemical reactivity. Mulliken discussed some improvements on his "magic formula" for the semiempirical calculation of bond energies, and Coulson discussed the problems that arose in extending the concept of bond order to heteronuclear bonds. Coulson also showed maps of electron density in benzene. The quantum mechanical calculation of the structures of activated complexes was considered one of the greatest rewards of valence theory. The papers of the third group addressed forces between molecules and nonbonded atoms. The papers of the fourth group were concerned with the problem of configuration interaction or correlation energy. The papers of the fifth group dealt with mathematical developments. Numerical integration schemes for Hartree–Fock equations were reviewed. Slater recalled that IBM machines were already in use for the calculation of self-consistent fields for atoms but cautioned that a "direct attack on problems of electronic structure is not yet within the reach of automatic computing machines; the machines can do complex arithmetic, but the problem of the molecule has not yet been reduced to an arithmetic level" (Parr and Crawford 1952, 551). The papers of the sixth group dealt explicitly with integrals. The awareness that "completely theoretical" molecular calculations as contrasted to "semi-empirical" were capable of giving useful results led to an increased need for comprehensive tables of values of molecular integrals "unto the fifth center and the farthest neighbor." Roughly 1 year before the meeting took place, Parr, in a jointly authored paper with D. P. Craig and I. G. Ross (1950), introduced in the literature the designation *ab initio* to name exact calculations in which no parameters except for the fundamental constants were allowed in, and which promised for the first time to be within reach of diligent practitioners. The possibility of cooperative work and convergence of concerted effort by all groups involved arose much interest. There were already eight groups turning to computers for the calculation of molecular integrals. Three teams were located in the United States. They included Mulliken's group in Chicago, in which Roothaan and Klaus Rüdenberg were computing and producing tables of one- and two-center integrals, with the collaboration of Harrison Shull, then at Iowa State University, and the IBM group there. Besides this joint venture, Henry Eyring's group in Utah was also working on one- and two-center integrals. Two other groups were located in Great Britain and included Coulson's group at King's College, London, in which a scheme had been set up to compute desired integrals from a basic set of unit molecular integrals, and Samuel Francis Boys's group at Cambridge University where automatic computing machines were used. In Göttingen, Germany, H. -J. Kopineck was compiling tables of two-center integrals (Peyerimhoff 2002); in Tokyo, Japan, Kotani was extending his sets of tabulated two-center integrals; and, finally, in Uppsala, Sweden, Löwdin's group was using direct numerical integration to evaluate integrals. Finally, to avoid duplication and to speed up results and their circulation among community members, an informal integrals committee at Chicago was created with the aim to act as a clearing house for information on integrals.

Parr and Crawford (1952, 547) started their report by quoting from Dirac's 1929 paper, in which a typical reductionist program for chemistry was outlined, and compared it with the recent developments in quantum mechanics. Edwin Wilson, the editor of the *Proceedings of the National Academy of Sciences of the United States of America*, believed this to be insulting the great man. He could not understand the relevance of having this quote. "Is this supposed to be a slam at Dirac or is it supposed to represent the conclusions of the Conference?" He thought that if the conference was far ahead of what Dirac had projected, then it "might be kinder not to quote him at all." Or one could say that Dirac in 1929 made this statement and "go on to point out that this is still the situation as it stood in the minds of the conferees at the end of the Conference—if that be true."[39] Mulliken did not intervene in this matter and thought that Parr and Crawford should deal with this issue the way they saw fit themselves and believed that the quotation was quite appropriate, urging them to make the connection with the conference more direct.[40] For the published version, they included a conclusion commenting that Dirac's statement had long been the hope and despair of quantum chemists. The hope because the problem could be solved "in principle," and the despair because the equations were "much too complicated to be soluble." On a very optimistic note, their last sentence was that "a frontal attack is at last being made on these equations—a hopeful, rather than desperate, attack" (Parr and Crawford 1952, 552). This conclusion in the final report substituted what they had in their draft[41]:

The session on Saturday afternoon [held at the local schoolhouse] began with Coulson, sitting at an eighth grade desk, picking up an eighth grade algebra book, and reading therefrom, in resonant tones, "the world is incurably mathematical." On this we all agreed. The encouraging note of the Conference is that the chemist is now squarely facing the need for mathematical development: he has turned from seeking nostrums for the incurable.

The editor had found this "another rather childish and undistinguished addition for a Journal like the *Proceedings*!" Two decades later, Parr (1975) was even more optimistic for he boasted that in the past 50 years, Dirac's claim had been replaced by the stronger claim that "we can calculate everything."

The 1953 Nikko Symposium and Slater's Solid-State and Molecular Theory Group

After Kotani's European tour, the Science Council of Japan held an International Conference on Theoretical Physics in Tokyo and Kyoto in September 1953, which was sponsored by the International Union of Pure and Applied Physics. It was "the first international conference in pure science held in the Far East" (Kotani, Kakiuchi, and Araki 1954, 1). Slater, Löwdin, and Mulliken delivered lectures at the session on molecules. Notably, they had all convened right before, on September 11 and 12, in another gathering especially dedicated to molecular physics, organized also by Kotani in Nikko, a national park renowned for its natural and architectural beauty (Kotani, Kakiuchi, and Araki 1954, 1). Despite the geographical distance, 17 scientists coming from Europe (England, France, Germany, Netherlands, Sweden), the United States, and India attended this meeting, in addition to about 50 Japanese scientists. Among the foreign attendees were Coulson, L. Néel, Löwdin, I. Waller, P. W. Anderson, Mayer, Mulliken, Slater, Van Vleck, and C. H. Townes. After the 1953 meetings, relations between members of Kotani's group and of other research groups in Europe and the United States were greatly enhanced, and exchange of annual reports became common practice.

At Nikko, presentations and ensuing discussions centered on the refinement of approximations used in molecular problems, the extension of methods to deal with larger molecules, crystals, and solids, and the assessment of the state of calculations of molecular integrals after the Shelter Island Conference.

Slater took the opportunity to summarize work done in his group and specifically to explain the rationale behind Alvin Meckler's work on the O_2 molecule, an exemplary case that explored possible relations between molecular theory and solid-state theory, in which configuration interaction was introduced as a refinement of the molecular orbital method, and which used the best available computer technology of the time (Meckler 1953).

In 1948, Compton stepped down as President of MIT, and in 1951 Slater accepted an offer for a newly created institute professorship, resigning from the chairmanship of the Department of Physics. In fall 1950, Slater managed to create a Solid-State and Molecular Theory Group, which was initially housed in the premises of the newly created Research Laboratory of Electronics. A few months later, Slater had already secured funds from the Office for Naval Research, the first quarterly progress reports were issued, and the possibility of using during night hours the electronic calculating machine Whirlwind I, being developed by MIT electrical engineers, became a reality (Slater 1975, 226–234, 237–242, 255–267). Slater chose to swim against the tide, insisting on the importance of solid-state theory and the potential of its applications. In the end, Slater's group became the core of the new Center for Material Science and Engineering (1962) (Bensaude-Vincent 2001). Slater had not been working on problems of quantum chemistry for some time. But now there were new opportunities, especially as quantum chemists were making ample use of computers. The analogy of crystals to big molecules, which had always helped his reasoning, made quantum chemistry of particular interest for solid-state theory. Meeting Löwdin at the Shelter Island Conference of 1951 reinforced this viewpoint. Their personalities, styles, and common aims facilitated their joining forces in the articulation of an American–Swedish collaboration in quantum chemistry. It is, thus, not surprising that the contributions of Slater's group to quantum chemistry were guided not so much by a specific interest in investigating molecules of chemical relevance as by his continuous attempts to investigate the possibilities for the application of quantum mechanics to problems of molecular structure.

Reasoning by analogy guided theorists in going from simple to more complicated cases. Meckler took into consideration nine configurations for the triplet state of the oxygen molecule and 12 configurations in the singlet state and solved the secular equation for several internuclear distances. He found that at infinite distance, the energy of the singlet and the triplet states was the same, and that the calculated values of dissociation energy, internuclear energy, and vibration frequencies agreed with experimental values. No commercial computer could help in such a gargantuan enterprise. It was the Whirlwind I—an advanced research computer, being developed to test ideas on computer design, which Meckler used during nighttime since spring 1952—that made calculations possible. Some extra approximations were also used: Meckler was one of the first to use Boys's suggestion to substitute Slater type orbitals for minimal basis sets of Gaussian orbitals, a change that facilitated substantially the calculation of molecular integrals. Slater called them later "a concession in the matter of accuracy" (Slater 1975, 259). As always, for Slater's "physical" soul, the correct mathematical handling of the molecular problems was exceedingly important, and, hence, his warning: "we feel that quantum chemistry has still not advanced far enough so that we even know what are the best approximate methods to use for molecules, and can estimate their accuracy" (Slater 1954, 4).

Löwdin and Slater were once again in syntony. Assuming as Slater always did an analogy between crystals and molecules, Löwdin (1954) also explored the configuration interaction in order to account for the correct asymptotic behavior of molecules in the separated atoms situation without having to deal with an intractable mathematical problem; that is, without solving secular equations of high orders. He introduced the alternant molecular orbital method as a way to handle the correlation problem. In an alternant system, the electrons with different spins may accumulate on different subsystems, and their separation may be regulated by parameters to be determined by the application of the variational principle. As Slater, he was after accurate but practical refinements that could account for cohesive energies and magnetic properties of both molecules and crystals.

Coulson's contribution revolved around issues of disciplinary assessment. The past 3 years had firmly established his reputation as a leading member of the scientific community and a key player in quantum chemistry. Elected as a Fellow of the Royal Society, he now held the Rouse Ball Chair of Applied Mathematics at the University of Oxford, had just created the Mathematical Institute, and had written his textbook *Valence*. In this new period of his scientific life, he became keen on offering his considerations on the status and future of quantum chemistry. His views turned out to be highly regarded. He discussed how Hückel's method could be refined to handle larger aromatic and conjugated molecules for which Hückel's initial assumption of σ–π separation did not hold. Although Coulson and his group were contributing to the computation of difficult integrals, Coulson reminded that one could choose between two possible approaches, which he dubbed the "easy" and "hard" methods or, alternatively, the empirical or the nonempirical methods. He believed that the discovery of the essential elements in a phenomenon could result from relatively accurate but simple models. Despite the availability of ever more accurate calculations, his reinforced conviction lay still in the power of the "easy" method, when correctly handled. Echoing Van Vleck and Sherman's old distinction between the optimists and the pessimists, he ended his talk stating that "the cynic may despair, but the optimist will continue to rejoice" (Coulson 1954, 32).

The Nikko Conference was also important for the opportunity to assess progress accomplished on the calculations of molecular integrals since the Shelter Island Conference. Two informal meetings were held during and after the symposium in which contributions from Kotani's Tokyo–Kyoto group, Mulliken's Chicago group, Coulson's Oxford and King's College, London, group, and Löwdin's Uppsala group were discussed.[12] The importance of progress reports and standardization of nomenclature was reiterated. And, again, a collective strategy for handling future calculations was delineated.

Quantum Chemistry as a Lifestyle

Löwdin's Quantum Chemistry Group in Uppsala

After his travel to the United States during the academic year 1950–1951, which culminated with attendance at the Shelter Island Conference and with visits to Mulliken's and Slater's groups (Ohno 1976, 1–11, 1976a, 13–23; Jansen 1977; Manne 1976, 25–31), during the next 5 years Löwdin (figure 4.2) visited the United States yearly, spending half the year at various American universities. It was during his stay at MIT that he worked and published his trilogy on the "Quantum Theory of Many-Particle Systems" (Löwdin 1955, 1955a, 1955b). He advocated the use of reduced density matrices in the analysis of general wave functions, provided detailed formulas for non-orthogonalized orbitals, introduced the notion of natural spin orbitals conducive to a configuration interaction expansion of more rapid convergence, and proposed extensions of the Hartree–Fock method. This series illustrated his concern

Figure 4.2
Per-Olov Löwdin.
Source: AIP Emilio Segré Visual Archives.

for introducing and discussing novel concepts and methods to be applied to the many-body problem that underlies the study of atoms, molecules, and crystals. His emphasis on mathematical rigor and search for methods conducive to numerical applications became the core of Löwdin's contributions to quantum chemistry.[43] His ability to establish and strengthen an extensive network of connections became also central to his role as a leading participant in discipline building. It is no surprise that his agenda encompassed an articulate plan to create an active research group at Uppsala, much like those of Mulliken and Slater.

In February 1955, Löwdin drafted a memo that was presented to two different institutions—the King Gustav VI Adolf 70 Years Fund for Swedish Culture, and the Knut and Alice Wallenberg Foundation, which eventually secured the funding of the group during its initial stage. He formulated a detailed plan to build a group in quantum chemistry where among the material resources he sought were the desk calculators (FACIT ESA-0).[44] In March 1955, he organized, together with Fischer, a well-attended international meeting that took place in Stockholm and Uppsala and that reflected the recognition he had secured for himself in the meantime as one of the prime movers in the discipline. Coulson, Mulliken, Slater, and Kotani supported his goal of creating a new research group in Sweden with enthusiastic letters of recommendation (Fröman and Lindenberg 2007, 14–15). In fact, his appointment as docent at Uppsala was coming to an end—the contract lasted for 5 to 7 years and could not be renewed. Löwdin was being lured to go to the United States, but in the end he stayed in Uppsala. The King of Sweden—Gustav VI Adolf—gave him the opportunity to carry out his vision. On July 1, 1955, the Quantum Chemistry Group was officially founded, and he became its first director, with a position at the Swedish Natural Science Research Council as well. The Quantum Chemistry Group was an autonomous unit within the university, administratively independent and pursuing its own research goals. Among several Swedish–American organizations contacted, the Texas Swedish Cultural Foundation covered subsistence expenses for foreign visitors in Uppsala. In fact, the group was international in constitution since its very beginning: Klaus Appel, Jean Pierre Calais, Jan Lindenberg, Anders Fröman, Hall, Shull, J. O. Hirschfelder, Bernard Pullman, Joseph and Maria-Goeppert Mayer, K. Ohno, were among those who visited the Quantum Chemistry Group for extended periods of time (Fröman and Lindenberg 2007, 16).

Löwdin had no teaching obligations but he was an inspiring teacher, eloquent and captivating, who was equally comfortable addressing specialized or broad audiences, and someone whose scientific agenda included teaching quantum chemistry to fellow chemists, physicists, as well as research students (Fröman and Lindenberg 2007, 16–17). Lectures series open to the public became a characteristic of the group, and group seminars organized on a frequent basis were attended by group members as well as interested outsiders.

By mid-1956, Löwdin was invited by the U. S. Secretary of the Air Force to deliver a set of lectures to the personnel of the Aeronautical Research Laboratory in the Wright Air Development Center at the Illinois Institute of Technology. As it had happened with Mulliken's and Slater's groups, American military agencies were eager to keep pace and even support research of no immediate military application. Löwdin's techniques to attack electronic structure problems did not go unnoticed, and starting in 1957, the U. S. Air Force through its European Office of Air Research and Development Command (EOARDC), located in Brussels, became one of the major sponsors of Löwdin's group (Fröman and Lindenberg 2007, 19–22). The visit to the United States was also used to discuss computations and get advice on a mid-sized computer to be acquired by the Quantum Chemistry Group. A Burroughs E-102, which had convenient programming features but was not as fast as larger computing machines, was examined; it was decided instead to acquire an ALWAC IIIE, built with vacuum tube technology and a magnetic drum as memory. The acquisition of such a machine, which turned the Quantum Chemistry Group into one of the best equipped centers in the world, necessitated the construction of a laboratory and the move to another building. The new structure required a much tighter organization, every member having assigned tasks, capacity for independent decision making, and regular discussions with the group's leader. A formal inauguration took place on April 23, 1958, the date of the centenary anniversary of Planck's birth (Fröman and Lindenberg 2007, 24–27).

By the time of the laboratory's official inauguration, Löwdin's academic situation and future prospects were unclear. The Swedish Natural Science Research Council was called to take a decision on his future. Löwdin had many supporters within the Faculty of Sciences, including Arne Tiselius, the Nobel laureate biochemist, and many supporters from abroad, who since 1957 believed that a personal chair should be created for him. But academic politics is never linear. Uppsala's university milieu was undersized and peripheral, and Löwdin had for a long time the opposition of Ivan Waller, his former Ph.D. dissertation advisor. A committee, which included among others Waller and Oskar Klein, was formed to take a final decision on the matter. Divided in their viewpoints, the internationally renowned Swedish theoretical physicists Lamek Hulthén and Klein agreeing with foreign experts and opposing their colleague, a compromise was reached calling for the creation of a regular chair in quantum chemistry, to which Löwdin could apply on an equal footing with any other candidates. In the meantime, Löwdin had two pending offers in the United States: the Fritz London Chair at Duke University, North Carolina, and a senior level research professorship at the University of Florida in Gainesville. The EOARDC also voiced its opinion on the matter. The agency was eager to continue and even enlarge its support to Löwdin's group if it was matched by Swedish agencies and considered it crucial to have a strong group doing quantum chemistry outside the United States. As usual in these matters, the administrative procedures leading to the creation of a new chair took time, and it was only in 1959 that Uppsala University included in its budget the funding for the new chair. The new chair-holder should be appointed by July 1, 1960. On May 13, the King of Sweden signed a resolution granting to Löwdin the privileges and duties of the first professor of quantum chemistry at the University of Uppsala, and possibly in the world (Fröman and Lindenberg 2007, 31–32, 39). Five years had passed since the creation of the Quantum Chemistry Group. Five bustling years had put Sweden on the map of quantum chemistry and shaped Löwdin into a new disciplinary guru.

The 1958 Valadalen Summer School

Circulation and training were an integral part of Löwdin's agenda. His goal was not only to strengthen relations between chemists and quantum chemists but also to enlarge and consolidate the community of quantum chemists and to train younger students. Building on Coulson's experience of organizing summer schools in Oxford since 1955, the idea of summer schools especially devoted to the discussion of methods, concepts, and results in quantum chemistry was implemented by Löwdin. They became famous for their contribution to "the removal of both national and scientific language barriers" (Gold 1961, 40).

The first such initiative took place in Valadalen in the Swedish mountains during summer 1958, from July 26 to August 30 (Fröman and Lindenberg 2007, 33–37). Far from worldly distractions, an intensive program of lectures and problem-solving sessions was interlaced with mountain hiking, soccer games, and swimming in chilly waters. Participants worked 5 days a week, and each day had six lectures and two exercise hours. During the last week, the number of lectures diminished so that more problem-solving hours and informal discussions on current topics in quantum chemistry could be accommodated. Participants were supposed to have read the first half of Pauling and Wilson's *Introduction to Quantum Mechanics* and were encouraged to bring with them textbooks such as Eyring, Walter, and Kimball's *Quantum Chemistry* (Fröman and Lindenberg 2007, 45–46). The summer school lasted for 4 weeks and was capped with a week-long symposium on "Correspondence between Concepts in Chemistry and Quantum Chemistry" for which well-known quantum chemists were invited, including Pauling and Mulliken as special lecturers. Presentations were followed by lively discussions, highlighted by the known conflicting viewpoints and styles of both founding fathers of quantum chemistry.[45]

Having among his audience many chemists, and always eager to clarify the relations of chemistry to quantum chemistry and to establish quantum chemistry on secure foundations within quantum science, Löwdin (1957) took the opportunity to come back to some of the considerations he outlined in a review paper on the "Present Situation of Quantum Chemistry." In this review, Löwdin summarized the basic principles of quantum mechanics in order to address the development of quantum chemistry and its goals. He offered diagrammatic representations of the various methods used in quantum chemistry for the solution of Schrödinger's equation as well as the various mathematical steps needed in solving the many-electron Schrödinger equation for a molecular system, which he explained carefully throughout the paper. He did the same in his lectures at the summer school. In addressing the development of quantum chemistry, Löwdin discussed its relations to quantum science, pointing to four different ways of establishing this connection. He repeated these considerations before the audience at Valadalen. In one of the methods, chemists themselves translated and adapted ordinary chemical concepts to the language and ideas of quantum science. Pauling's resonance theory was such an example. His opinion was that "it is very hard to decide whether the electronic interpretations given have a real background in nature or not" (Löwdin 1957, 57). In another, the unification of chemical and quantum mechanical ideas was accomplished by means of semiempirical approximations taken as devices correlating one set of experimental chemical data with another, offering the advantage of simplicity, and enabling one to make some quantitative predictions if not pushed too far. The other two methods of establishing connecting links between chemistry and quantum science started by the quantum theory of many-electron systems and attempted to derive solutions of Schrödinger's equation, either approximate or highly accurate. This last case was the 1927 Heitler and London treatment of the ground state of H_2 perfected by Hubert M. James and Albert S. Coolidge, working at the Mallinckrodt Chemical Laboratory, Harvard University, in 1933. Löwdin concluded that the link between chemistry and Schrödinger's equation was still weak if one was only seeking quantitative results. For him, the goal was to work on an ever more reliable basis for the theory itself and for other approaches built by analogy, based on the exploration of mathematical and numerical techniques. In this framework, the development of modern electronic computers was of "almost revolutionary importance" (Fröman and Goscinski 1976, 38), as small molecules could already be treated in a theoretically satisfactory manner.

For Löwdin, there was still a long way to go before the ultimate goal of quantum chemistry could be reached. In the review, he had no qualms in stating that this goal was the accurate prediction of the properties of a hypothetical polyatomic molecule before its laboratory synthesis. And he added: "The aim is also to obtain such knowledge of the electronic structure of matter that one can construct new substances having properties of particular value to mankind. To learn to think in terms of electrons and their quantum mechanical behaviour is probably of greater technical importance than we can now anticipate" (Löwdin 1957, 68). These statements were a particularly characteristic expression of his agenda for quantum chemistry: to turn quantum chemistry into an exact and predictive discipline established on secure quantum mechanical foundations.

In the discussions at Valadalen, the goals of quantum chemistry were schematically expressed in ways that underlined their utilitarian role (Löwdin 1986, 22). Löwdin was fond of visual representations, and he knew that chemists were also particularly fond of them. He depicted a simple little diagram. The horizontal axis represented the refinement of theory, and the vertical axis indicated agreement between theory and experiment. The wave-like curve approaching an asymptotic limit was meant to show that "an elementary theory may often be brought to give excellent agreement with the experimental results, whereas most refinements of the theory will disturb the nice situation and cause disagreements" (Fröman and Goscinski 1976, 39). He called the first peak "Pauling point" and the first minimum the "Ph.D. point" to call attention to the sort of problems given to Ph.D. students. Electronic computers enabled the theory of small systems to overcome the first maxima and minima on the agreement curve, but in the case of more complex systems it was not even clear where theory stood at the time. The diagram purported to show that agreement between theory and experiment was a necessary but not a sufficient condition for the validity of theory. And that the aim of quantum chemists should be to go beyond the Pauling point in order to test the final outcome of theory. Despite the diverging views expressed in these discussions, there was a deep commitment and collective concern for the future of quantum chemistry (Fröman and Goscinski 1976, 47–51).

A Fruitful Swedish–American Connection: The Quantum Theory Project and the Sanibel Island Conferences

Individual and institutional ties with American scientists, universities, and organizations were a distinctive feature of Löwdin's career: Slater acted as his mentor; the organization of both Mulliken's Laboratory of Molecular Structure and Spectra and Slater's Solid-State and Molecular Theory Group inspired the creation and organization of the Quantum Chemistry Group. American agencies sponsored the group's activities.

While the process leading to the establishment of the chair of quantum chemistry was slowly taking place, negotiations with the University of Florida at Gainesville, involving the dean of the Graduate School and the heads of the Departments of Chemistry and Physics, led to the creation of the Florida Quantum Theory Project, a project very much like the one in Uppsala (Löwdin 1977). It acted as a bridge between quantum chemistry on both sides of the Atlantic, as a bridge between the Department of Chemistry and the Department of Physics at Florida, and finally as an institution to attract students from Latin America. It was agreed that Löwdin would spend one third of the year at Florida and bring with him some Swedish collaborators for 1- to 2-year stays in order to run the program during Löwdin's absence (Fröman and Lindenberg 2007, 38; Löwdin 1986).

Soon Löwdin and his group were joined by Mulliken and Slater. Mulliken was first invited to Tallahassee, Florida, in 1964 to spend a month at the Institute of Molecular Biophysics directed by Michael Kasha, his former student at Chicago, and he was appointed Distinguished Professor of Chemical Physics at Florida State University. After retiring from MIT, Slater joined Löwdin's Quantum Theory Project in 1964 and proceeded to set up something much like his MIT Solid-State and Molecular Theory Group in this new context. Together they were able to build a powerful center for high-level research in "Quantum Chemistry, Solid-State Theory and Quantum Biology" in Florida.

Starting in December 1960, winter schools were also established in Florida. They were planned to last from 4 to 5 weeks and to be sponsored by the National Science Foundation. These symposia came to be known as the Sanibel Symposia as they were organized for the first 15 years at Sanibel Island in the Gulf of Mexico outside Fort Myers, Florida. As Mulliken (1989, 183–184) reminisced in his autobiography, the site of the conference was a particularly attractive location, especially notable for the great variety of sea shells that could be found on the beaches. The idea behind the Sanibel Symposia was to organize forums for the discussion of the main lines of research, methods, and problems dominating the discipline.[46] The odd-year symposia were arranged in honor of the great pioneers in the "Quantum Theory of Atoms, Molecules and Solid-state," starting with Hylleraas in 1963, Mulliken in 1965, Slater in 1967, Eyring in 1969, Van Vleck in 1971, E. U. Condon in 1973, and L. Thomas in 1975.[47] Löwdin's agenda for quantum chemistry included as one of its central items the discussion of the importance conceded to disciplinary history, and the odd-year Sanibel Symposia fell into this category. In fact, since the late 1950s, Löwdin expressed his concern for the oblivion in which the first decades of quantum chemistry were enveloped, and he decided to remind quantum chemists of many of the events and personalities involved in the founding of the discipline, being certain that such knowledge could guide future developments. The shaping of disciplinary history due to the impact of computers was also discussed at the Sanibel meetings. One such example is provided by the concerns voiced by Enrico Clementi (born 1931), one of Mulliken's former students, then working at the IBM Research Laboratory, who appropriately chose to speak about "Chemistry and Computers" in the 1967 meeting honoring Slater.

Clementi was very assertive in claiming that computers could be extremely useful in the future if, and only if, one departed from the then-current trend in computational chemistry, which pointed "toward the formation of an enormous library of wave functions with little attention to chemistry as such. This, of course, will lead to chemistry but only if we compute a very significant fraction of the possible molecules. Such a goal seems most unrealistic" (Clementi 1967, 308). He reacted against the increasing "computation" of the discipline if "computation" implied its seclusion from chemical problems. Quantum chemistry without chemistry seemed to be pointless. For him, the only meaningful way to use computers was to write computer programs able to cope with realistic chemical problems such as those occurring in nature. The mathematical model behind such an endeavor was, of course, quantum mechanics with as many approximations as a chemical problem could afford to sustain "before becoming an irrational 'soup' of floating numbers of questionable physical meaning" (Clementi 1967, 308). Then if the computer program was meant to solve a "synthetic chemistry problem," it should be able to start from the component atoms and arrive at the final molecule, in the process elucidating intermediate chemical steps and reaction mechanisms. If the program was written to solve a "spectroscopic problem," it should give the basic spectroscopic constants. If the problem was a "structural problem," the computer program should give internuclear distances and electronic density mappings. In a few years, Alberte Pullman (1971) would be voicing the same concerns about quantum chemistry: that more computation and more successful results should not mean less chemistry.

The *International Journal of Quantum Chemistry*

The proceedings of the two Sanibel meetings, commemorating the contributions of Hylleraas and Mulliken, were published in the *Reviews of Modern Physics* and in the *Journal of Chemical Physics*, respectively. According to Löwdin's recollections, the American Institute of Physics seemed worried that these activities were taking too much space in its publication outlets, a situation that created an extra incentive for the industrious Löwdin to create a new journal (Löwdin 1986, 21). The *International Journal of Quantum Chemistry* was founded in 1967 and published by Interscience Publishers, a division of John Wiley & Sons. Clementi's talk and a selection of other communications delivered at the Sanibel meeting honoring Slater were included in a supplement to its first volume.[48]

In his characteristic style, Löwdin took the opportunity of introducing the new journal by writing a concise but sharp editorial manifesto (as we noted in the introduction), which embodied both a reflection on the discipline's past and an announcement of its more promising goals.[49] Written at a time in which quantum chemistry was experiencing intense networking and growing internationalization and was exploring the potential of the electronic digital computer while at the same time extending its domain to molecules of biological interest, it called attention to a number of specific features of the subject matter of quantum chemistry—the elucidation of the electronic makeup of atoms, molecules, and aggregates of molecules; the interplay of theory, experiment, mathematics, and computational algorithms in forming the methodological framework of quantum chemistry; its relationship with mathematics, physics, and biology; and finally the assessment of the role of quantum mechanics in providing a unifying framework for the natural sciences and eventually for the life sciences. For honorary editors Löwdin chose Heitler, Mulliken, and Slater. On the editorial board Coulson, Raymond Daudel, Kotani, McWeeny, Roothaan, Shull, J. O. Hirschfelder, L. Jansen, and R. Pauncz worked jointly with him. Obviously, the selection of personalities, and those that were not chosen, reflect Löwdin's disciplinary allegiances.

The main fields to be addressed in the journal reflect how he planned to influence the discipline's future: fundamental concepts and mathematical structure of quantum chemistry, applications to atoms, molecules, crystals, and molecular biology, and computational methods in quantum chemistry outlined the main roads to be explored. Past, present, and future were interlaced in the editorial program. A reflection on the "Nature of quantum chemistry" completed the picture. Disguised as straightforward guidelines to the journal's contributors, it really offered Löwdin's perceptive considerations on the nature, methods, and tools of his discipline by delineating the "ideal form of a theoretical paper," by calling attention to the interplay of experiment and theory in any "science," by framing the role of interpretations as rules to go from one to the other, and by considering that quantum chemistry "could boast more of its conceptual framework than of its numerical achievements" (Löwdin 1967a, 8), Löwdin emphasized that the future of the discipline was tied to numerical computations, numerical analysis, and the use of potent and fast computers. But he also emphasized that "various types of theories are constructed for different purposes" (Löwdin 1967a, 9), so that ab initio, semiempirical theories or any other sort of theory have a role to play in any phase of disciplinary development, to such an extent that they should be explored in parallel having always in mind their respective domains of applicability. For Löwdin (1967a, 10), the construction of "meaningful semiempirical theories" continued to be one of the most important future goals for applied quantum theory. Despite Pauling's estrangement from the discipline's post-war developments and his absence from the list of honorary editors chosen by Löwdin, his legacy continued to play a role in disciplinary inroads.

Old Contexts, New Agendas: Quantum Chemistry as a Quasi Laboratory Science

Mulliken's Laboratory of Molecular Structure and Spectra

Resuming activities after the end of the war was not easy for anybody, and certainly not for Mulliken and his group. New hierarchies appeared to be created, and nuclear and particle physics had been accorded the status of scientific stardom, high in the consideration of public opinion and government funding. Members of the scientific community had to rethink their agendas and priorities in view of subdisciplinary rearrangements. Before the war, the moves of the American physics community to opt for molecules, "leaving" atoms for their European colleagues, enabled American physics to "come of age" (Van Vleck 1964). Now, molecular scientists were not anymore on center stage. To stick to his pre-war molecular program meant to revitalize a research group and fight for government and federal support under very different constraints. Nevertheless, Mulliken was able to secure funding, initially from the Office of Naval Research, which did not commit the group to any research of immediate military application but only to pure research of possible future use. Additional support was also secured from the Army Research Office and the Air Force Cambridge Research Center, besides grants from the National Science Foundation (Mulliken 1989, 117).

Adjusting to the new funding system proved to be easier than coping with the institutional rearrangements and different policies within the post-war Department of Physics in Chicago. Contrary to expectations and despite his high status, Mulliken never managed to create at the university an Institute of Molecular Physics mirroring the Institute for Nuclear Studies and other interdepartmental institutes such as the Institute for the Study of Metals or the Institute of Radiobiology and Radiophysics (Mulliken 1989; Butler 1994; Park 1999a). His small-scale institute, known after 1952 as the Laboratory of Molecular Structure and Spectra (LMSS), occupied two different locations, one for experiments and one for theory. From 1947 to 1970, the group issued yearly reports, known as Red Reports, which were delivered to supporting agencies and were circulated worldwide in an effort to speed up communication among molecular scientists and quantum chemists. Besides Mulliken, the group included initially J. R. Platt and C. C. J. Roothaan as faculty members, some people who worked exclusively with calculations, a few postdoctoral research associates, visiting professors, and graduate students (Mulliken 1989, 117–359). Mulliken and his group concentrated on converting molecular orbital theory from a descriptive and semiempirical theory into a more quantitative theory. Two lines of research were developed in parallel: improvements of Hückel's method and the search for other options. Together with other young collaborators, Roothaan pushed forward the second front, selecting as their main target the difficult research on electron-repulsion and overlap integrals.

C. C. J. Roothaan (born 1918) joined Mulliken's group in 1946, after enduring "the most traumatic experiences" in his life (Roothaan 1991, 1). He was a young man from the Netherlands with a background in engineering and physics at the University of Delft, where he attended classes by R. Kronig and H. A. Kramers. He was caught in the dark webs of war and taken prisoner by the German occupiers in April 1943, together with a brother, who did not survive the concentration camps.[50] They were held prisoners first at the police quarters, then moved to a concentration camp at Vught, close to Eindhoven, where they stayed until September 1944. Roothaan managed to work in the "Philips Commando," the name given to the manufacturing facilities set up inside the camp by Philips managers in order to take advantage of prisoners' labor. He also belonged to the "Computation Chamber," another Philips creation, which used a selected few academically inclined prisoners. The work done in this context on the calculation of elastic constants in a classical crystal brought him closer to theoretical quantum science and was accepted after the war as his diploma of "Ingenieur," roughly equivalent to an M. Sc. degree, by the University of Delft on October 15, 1945. As the pressure from the Allied forces grew, prisoners were moved to a camp in Oranienburg, in the Berlin suburbs, a camp where the death rate was similar to that of Dachau and Buchenwald.

Returning to the Netherlands, Roothaan immediately applied to a postgraduate position at the University of Chicago where he arrived in early January 1946. He felt certainly like fleeing from hell to heaven: "Chicago was the most exciting place to be for a young physicist" (Roothaan 1991, 8). He did not opt for nuclear physics, feeling that as an "alien" he had few chances to be given clearance and an interesting topic to work on. He chose instead to work toward his Ph.D. on molecular structure and spectra with Mulliken. His initial topic dealt with semiempirical molecular orbital calculations on substituted benzenes using Hückel's approximation. For 2 years, from 1947 to 1949, he worked toward his Ph.D. while holding a job as a physics instructor at the Catholic University of America in Washington D. C. , where he became a member of Karl Herzfeld's group. Right at the beginning of his doctoral work, Roothaan became dissatisfied with the then-current Hückel approach to molecular orbital theory, which started by adopting "a mystical one-electron Hamiltonian" (Roothaan 1991, 9), of which he could not find any definition in the literature. In June 1948, Roothaan found a solution to his queries by the method of Hartree and Fock. He returned to Chicago in 1949 to receive his Ph.D. in 1950, and then joined the Department of Physics of the University of Chicago. From 1962 to 1968 he was the director of the University of Chicago Computation Center. His career pattern exemplifies a constant alternation between "physics, quantum chemistry and computer development" (Roothaan 1991,12), a pattern that was common for many of the new generation involved in post-war quantum chemistry.

Computers and Ab Initio Computations

Four years after the Shelter Island Conference and 1 year after the Nikko Symposium, it had become clear that commercially developed computers could handle programs for large computations in the foreseeable future.[51] But although computer technology was making big strides, the cost for using them remained forbidding.

The quantum chemical community stood in direct competition with experimental chemists for common funding resources, and in what relates to equipment, another group—the crystallographers—relied on automated computations and was also fighting for computer access. Some quantum chemists decided to take immediate action, and at the 1955 Molecular Quantum Mechanics Conference, organized at the University of Texas, Austin, by F. A. Matsen, conference participants passed a recommendation urging "governments, industries, foundations, and private philanthropists (to) give special attention to the problem of providing more high-speed computing facilities for use in molecular problems" (*Molecular Quantum Mechanics Conference* 1956, 60).

Some universities responded positively to this call, developing such installations and making them available to the various departments. Researchers at the University of Chicago, MIT, and the University of Cambridge had appreciable means at their disposal for calculating molecular wave functions, especially for diatomic molecules. Nevertheless, the different ways practitioners envisioned the development of computers' hardware and software and their views concerning progress of both semiempirical and ab initio computations contributed to the further entrenchment of the two distinct cultures for doing quantum chemistry.

After the deadlock that followed the Heitler and London 1927 paper and the subsequent early attempts at ab initio computations applied to the hydrogen molecule, from those by Sugiura (1927) to those of Hubert M. James and Albert S. Coolidge (1933),[52] the first all-electron ab initio calculation in a molecule larger than H_2—the N_2 molecule—using a minimal basis set of Slater type orbitals and dealing with the ground state and several excited states was performed by Mulliken's group at Chicago. The prowess was due to W. C. Scherr (1955), one of Roothaan's students, and was part of his Ph.D. work. The calculation was done on desk calculators (Marchants, Fridens, and Monroes) with the help of some assistants and took 2 years to complete. In 11 years' time, the same computation could be done in 2 minutes with the largest available computers, provided the machine program was already written (Mulliken 1966, 152; Bolcer and Herman 1994, 8)!

The availability of a method such as the LCAO-MO-SCF (linear combination of atomic orbitals–molecular orbitals–self-consistent field) to calculate atomic and molecular integrals, which was well suited to the logical sequencing of computer programs, increased the demand for their computation and forced computer programs to be developed (Roothaan 1991, 11). In fact, with rapid advances in computer technology, the need for tables of molecular integrals from which required values were painstakingly interpolated became obsolete. Integrals and everything else needed for each molecule were directly provided by a variety of computer programs written by quantum chemists themselves. This task was undertaken by Roothaan himself, together with Platt and Rüdenberg, who arrived from Zürich with his mentor Gregor Wentzel in August 1950. In 1956, Bernard J. Ransil joined them.

Bernard J. Ransil (born 1929) obtained his Ph.D. in 1955 at the Catholic University in Washington, D. C. , for work on the LCAO-MO-SCF treatment of the H_3 molecule. One year later, he became a research associate at the Laboratory of Molecular Structure and Spectra, replacing Roothaan who was on leave in Europe on a Fulbright Scholarship and doing work that led to the first generation of machine programs. A group of graduate students and assistants helped in the design and test of the first computer program able to generate diatomic wave functions. They performed minimal orbital LCAO-MO-SCF approximate calculations for all diatomic molecules and hydrides of the first row of the periodic table and three additional heteropolar molecules (BF, CO, and LiF). The program was written in machine language for the UNIVAC (Remington-Rand 1103) as computer languages such as Fortran were not yet available. Despite the fact that there was no such computer facility at the University of Chicago, Mulliken and Roothaan had secured a contract for the use of excess computer time on the UNIVAC at Wright Field Air Force Base in Dayton, Ohio (Mulliken 1989, 156–163; Bolcer and Herman 1994, 8–10). For 18 months, every 2 or 3 weeks, members of the group stayed for a few days in Dayton, using the computers in the evenings or through the nights.

Luckier than most other groups for having access to a computer, their situation illustrates the difficulties encountered by those whose work was increasingly dependent on the use of computers. They had to be frequent travelers and their own software engineers, able to adapt computer programs to new computer designs whenever necessary. In fact, when the process was nearing production-run status, the Air Force base switched to a 1103A computer, entailing extensive program rewriting. In the meantime, Roothaan returned from Europe, and the diatomic project was completed in 1958–1959: the program became operational in spring–summer 1958, and minimal orbital calculations on all first-row diatomic molecules, hydrides, and the three heteropolar molecules mentioned above were completed by the next winter. The results were encouraging. No wonder that Ransil (1960, 1960a) used the opportunity provided by the upcoming Boulder Conference to present the results of the diatomic project, discussing its advantages and shortcomings. For the 12 diatomics studied, the program underestimated energies to 1% or less; gave correct sign and order of magnitude for dipole moments; estimated ionization potentials to one-figure accuracy; and gave correct order of magnitudes and one- to two-figure agreement for spectroscopic constants in molecules for which those experimental values existed (Mulliken 1989, 136–161; Bolcer and Herman 1994, 9). No wonder that, from 1959 onward, Mulliken dubbed the success of the first computer program as announcing a new era.[53] In his autobiography, Mulliken (1989, 159) recollected that as both computer speed and memory capacity improved, the main difficulties for obtaining accurate analytical wave functions and energies for small molecules would be "how to give accurate analytical representation to electron correlation. All other considerations . . . were technical problems that would yield to the inexorable advance of computer technology."

Computers and Semiempirical Approximations
By the end of the 1950s, the first successful inroads into exact computations using the various new computer machines provided a realistic prospect to Dirac's dictum: though

analytic calculations were impossible, numerical approximations could become feasible to arbitrary accuracies. Furthermore, the availability of computers guided the development of new semiempirical techniques of increasing sophistication and precision that could be applied to large molecules. Such was the case, for example, of the Pariser–Parr–Pople approximation. Worked out jointly by Parr and Rudolph Pariser, and independently by John Pople (1925–2004), it was a more elaborate self-consistent field form of Hückel's π-electron theory in which electron repulsion integrals and exchange integrals were explicitly taken into account, and the wave function was properly antisymmetrized, but which nonetheless introduced a simplification in the LCAO-MO-SCF approach by assuming "zero differential overlap."

Pariser was employed as a physical chemist at the Jackson Laboratory of E. I. du Pont de Nemours & Co. when he was assigned the task of characterizing the dyes being synthesized in the laboratory. Limitations of instrumentation and the overwhelming number of dyes under scrutiny made him decide to approach the subject from a theoretical viewpoint, trying to understand the relation between structure and properties, especially color, in the case of molecules of complex organic dyes. Having been an undergraduate student of Parr at University of Minnesota, he sought his advice in July 1951. Parr was then an associate professor at the Carnegie Institute of Technology, Pittsburgh. In this way, a rather unique collaboration between academia and industry was inaugurated. They both agreed that any theory with the ability to make quantitative predictions had to include electronic repulsions, although they were also aware that this was practically impossible for molecules larger than benzene, having in consideration the limited computer capabilities of the early 1950s.

In early November, Parr realized that electron-repulsion integrals involving products of two charge distributions associated with orbitals on different centers could be discarded, and one was left with electron-repulsion integrals of purely Coulomb type. By using an approximate representation of $2p$ π-orbital densities in the remaining two-center repulsion integrals, they were able to apply LCAO-MO theory, including configuration interaction, without being "held up by the awful problem of integrals" (Löwdin 1990; Parr 1990, 329). The first calculations, done with a mechanical desk calculator by Pariser on benzene and with difficulty on naphthalene, agreed with other theoretical calculations but were still far from reproducing the experimental results. In the next step, the training of Pariser as an experimentalist helped him in selecting "experimental" values for certain integrals for which they were able to offer a convincing theoretical justification (Pariser and Parr 1953, 1953a). Their approximation, which was probably baptized by Rüdenberg in August 1952 as the zero differential overlap approximation (ZDO or NDO), was justified by Löwdin 3 months later as resulting from the transformation to symmetrically orthogonalized atomic orbitals (Parr 1990, 329, 343). To apply the ZDO approximation with the new empirical integrals to larger molecules required the modern solid-state high-speed computer. They remembered "frustrating experiences with vacuum-tube based and card-programmed computers" (Pariser 1990, 320), followed by the excitement of using an IBM 701, for which Pariser himself developed a program in machine language for their approximation method. Independently from Parr and Pariser, Pople (1953, 1990) arrived at the same results. As he recollected later, during this period he believed in the subsidiary role of theoretical chemistry as handmaiden to other developing branches of chemistry, to such an extent that applications of theory should use all the necessary approximations in order to be immediately useful (Handy, Pople, and Shavitt 1996, 6014). He was not yet seeking to develop the exact computations for which he later became famous, becoming one of the recipients of the Nobel Prize in Chemistry for 1988 (Pople 1998; Radom 2008).

Developed in a period in which the reigning principle of structural chemistry was still Pauling's principle of maximum overlap, the application of the NDO approximation enabled the reassessment of an assumption as old as quantum chemistry itself, and furthermore showed that semiempirical methods could be applied to the quantitative prediction of the electronic structure and spectra of molecules of real interest to chemists. In fact, during the period 1961–1977, the papers of Parr and Pariser were among the five most cited publications of the 1950s in chemistry and physics (Pariser 1990, 322). Quantum chemistry was becoming an increasingly valuable tool for chemists—and computers played an indispensable part in this new trend.

Computers and Cultures of Quantum Chemistry

While everywhere, in the United States, the United Kingdom, France, Germany, Japan, and Sweden, quantum chemistry groups were giving a prominent place to computers and computations, the outlooks of Mulliken's group in Chicago and Slater's group at MIT expressed appreciable differences with the outlooks of Lennard-Jones's (and later of Longuet-Higgins's) group in Cambridge and those of Coulson's group in Oxford, the two main computer centers in the United Kingdom during the 1950s.

Despite considering himself a "middle-man between experiment and theory," whose far-reaching contributions stemmed from the qualitative interpretation of molecular spectra, Mulliken wished to strengthen theoretical endeavors. Though he never really became himself an expert on the use of computers, Mulliken foresaw their potential and fostered the exploration of computers in his laboratory. From tables of molecular integrals to programming, his vision of quantum chemistry was tied with exact ab initio calculations. The same meticulous personality who was professionally thrilled by ordering molecular spectra, and whose preferred hobby was the identification of countryside plants, found a strong appeal in the tabulation of molecular integrals and in the calculation of molecular wave functions. Classification, nomenclature, and organization were always at the forefront of his interests, professionally or otherwise. Mulliken's interests tipped without qualms toward ab initio computations, but he still attributed much relevance to the exploration of semiempirical approximations by discussing at the Nikko Symposium the recent refinement of Hückel's method that came to be known as the Parr–Pariser–Pople approximation (Mulliken 1954).[54] Slater partook of Mulliken's appeal for computers and ab initio computations and also showed no reservations in their applications to quantum chemistry and solid-state physics, never overcoming his suspicions in the use of approximations that could not be justified theoretically, something that he expressed in his talk at the Nikko Symposium.

A different way of looking at these problems was exemplified by Coulson's intervention at the Nikko Symposium, in which he contrasted the "easy" with the "hard" methods in doing quantum chemistry, and which followed smoothly his many considerations steadfastly voiced since the time he became the Rouse Ball Professor of Applied Mathematics at Oxford in 1951 (figure 4.3), and which culminated in the after-dinner address at the 1959 Boulder Conference. More than anyone else he was a stubborn and committed advocate of methodological pluralism, of the possibilities for exploring different approaches in different problems, always eager to compare and

Figure 4.3
Charles Alfred Coulson teaching at the Mathematical Institute, 1959.
Source: Courtesy of the Coulson family.

contrast them, to foster semiempirical calculations while at the same time exploring the potential of ever more potent computers, all within the overarching view that privileged conceptual understanding over numerical accuracy.

The ambivalent reaction to the extensive use of computers in ab initio computations and their role in reshaping quantum chemistry may well be illustrated by the difficulty in understanding and supporting Boys's work by such prominent personalities as Lennard-Jones and Longuet-Higgins, the two successive presidents of the Department of Theoretical Chemistry at the University of Cambridge to which Boys belonged.[55] Some of Boys's former students and colleagues referred to him as "the odd man out" (Handy, Pople, and Shavitt 1996, 6008), in the sense of a visionary, as someone able to predict the potential of ab initio *avant la lettre*. And Pople recollected that in the early 1950s, Boys took the "long-term view" while Pople and many others were after new and more exact approximations and very "sceptical" about this "long-term view" (Handy, Pople, and Shavitt 1996, 6014). In another biographical memoir, Coulson (1973a, 111–112) called attention to Boys's personality, stubbornly undisturbed by scientific fashions and steadfastly implementing his agenda, in order to account both for his unsuccessful career and the inability to understand his agenda revealed by some fellow quantum chemists. Notwithstanding the role played by personality characteristics or scientific incompatibilities, we would like to argue that his case offers a clear instantiation of how the culture of applied mathematics informed the work of many British quantum chemists and accounted for their specific allegiances.

Samuel Francis Boys (1911–1972) earned an undergraduate degree in chemistry with first-class honors at Imperial College in 1932 and spent the next 3 years taking extra degree courses in mathematics. In 1935, he moved to the University of Cambridge to start his Ph.D. work under the supervision of T. M. Lowry, but upon Lowry's death he became Lennard-Jones's student. This change reinforced his early interest for the electronic structure of molecules. The war slowed down his first incursions into quantum chemistry, as he concentrated on rocket propellants in the Ballistic Branch of the Armaments Research Department, having again Lennard-Jones as one of his superiors. After the war, he was a recipient of the new I. C. I. fellowships at Imperial College, where his lifetime immersion in ab initio electronic structure calculations began. He returned to Cambridge in 1948 as a lecturer in theoretical chemistry, staying there until his death.

His series of papers on the "Electronic Wave Functions" (1950–1954) dates from this period (Boys 1950, 1950a, 1951, 1951a, 1951b, 1952, 1952a, 1952b; Bernal and Boys 1952, 1952a; Boys and Price 1954; Boys and Sahni 1954). In the first paper of the series, Boys (1950) introduced Gaussian basis functions in quantum chemistry calculations. They had been used previously to solve the harmonic oscillator in quantum mechanics, and McWeeny used them in his 1948 dissertation with Coulson.

Most probably Boys was using them since 1942 or earlier, as witnessed in his notebooks (Handy, Pople, and Shavitt 1996, 6009), but it took him some years until he introduced them in the literature to replace Slater type orbitals (STO). They offered the advantage over exponential functions of greatly simplifying the calculation of multicenter integrals. The first tests of Gaussian basis sets were carried out by Boys's Ph.D. student Robert K. Nesbet on the ground state of methane in 1954 and at about the same time by Meckler in Slater's MIT group.

Boys and his collaborators initially dealt with electronic structure computations for Be, B, C, F-, and Na+, including both the ground state and some excited states, using desk calculators (Boys 1950a, 1952a, 1952b; Bernal and Boys 1952a). They, then, presented results already obtained by using the EDSAC, the first electronic computer designed at the University Mathematical Laboratory by Maurice V. Wilkes (Boys and Price 1954; Boys and Sahni 1954).[56] Similar to the American Electronic Discreet Variable Automatic Computer (EDVAC), it became operational in mid-1949. When his group moved into molecular electronic structure calculations, all the computational steps and derivation of results were in the process of being completely automated. Still later, in 1956, when the EDSAC II, the successor to EDSAC, which depended on transistors instead of vacuum tubes, became operational, Boys was able to have the computer for himself for some weeks, as he was the only one who could write programs and subroutines for the new machine. By then, he was convinced that it had become possible to predict molecular structure and properties with "unlimited accuracy" using Schrödinger's equation and the numbers and properties of electrons and nuclei making up each molecule. And so he believed that a short manifesto for computational quantum chemistry was a pressing need—to discuss the complicated mathematical formulations, the involved intermediate problems and the extent of the numerical operations, and to stress that computing machines not only gave the opportunity for getting numerical results but also "for carrying out much of the mathematical analysis of the most formal type" (Boys et al. 1956, 1207).

Boys believed strongly that advanced computers would be the clue for attaining many results that many people thought to be unattainable. Wilkes's recollection corroborate this view: "Boys' real trouble was that he was trying to operate on a scale that was beyond the means available at the time. Later, when machines more powerful than the EDSAC became available, the full extent of his vision became apparent" (Smith and Sutcliffe 1997, 279). A few years later, Pople considered that Boys was "the hero" of the 1959 Boulder Conference. His presentation was extraordinary. An early advocate of complete program packages, which he wrote himself, while reporting on his work "he produced a paper tape of his whole computer program and unrolled it along the length of the chemical lecture bench. There, in one roll, was something, of which one could ask a chemical question at one end and it would produce an answer at the other! . . . most of the audience probably thought the demonstration bizarre. But it was prescient" (Handy, Pople, and Shavitt 1996, 6015). And Pople confessed that he was only converted relatively late to ab initio calculations, in 1967, adding that the first program his group developed was, in effect, "pure Boys" (Handy, Pople, and Shavitt 1996, 6015).

The prevalence of the so-called "spirit of applied mathematics," to which Coulson referred in his inaugural lecture as professor of applied mathematics by discussing fundamental aspects of the British tradition of applied mathematics, expressed, in a way, a particular outlook toward computers. Within this culture, the exploration of different methods was cherished. The accommodation of diversity was considered an asset. And computers, he thought, should reinforce this trend; they should never downplay it. This culture accommodated initially both the VB and MO approaches, and later the semiempirical and ab initio approaches. Decades later, during the reigning era of MO theory, McWeeny, one of Coulson's Ph.D. students, a former collaborator and the editor of the 1979 edition of *Coulson's Valence* (1979), worked both on MO and on VB, and tried to counteract the received view according to which VB was not conducive to ab initio approaches (McWeeny 1954, 1954a, 1955, 1986, 1989, 1990). No wonder that McWeeny often recalled Coulson's concern for the abuse of "bastard mathematics," an accrued imminent risk if computers were not used properly. He also subscribed to Coulson's unending search for "primitive patterns of understanding" (McWeeny 1995, 20). Once again, computers should be used to help in this quest, never to obliterate it.

The same concern was voiced by G. G. Hall, a former Cambridge graduate student. According to him, the distinctive feature of the British culture of applied mathematics laid in model building (Ford and Hall 1970; Hall 1972), and the Cambridge emphasis on theoretical model building and semiempirical calculations fitted like hand and glove to this grand schema. While this agenda was being implemented, Boys was not only alone in pursuing ab initio computations but foresaw the deep significance of electronic computers. As Hall (1996, 313) recalled "his mistrust of semi-empirical theories and his refusal to guess at results also isolated him from many colleagues. From time to time he was the object of biting attacks. He was promoted to a Readership but never became a Professor." For Hall and for Coulson, as computational quantum chemistry grew in importance, there was "the need for a stronger interaction between calculation and understanding" (Hall 1991a, 15). The insiders' viewpoint fitted with the standpoint of an outsider. For Bernard Pullman (1976, 134), the "prestigious school of English quantum chemistry has always combined an interest for fundamental theory with an equal interest for the elucidation of the properties of large molecular systems." In this way, he expressed how the ability to articulate apparently antagonistic trends in the practice of quantum chemistry was central to the British culture of applied mathematics. It seems that it, also, became by extension a distinctive feature of the culture of quantum chemistry. And, as we already discussed, the French appropriated central aspects of this specific culture as their own.

A New Era

The 1959 Boulder Conference and "the Hyperbola of Quantum Chemistry"

The Conference on Molecular Quantum Mechanics held at Boulder, Colorado, in June 1959 was the culmination of a succession of meetings convened after the Shelter Island Conference with the express purpose of dealing with computations of complicated integrals accompanying the move from human computers to desk calculators to electronic digital computers.

Organized by the National Science Foundation, its steering committee included Mulliken and Slater as representatives of the first generation of quantum chemists and strong believers in the promises of heavy computations. It also included some already well-known names of the younger generation such as Parr, who had been one of the official reporters of the Shelter Island Conference, and Pariser, both of who worked out the approximation that bore their names. Of the more than 100 participants, only 18 did not come from the United States. Non-American participants included Coulson, Boys, McWeeny, Daudel, the Pullmans, and, of course, Löwdin. They were good examples of how the field was expanding itself both by exploring different methods and by incorporating an international network of specialists, often with complementary or even antagonistic methodological commitments. The topics to be discussed in the various sessions covered old and new themes, illustrating the incursions of the field into big molecules, the test of new calculational methods and computer programs, and at the same time it highlighted the move from structure to molecular dynamics and the consideration of forces other than the chemical bond in playing a role in quantum chemistry (Parr 1960).[57] Furthermore, the discussion of the promises offered by quantum chemistry in understanding the basic processes of life turned biochemistry into an exciting emerging field where molecular quantum mechanics found many applications (B. Pullman and A. Pullman 1960).

What made the Boulder Conference a historical event was that it marked, in no uncertain terms, the transition from the founding generation of quantum chemists to a generation whose success would be dependent on the way they would make use of the electronic computers. During the conference, the promising prospects of electronic computers was discussed together with the dangers that these prospects had for the change of the character of quantum chemistry as it had been articulated since the Heitler–London paper of 1927. Everyone was convinced that the improving calculational techniques and electronic hardware would bring forth many and new results. Not everyone agreed on the extent to which the new practices would distort accepted norms, thus reconfiguring quantum chemistry (almost) beyond recognition. Furthermore, the Boulder Conference is situated between the Pariser–Parr–Pople suggestions of 1953 and the year Pople started to work on the approximation procedures that were to be expressed in the form of computer programs that would lead to his seminal contributions a few years later, in 1964, for which he was awarded the Nobel Prize in Chemistry for 1988.

Perhaps Ransil's thoughts were rather indicative of the "climate" at the meeting. Ransil (1960, 1960a) introduced his extended discussion of the advantages and shortcomings of the diatomic project being developed at Mulliken's Laboratory of Molecular Structure and Spectra by general considerations on the attitudes one could adopt in relation to computers. His seminal long introductory paragraph is quite illuminating:

The coming of age of the digital computer and its impact on the field of molecular structure has recently been variously characterized as "disastrous to theoretical chemistry" and as "the means which will enable modern structural chemistry to become less of an art and more of a science." Insofar as the digital computer provides the means for critical calculations upon which theoretical concepts may be justified, tested, or based, the author is inclined toward the latter point of view; insofar as the use of a digital computer might blunt one's critical faculties and stunt the free play of his scientific imagination, reducing his research to little more than calculations for the sake of calculations, he agrees with the former estimate. Obviously a wide middle ground exists where the digital computer, intelligently used as a research instrument, can quickly provide the theoretical chemists with accurate results to an illuminating but complex critical calculation. Properly used, the numerical experiment can be as much of an aid and stimulus to the theoretical chemist as a well thought out and executed physical experiment. (Ransil 1960, 239)

As it is clear from this initial statement, Ransil quotes views without acknowledging the sources, thus we can surmise that these views were widely circulating and were, in fact, characteristic of the shop talk of the community. These views expressed the core of a wide spectrum of opinions, which were no doubt expressed in the soul-searching discussions during the conference. Notably, he did not uncritically embrace all the promises for a golden future. But he emphasized that a number of household words for the quantum chemist, especially the present-day understanding of concepts such as bond order, bond length, charge density, conjugation, hyperconjugation, and resonance, would "benefit from a reevaluation based upon accurate *a priori* quantum mechanical calculations" (Ransil 1960, 239).

Ransil encapsulated the theoretical agenda of his group in the following command: "go as far by calculation as is reasonably possible, guided by chemical and physical intuition and computer economics, without introducing empirical schemes or data" (Ransil 1960, 239). It is, therefore, no wonder that a few months before the meeting, Mulliken was boasting to whoever wanted to listen that we "are standing on the threshold of a new era" (Mulliken and Roothaan 1959, 398).

Löwdin's contribution, alternately, exemplifies all the characteristics of quantum chemistry, from its beginning to the 1970s. He stressed the importance given to mathematization and to the rigorous quantum mechanical expression of different quantum mechanical schemes he and his group pushed to new horizons. But Löwdin (1960, 334) also emphasized the perennial opposition between "physical and chemical visuality" and "extremely cumbersome procedures" in order to argue for a balanced whole involving both visual imagery and heavy computations.

Coulson aired his worries in no uncertain terms. His pessimism about the future was evident, not about the prospects of the new techniques but about the effects these would have in bringing about the metamorphosis of the practices of quantum chemists. Despite Coulson's own contributions and those of his research associates to the calculation of molecular integrals using ever more elaborate computer programs, Coulson was never oblivious to the major shortcomings of their indiscriminate use. By the time of the conference, he realized that in the process of expansion of the field, deep changes had occurred within the community of quantum chemists. Coulson gave the after-dinner speech, summing the main trends of the meeting in his characteristically meticulous way and listing the problems he believed were to occupy the chemists in the years to come. But in this speech, one senses a very worried Coulson, a Coulson who realized that there are now deep and perhaps irreconcilable—divisions in the community of quantum chemists. These were divisions that he believed were absolutely detrimental to the discipline and, for the only time in his published work, one finds Coulson not preaching tolerance but advocating partisanship.

In discussing the major conclusions from the conference he noted: "there is one of these [conclusions] about which I feel very strongly, and because it is of such great importance for any future conferences on molecular structure, I make no apology for coming straight to it. It seems to me that the whole group of theoretical chemists is on the point of splitting into parts . . . almost alien to each other" (Coulson 1960, 172). The splitting was the result of the different views concerning the large-scale use of electronic computers—but there could even be a deeper reason than that. During the week of the conference, he had heard more than once the phrase "Oh, but you're not doing quantum chemistry." The occasions that gave rise to such assessments were the computational techniques presented for calculating energy values for atomic helium and molecular hydrogen, the calculations of a "highly empirical" kind to estimate energy levels and charge distributions of heteronuclear aromatic molecules, and the tabulation and interpretation of barriers to internal rotation in substituted ethane type molecules. His view was that these three situations represented quite distinct aspects of quantum chemistry, as they differed considerably in their underlying assumptions. But each group thought that what the others did was not quantum chemistry. "The situation is indeed serious. For my own part, I am very far from laughing at it, and I want us to look at it openly and as dispassionately as possible.

The questions that we are really asking concern the very nature of quantum chemistry, what relation it has to experiment, what function we expect it to fulfill, what kind of questions we would like it to answer. I believe we are divided in our own answers to these questions" (Coulson 1960, 172). Coulson believed that the problems culminating in the then-present deadlock could be traced to the recommendations made at the 1955 Molecular Quantum Mechanics Conference, organized at the University of Texas (see the previous section "Computers and Ab Initio Computations"). He was uncharacteristically persistent: "It is in no small measure due to the success of these programs that quantum chemistry is in its present predicament" (Coulson 1960, 172).

The splitting, he thought, in the community resulted from the antagonism of two extreme groups. The first group possessed great computational skills and advocated that there are a number of problems that a dispute can only settle by computation because experiments are too difficult. Examples of this were the absorption of hydrogen as a function of wavelength (very important for the astrophysical study of solar radiation) and the shape of the ground state of the methylene radical. This kind of work must have great accuracy and involved much use of electronic computers. To many people, this group of chemists appeared to be moving away from the conventional concepts of chemistry, such as bonds, orbitals, and overlapping hybrids "as to carry the work itself out of the sphere of real quantum chemistry" (Coulson 1960, 172). On the other extreme were calculations with very rough approximations for biological molecules. These calculations gave quite interesting results, but the approximations put forward would be greatly upsetting to the people who used computers extensively.

"Where, in all this, does 'real' quantum chemistry lie?" Coulson wondered. The possibilities offered by the electronic computers enabled one to distinguish three levels of activity—a distinction with which most of the exponents of computing at the conference agreed.

First, there were the molecules or atomic systems of one to six electrons, for which one could effectively calculate energies as accurately as they can be measured. Second, the all too realistic prospects for faster computers allowed one to extend the range of molecules for which it would become possible to have effectively exact solutions to those with 6 to 20 electrons. Nevertheless, accurate results for these cases were achieved at the expense of visualizability: even in the five-term James Coolidge function—which Coulson believed to be "the best compromise between accuracy and simplicity"—there was nothing easily visualized about the wave function, and it required a further numerical integration on an electronic computer to derive from the full 13-term wave function the electronic charge density of the electron.

Coulson thought it was not very probable—and also not particularly desirable—to deal in such a manner with molecules of more than 20 electrons. He was reminded of Hartree's remark that if one had to print the wave function values of the ground state of the iron atom with sufficiently small intervals in all the electronic coordinates, then a whole library was required to house the printed results and that there were not enough atoms in the solar system to make the paper and ink necessary to do the same thing for the uranium atom! It appeared that 20 electrons may be a criterion for the upper limit to the size of a molecule for which accurate calculations are expected to become practicable. Of course, one should keep in mind that the 1 to 20 range had many interesting cases—but it surely left out a lot.

Coulson believed there was such a deep distinction between those chemists whose main interest laid in the 1 to 20 range, and thus thought in terms of full electronic computation, and those who did not think in these terms that the two groups deserved distinct names—Group I (the electronic computers, or ab initionists as some called them) and Group II (the non-electronic computers, or a posteriorists): "I cannot help thinking that the gap between the two groups is so large that there is now little point in bringing them together. This is probably the last conference of the old kind. In future we should either have two distinct conferences or be prepared to plan parallel sessions for Group I and II enthusiasts" (Coulson 1960, 173). Had Coulson predicted that in 30 years computers would run 100,000 times faster, he might have revised the number of electrons making up the boundary between the two groups! However, at the time of the conference, Coulson thought that it would be an oversimplification to think that the difference is only a difference having to do with the use of electronic computers. In their desire for complete accuracy, Group I appeared to be prepared to "abandon all conventional chemical concepts and simple pictorial quality in their results." Against this, the exponents of Group II argued that chemistry is an experimental subject, whose results are built into a pattern around quite elementary concepts. He did not make any effort to conceal that his sympathies lay with the latter and re-emphasized that the role of quantum chemistry is to understand these concepts and to show the essential features in chemical behavior.

Coulson believed that it would be a great disaster if quantum chemistry were limited to either the "very deep" or the "shallow" level. And certainly it would be a serious loss if it did not maintain a close link with experiment and with conventional thought forms of chemistry. He believed strongly that there was a danger that Group I people would forget that chemistry is associated with the real world. He believed that for Group I mathematically a bond was an impossible concept and it was nearly never used by them, yet this was particularly disturbing because the bond is so very basic to all chemistry. He was rather pessimistic about the prospect of interactions between these two groups. "It is not surprising that the orientations of these two groups of quantum chemists are so different that cross fertilization has now become much less frequent than in earlier days . . . and members of both groups display an undesirable lack of sympathy for each other's work" (Coulson 1960, 174).

The division analyzed by Coulson for the first time during the Boulder Conference became a concern for many practitioners. As for Coulson, he became mellower with the passing of time. His first assessment of an irreconcilable schism within the community of quantum chemists gave way to a more conciliatory mood, much more consonant with the views he held for many years about his colleagues and the diversification of their practices. He lived to witness that the speed increase in computer performance enabled the successful parallel development of both ab initio and semiempirical approaches, to such an extent that it became increasingly misplaced to refer to Group II as the "non-electronic computers." Future developments showed that the divergent trends in the quantum chemical community that haunted the perceptive Coulson in time converged into a peaceful cohabitation and eventually into a synthesis of these two different cultures. It was, of course, the case that the community of quantum chemists was rather used to accommodating members with diverging approaches to their discipline: there was the period with those who favored the valence bond theory and those who had adopted the molecular orbital theory; then there were those eager to stick to semiempirical approaches and those prone to explore ab initio methods; then, among the ab initionists there were those preferring the Slater type orbitals (STO) or those opting for Gaussian type orbitals (GTO) as the preferred components of basis sets.

If Coulson wanted to raise walls against the threatening prospects, Ransil wanted to be the welcoming host. If Coulson felt intimidated, Ransil expressed his belief that awareness of the dangers is the only way of avoiding their negative effects. Coulson echoed a whole generation of physicists, chemists, and mathematicians who had become quantum chemists. Ransil presented evidence that a new generation of physicists, chemists, and mathematicians, highly versed in physical chemistry, chemical physics, theoretical chemistry, computational chemistry, and computer programming, was in a position to start churning out numbers about almost all the experimentally measurable parameters of a hitherto unimaginable number and kinds of molecules. Coulson spoke as a member of a community who did much to legitimate quantum chemistry at many levels. Ransil spoke as a member of a community respecting tradition but indifferent as to whether the threats from the radical changes in the practices of quantum chemists would bring about a reassessment of their collective identity.

In a way, their views symbolized two diverging modes of thought and action that had accompanied the evolution of the discipline since its very beginning and that had been identified already by Van Vleck and Sherman in their 1935 review paper back in the pre-computer context. One such mode of thought embodied the tradition of applied mathematics, represented by Coulson and Longuet-Higgins, who looked askance at computers as substitutes for other modes of thought. The other was the establishment of large groups now headed by Mulliken and Slater, who were eager to take all risks involved in giving computers central stage.

Coulson's splitting of the community into Group I and Group II became the object of a graphical analysis put forward by Pople (1965) during the Sanibel Island Symposium dedicated to Mulliken, and which came later to be known as the "hyperbola of quantum chemistry" (figure 4.4).[58] He depicted the inverse relationship between the size of the molecules under study measured by the number of electrons (horizontal coordinate) and the sophistication of computational methods, the accuracy of their approximations, and the number of features of electron distribution correctly reproduced (vertical coordinate).

Toward the middle of the vertical axis, he placed the minimal basis set LCAO-SCF method developed by Roothaan in the 1950s in Chicago. Above were calculations with larger basis sets approaching the Hartree–Fock approximation. Still above there were the improvements over single determinant wave functions using, for example, alternant molecular orbitals or configuration interaction. At the very top, he depicted exact solutions of Schrödinger's equation, which were just possible for one-electron systems. Below the middle of the vertical coordinate, occupied by the LCAO-MO method, were ordered semiempirical methods including different sorts of mathematical approximations. At the bottom laid those methods that treated electrons as effec-

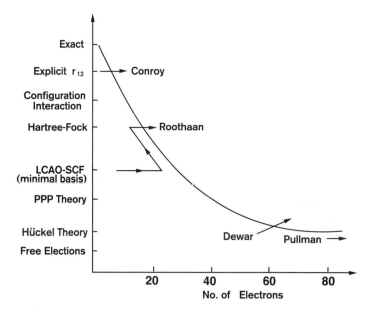

Figure 4.4
The hyperbola of quantum chemistry.
Source: Reprinted with permission from John Pople, "Two-dimensional chart of quantum chemistry," *Journal of Chemical Physics* 43 (1965): S229–S230 (on p. S229). Copyright © 1965, American Institute of Physics.

tively independent such as Hückel theory and free-electron theories. Above them were those methods, such as the Pariser–Parr–Pople method, which allowed for electron interaction but in a relatively rudimentary way.

In Pople's graphical representation, Group I people appeared on the top left and Group II appeared at the bottom right. The hyperbolic type curve represented the boundary separating what was already possible from what was still ahead for quantum chemists. The estrangement of the two groups was the result of Group I moving vertically while Group II moved horizontally, so that they "move continually further apart" (Pople 1965, on S229–S230). But this was not the only way to go. Progress could also be achieved if Group I people moved horizontally from left to right, applying a known method to larger molecules of greater chemical interest, or if Group II people moved vertically from bottom to top, exploring more exact methods for a given molecule. In this situation, the two groups would eventually "approach common ground in the center of the diagram to the benefit of the whole subject" (Pople 1965, on S230). Pople was looking after ways to ensure the convergence of the two groups. He did not consider their divergence an ineluctable result of the impact of computers, which was increasingly at the center of debate among quantum chemistry practitioners.

The 1970 Conference on Computational Support for Theoretical Chemistry

In 1970, the Conference on Computational Support for Theoretical Chemistry took place at the National Institutes of Health in Bethesda, Maryland, under the auspices of the National Academy of Sciences. The conference recommended the establishment of a national center for quantum chemistry, which would be built around a major computing facility. This facility—at a cost of approximately $10 million—would serve as a repository of computer programs as well as a place for the production of molecular calculations. Though this was unanimously agreed to by the conference attendees (quantum chemists, physical chemists, applied mathematicians, physicists, and computer experts), the academy's Committee on Science and Public Policy, chaired by the professor of applied physics Harvey Brooks, did not believe that the conference recommendations were "fully persuasive" (National Academy of Sciences 1971, iii), especially for the establishment of a national center. The committee's reservations had to do with uncertainties concerning the overall strategy for the development of large-scale computer installations. There did not appear to be a well thought out analysis of whether, for example, one could combine calculations involving nuclear structure physics, solid-state band theory, and quantum chemistry within the envisaged single-class computer, or whether one should be thinking in terms of a center for large-scale numerical analysis and "number crunching." The attendees of the conference agreed that the following characterized quantum chemistry in the post-Boulder period:

1. The development of several alternative methods of evaluating quantum mechanical integrals for diatomic and polyatomic molecules, both for Gaussian and for Slater type orbitals.

2. The emergence of complete systems for self-consistent-field (SCF) calculations on small and medium-sized closed-shell molecules.
3. Restricted-basis-set Gaussian-orbital SCF calculations on larger molecules.
4. Great improvement in semiempirical methods: inclusion of σ electrons, inclusion of many more integrals, closer correspondence with rigorous theory, study of the effect of approximations.
5. Scattering calculations for atoms and electrons.
6. The development of methods for understanding and calculating correlation energy; preliminary studies of the systematics of correlation energies and the effectiveness of procedures for estimating them.
7. Substantial developments in perturbation theory, of both low and infinite order, extending its sophistication and applicability (National Academy of Sciences 1971, 5).

It should be noted that these developments were not only due to the inertia of the practices of the first period, but also had been greatly facilitated by the advent of computers, and the extensive use made of them by the first post–Second World War generation. By the early 1970s, computational techniques were at a stage when it was becoming possible to have reliable a priori calculations and predict various parameters related to molecular structure and reaction kinetics in a more detailed manner than by the experimental results. In fact, in case the new techniques proved reliable, then it would have become possible to predict values of parameters that were exceedingly cumbersome and costly to measure experimentally, especially for large molecules, which were the ones of interest for application. Though the academy's Committee on Science and Public Policy rejected the recommendations of the conference, those who attended it reached unanimous conclusions on a number of issues related to the future of quantum chemistry. Among these issues and apart from the establishment of a national computation center for quantum chemistry and perhaps for other fields having related computational requirements, there was the establishment of a quantum chemistry institute connected with the national computation center in order to maintain a program library, the development of new programs, methods, and theories, and the planning for new computers. It was believed that the new center could have an organization and administration resembling that of Brookhaven National Laboratory—the facility at Long Island, New York, where at the time new and exciting results were produced from experiments with elementary particles performed by a consortium of universities. Furthermore, they investigated the possibility for remote capability based on financial support, noting that such capability must be of a character that will not degrade the computer from its primary purpose and must not constitute a major cost, proposing that in order to achieve resource sharing the Advanced Research Projects Agency (ARPA) network might serve as a model and therefore should be studied. Finally, they entertained the thought of a possible use (and financial support) of the center by industrial organizations. There were, of course, disagreements. The question of whether those who use the center for production runs should be required to make their programs available to the center was discussed but not resolved (National Academy of Sciences 1971, 4).

What, however, is particularly interesting is the views expressed at the conference about quantum chemistry and the ways it envisaged to influence the development of other branches of chemistry. It was noted that many chemists "use tools provided by quantum chemistry" (National Academy of Sciences 1971, 2). Furthermore, the needs in computer time and facilities were continuously on the increase among the quantum chemists and users of quantum chemical techniques and, hence, (quantum) chemists could not meet "their needs . . . to develop quantum chemistry further as a powerful tool in the service of chemistry and society" (National Academy of Sciences 1971, 3). It was also stressed that the computational techniques of quantum chemistry and theoretical chemistry, in general, should be regarded as an "instrument for solving chemical problems, and hence as an adjunct to other instrumentation" (National Academy of Sciences 1971, 13).

In the summary of the conference proceedings, it is rather surprising to read that due to the developments in the use of computers, quantum chemistry was no longer considered to be simply a curiosity but a specialty contributing to the mainstream of chemistry (National Academy of Sciences 1971, 1). It was believed that a close cooperation had developed between the experimentalist and the quantum chemist, with great benefit to both. The quantum chemist could often make predictions more accurate than the experimentalist's measurements. The experimentalist, in turn, could identify significant problems for the quantum chemist. In summary, quantum chemistry can now be regarded as a highly refined instrument on a par with, or even superior to, the finest laboratory instruments.

This is a rather remarkable statement coming after a period when the conceptual problems and the legitimization of the overall discourse of quantum chemistry was almost the only item on the agenda of quantum chemists. But by the 1970s, ab initio calculations with accuracies to within 0.1 electron volt brought to the fore the realization that quantum chemistry could be something more than the application of quantum mechanics to chemical problems, as well as the discussion of some esoteric points concerning the conceptual compatibility of the various theoretical schemata being proposed. The development of computers and the parallel developments of computer programs and general methods for calculations in quantum chemistry brought quantum chemistry to the same "level" with other "more respectable" branches such as molecular biology, solid-state physics, and atomic physics. In fact, those who attended the conference expressed their wish that the hardware configuration and software systems should be selected by "quantum chemists with computer experience, not by computer scientists or others unfamiliar with quantum-chemical research" (National Academy of Sciences 1971, 7). The possibilities provided by calculations of more or less arbitrary accuracy made quantum chemistry "relevant" not only because it showed in a practical manner that quantum mechanics can indeed be applied to chemical problems, but also because such an application entailed the development of specific kinds of hardware and software that could have a rather general use. Those who were pushing forward the agenda related to a national center for quantum chemistry were also aware that they should not develop an argument implying a "narrow utilitarianism, but to place quantum chemistry in its proper setting as a branch of chemistry" (National Academy of Sciences 1971, 12). The development of computers allowed for the first time a large-scale information storage and retrieval, data logging, data reduction and control, simulation and modeling of chemical systems. The main need at the time was that of rapid storage and retrieval of data, as quantum mechanical computations produced vast amounts of intermediate data.

All these developments were the result of the use of large-scale computation equipment in "one of the most extensive and sophisticated areas of computer application" (National Academy of Sciences 1971, 5), though the scarcity of resources hindered their further development. This had not been the case in many other branches of chemistry. Nevertheless, there were still many shortcomings, and for many problems "no substantial agreement yet exists as to how best to handle them" (National Academy of Sciences 1971, 5).

It was progressively becoming clear that the availability of large-scale computing facilities would enable elucidation of many problems and allow quantum chemistry to contribute to many other areas of research. Surely, it would facilitate the development of further understanding of electron correlation and of more complete methods for its calculation together with the development of more self-contained and reliable programs for making complete calculations. Odd-electron molecules and reaction intermediates would be treated in an effective way by the development of open-shell calculational systems. Large-scale facilities would facilitate the production of calculations for entire reaction-potential surfaces not limited to specified geometric configurations and the improvement of techniques for computing integrals. One of the points was related to the further insight one would be getting on how to "extract chemical concepts from wavefunctions" (National Academy of Sciences 1971, 6). Concerning the further development and use of quantum chemical techniques, there was consensus that it could be realized not only if some needs in equipment were met, but also if communication among quantum chemists as well as with other chemists was enhanced.

Notably, an issue discussed was the ways of fostering communication among (quantum) chemists—something rather odd, considering the great advances in this area compared with the early days of the discipline. Apparently, it was believed that neither the conferences nor the technical journals provided the forums for "the exchange of detailed information describing methods for carrying out quantum-chemical computations" (National Academy of Sciences 1971, 7). Though there was a rather widespread interchange of computer programs, an increasing number of researchers were not willing to share their programs with others. Such detailed communication had thus far consisted mainly of the interchange of computer programs: credits were not properly given all the time, and people had to spend a lot of time answering questions raised by those who were planning to use the programs. Not least of all were worries about the possible profits to be obtained by the commercial use of programs, even trade suits from competitors. There was thus a suggestion that the descriptive and explanatory material issued by the Quantum Chemistry Program Exchange (QCPE)[59] could turn into a journal of reference. Chemists who were working in other areas needed to be informed about the possibilities provided by quantum mechanical calculations for their research—in fact, most frequently requested QCPE programs had been for semiempirical calculations, and they had come primarily from non-quantum chemists. By 1970, there were about 1700 users of such programs throughout the world. The ratio was 1 to 7, in favor of non-quantum chemists. Of course, many good programs were not available to QCPE, and many programs were machine dependent.

Chemists were basically ignorant of the reliability of the methods, and this was particularly significant, as many calculations were of the same degree of reliability as experiments, and others were the only way to get information about specific phenomena such as repulsive potential curves in adiabatic descriptions of atomic scattering and energy bands of polymers. However, participants were worried that in case quantum chemical problems were studied, then this was going to be financed out of chemistry budgets, and the results had to be shown to be "worthwhile to chemistry as a whole" in order to convince the chemists to bear the financial costs. The prospects were bright: "a situation is rapidly approaching in which very large calculations are going to be fully justified and should be funded. These included calculations of properties that could not be measured experimentally (for example, at extremely high temperatures) and calculations that would essentially replace experiments (for example, on the detailed courses of reactions) at great savings in cost" (National Academy of Sciences, 1971, 18).

Some Further Remarks

By the time of the 1970 Maryland Conference, which discussed how computational support for theoretical chemistry could be efficiently achieved, it was clear to all quantum chemists that many changes had occurred since the publication of the Heitler and London paper in 1927. Fostered by the organization of scientific meetings, the foundation of the new *International Journal of Quantum Chemistry*, the creation of summer and winter schools and of the Sanibel Island meetings, and the presence of computers and the developments in hardware and software, the practice of quantum chemistry was reshaped in unforeseeable ways. Acting initially as a bond tying together distant groups, the computer became a central and absolutely decisive incentive in reassessing the status of established traditions and in molding new ones.

Whereas the first decades of quantum chemistry induced many chemists to rethink the status of theory within chemistry, the establishment of mathematical laboratories and the wide use of computers compelled many to rethink the status of experiment. In fact, computers came to be seen as virtual laboratory sites in which new types of experiments, both virtual *and* actual, could be enacted to give answers to chemical questions. Virtual experiments and mathematical laboratories expanded the sites of quantum chemistry.

Coulson, perhaps more than any other theoretical chemist, contributed to the development of a host of numerical methods to deal with quantum mechanical problems in chemistry. From the start, he was among the protagonists who supported large-scale computations in quantum chemistry using the newly developing electronic computers and one of the scathing critics of the indiscriminate use of computers despite the fact that the new instruments provided results of amazing accuracy, and they opened vistas to a promised land unreachable by traditional experimental means.

Diversity of styles became an identity hallmark of quantum chemistry. What for many years was considered as incompatibility between the VB and MO approaches, or between the semiempirical and ab initio approaches, when contrasted with the practices of quantum chemists who either used both VB and MO approaches (although they were more the exception than the rule) or used computers in developing semiempirical approaches or, inversely, fed in at times parameters in ab initio calculations, eventually became an apparent incongruity resolved by the realization that the success of quantum chemistry arose from an acquired ideology to accommodate a confluence of diverging trends and reasoning styles. No wonder that even while expanding its domain to big molecules and macromolecules, practitioners recalled Van Vleck and Sherman's old contrast between the optimists and the pessimists and their plea for a middle-ground attitude. In the end, the changing character of quantum chemistry became its more permanent defining trait.

In this respect, the contributions of the first generation of French quantum chemists are quite striking. Embracing quantum chemistry as part of a local agenda for the reformation of French theoretical chemistry, they opted for seemingly incompatible aims: a frontal assault on big molecules with nearly exact methods, and the articulation of old chemical concepts adopted by quantum chemistry with the apparently irreconcilable constraints imposed by quantum mechanics. This epistemic strategy was reinforced by the establishment of individual and institutional ties with other quantum chemists such as Mulliken and Coulson and their groups, by the active participation in and organization of international meetings, and by a commitment for the pedagogical indoctrination of fellow chemists and young students centered on the writing of many textbooks.

Löwdin's agenda for quantum chemistry aimed first and foremost to establish it in secure foundations within quantum mechanics, together with the exploration of its connections with solid-state physics and material sciences. An untiring activist for whom quantum chemistry became a lifestyle, strongly inspired by his mentor Slater, his imprints were left at the epistemic, institutional, and computational levels, the whole interlaced with a strong emphasis on philosophical reflections, including the relation of theory to experiment, merits and shortcomings of the semiempirical versus the ab initio approaches, and the role of numerical and computational inputs in the new era marked by the appearance of the computer, that "post-war wonder" according to Clementi's apt expression.

Notably, throughout the successive phases of quantum chemistry there were always eloquent practitioners keen to connect the various disciplinary stages with a systematic concern for philosophical and methodological considerations. Lewis offered very cogent considerations on the role of theories in chemistry and the role played by visualizations when contrasted with their role in physical theories. After the foundational years marked by a turbulent assessment of various, and at times conflicting, contributions by Heitler, London, Pauling, Mulliken, Slater, and Van Vleck, both Coulson and Löwdin were able to contribute toward the construction of a consensual framework where conflicting views would not undermine the wealth of possibilities being realized for quantum chemistry. But this framework was catastrophically undermined by the intrusion of the computer, which came to shape quantum chemistry in such a way that technical details, empirical data, experimental results, numerical solutions, visual representations, institutional setups, group building, organization of meetings, textbook writing, and, last but not least, philosophical reflections were all (re)negotiated to define the changing character of quantum chemistry.

5 The Emergence of a Subdiscipline: Historiographical Considerations

The story of quantum chemistry has been a story with a happy ending: the happy ending of a tortuous journey, the beginning of which was marked by a self-negating realization that there could be no analytical solutions to almost all the problems of chemistry by using quantum mechanics, though in most of the cases the relevant equation(s) could be written down. But, the nightmare was punctuated by a dream of a dream world: A single instrument, the electronic computer, promised a boundless frontier of numerical solutions of arbitrary exactness. With it, however, as it often happens in dream worlds, came another unnerving realization: As the first pioneers were experiencing this new frontier, the attractions provided by the very instrument of salvation led many astray. There was concern that progressively less and less chemistry was involved in the work of quantum chemists, something dramatically encapsulated by the rallying cry of a latecomer to the field, the French Alberte Pullman, a senior researcher at the Centre National de la Recherche Scientifique (CNRS). In 1971, she urged theoretical chemists not to be carried away by the possibilities provided by the powerful calculational methods and "not to stay in the ivory tower of abstractions and applications with always the same standard compounds" but to explore "more and more *problems* of chemistry, of all chemistry" (A. Pullman 1971, 16). Alberte Pullman exhorted quantum chemists to reintroduce chemistry into their calculations and denounced the tendency on the part of many theoretical chemists to forget that quantum chemistry remained nonetheless *chemistry*, despite the possibility of increased accuracy in calculational standards due to the use of computers. The obsession for getting better and better values of parameters, integrals, or other quantities gave the impression that for some, quantum chemistry aimed solely at "the reproduction of known results by means of uncertain methods," contrary to the other sciences, which aimed at "using known methods to search for unknown results" (A. Pullman, 1971, 13).

Whether chemistry had been forgotten in the euphoria of the age of the computer is a debatable issue. What, however, is not debatable is that from the very beginning of the period when chemical problems were examined quantum mechanically, everyone involved in the subsequent developments tried to understand the *chemical character* of what was begotten in the encounter(s) of chemistry with quantum mechanics. Was quantum chemistry an application or use of quantum mechanics in chemical problems? Was quantum chemistry the totality of chemical problems formulated in the language of physics and which could be dealt with by a straightforward application of quantum mechanics with, of course, the ensuing conceptual readjustments? Or was it the case that chemical problems could be dealt with only through an intricate process of appropriation of quantum mechanics by the chemists' culture? Research papers, university lectures, textbooks, meetings, conferences, presidential addresses, inaugural lectures, even correspondence among chemists and physicists became the forum for the discussion of these questions. By attempting to provide answers to these seemingly pedantic, and often implicitly posed, questions, various individuals or groups of individuals attempted to legitimize methodological outlooks and define the status of quantum chemistry. They attempted, that is, to achieve a consensus about the degree of relative autonomy of quantum chemistry with respect to both physics and chemistry and, hence, about the extent of its nonreducibility to physics. Terminologically, it appeared that there was a consensus that quantum chemistry had always been a "branch" of chemistry. Its history, however, shows that what appeared to be nominally so was also the result of the failures of the different (sub)cultures (physics and applied mathematics) to appropriate it and the difficulties in convincing the chemical community at large that talk about quantum chemistry was, in fact, talk about chemistry.

We attempted to show throughout our narrative that the various developments that brought about the development of quantum chemistry revolved around the six clusters of issues we discussed in the introductory chapter: the epistemic aspects comprising mainly the conceptual framework and the calculational techniques that had been developed; the institutional developments that reflected the emergence of the subdiscipline; the contingent character of the various developments; the catalytic role of the digital computer, the philosophical issues related to quantum chemistry; and the role of styles. The structuring of our narrative was not so much a way of taking stock and tracing in the same form and manner how each of these clusters of issues was realized in each chapter as if we planned to fill in an accountant's sheet. Nor are we interested in assessing the extent to which each protagonist fulfilled these aspects. These six clusters of issues—and most importantly their multifarious interrelationships—composed a way to articulate the constitutive characteristics of the culture of quantum chemistry. None of the issues related to each of the six clusters can be understood independently of the way each one of them has been expressed through, influenced by, adapted to, and juxtaposed with all the other issues, eventually redetermining them in the arduous process of the formation of a "standard" mode of practice in quantum chemistry. And by discussing the complex of the issues related to each one of these clusters and their relationships, we attempted to substantiate our claim that the story of the emergence and establishment of quantum chemistry could be told as the emergence and establishment of a new culture progressively adopted and propagated by the ever increasing practitioners of this "in-between" discipline—some of whom started their careers as physicists, some as chemists, and some as mathematicians.

The Role of Theory in Chemistry

It appeared that developments in quantum chemistry inaugurated discussions on a cardinal issue: the status of theory in chemistry. Many quantum chemists became actively involved in clarifying what chemists (should) mean by theory and in what respects specific theories differed from those of physics. For generations, chemistry was identified as a laboratory science, and chemists were content with (empirical) rules. In ways that bear amazing similarities with the case of van't Hoff's chemical thermodynamics, quantum chemistry was enthusiastically embraced by some and was barely tolerated by most—yet, because it worked, those who ignored it could not do so for a long time. And, thus, understanding the character of "theory" in chemistry may, perhaps, be an intriguing historiographical challenge. Much of the history, and to a large extent the philosophy, of chemistry shies away from discussing the role of theory in chemistry—as opposed, of course, to the case of physics. In contradistinction to the physicists, chemists have been happy with expressing allegiance to more than one theory or theoretical schemata—an anathema for physicists. Chemists were always open in making a rather liberal use of empirically determined parameters in constructing their theoretical schemata and, often, their schemata appeared to be "propped up" expressions of the rules they had already devised. For physicists, the predictive strength of a theory was of paramount importance. Philosophers of science have attempted to understand the intricate balance between the descriptive, the explanatory, and predictive power of mainly the physicists' theories. And chemists were rather happy in trying to explain to their colleagues how they will be using the theories they were devising or "borrowing," often realizing that these were theories that the physicists would snub and most philosophers of science simply ignore.

It was Lewis who, already back in 1933, contrasted the different features of theories in chemistry and physics. He presented structural organic chemistry as the paradigm of a chemical theory, as an analytical theory in the sense it was grounded on a large body of experimental material from which the chemist attempted to deduce a body of simple laws that were consistent with the known phenomena. He called the paradigm of a physical theory a synthetic theory to stress that the mathematical physicist starts by postulating laws governing the mutual behavior of particles and then "attempts to synthesize an atom or a molecule" (Lewis 1933, 17). He maintained that an inaccuracy in a single fundamental postulate may completely invalidate the synthesis, whereas the results of the analytical method can never be far wrong, resting as they do upon such numerous experimental results.

But theories in chemistry needed a reappropriation of a number of concepts that had their origins in the physicists' *problematique*. Lewis's work in thermodynamics was indicative of the feasibility of such a process of reappropriation. His aim (and he was in tandem with van't Hoff) was to convince chemists of the deep significance of thermodynamics for the study of chemical systems, at a time when thermodynamic potentials were known mainly to physicists. The few chemists who had heard about them could hardly see how they could be applied to complex real chemical systems.

Let us be reminded that the formulation of chemical thermodynamics did not automatically lead to its adoption by the chemists. There ensued a stage of *adapting chemical thermodynamics to the exigencies of the chemical laboratory*. Chemical thermodynamics had to appeal to the chemists not only because it provided a theory for chemistry, but also because it formed a framework sufficiently flexible to include parameters that could be unambiguously determined in the laboratory. An aim shared by both van der Waals and Lewis was the definition of entities that could be of practical use to experimentalists by avoiding a direct reference to entropy in them. They both made efforts to propose *visualizable* entities, something that was not independent of the special relations of each with particular *laboratory practices*. For Lewis, thermodynamics could be assimilated in chemistry only if it became possible to work with parameters that could be unambiguously related to situations one meets in the laboratory, rather than seeking the extension of parameters, originally defined for ideal systems, to problems occurring in the laboratory. Thermodynamics could lose all its appeal to chemists if it remained a theory formulated in terms of parameters that could not be unambiguously measured in the laboratory. For example, it was notoriously difficult to determine exactly partial pressures and concentrations, which were the parameters in terms of which most of the equations of chemical thermodynamics were formulated.

Lewis proposed to base chemical thermodynamics on the notion of escaping tendency or *fugacity*, which he considered as being closer to the chemists' culture, more fundamental than partial pressure and concentration, as well as being exactly measurable. Lewis hoped that this new concept would become the expression for the tendency of a substance to go from one chemical phase to another. After discussing fugacity, whose experimental determination involved the difficult measurements of osmotic pressures, Lewis proposed to reformulate chemical thermodynamics in terms of the *activity* of a substance, which measured the tendency of substances to induce change in chemical systems and was defined as its fugacity divided by the product of the gas constant and the absolute temperature.

In 1907, Lewis published a paper titled "Outlines of a New System of Thermodynamics in Chemistry" where, among other things, he explicitly articulated his overall approach to chemical thermodynamics. He started by stating that there are, basically, two approaches in thermodynamics. The first makes use of entropy and the thermodynamic potentials and had been used by Willard Gibbs, Pierre Duhem, and Planck, and the second approach, where the cyclic process was applied to a series of problems, had been used by van't Hoff, Wilhelm Ostwald, Walther Nernst, and Svante Arrhenius. The first method was rigorous and exact and had been, mainly, used by physicists, whereas chemists preferred the second. According to Lewis, the main reason for the chemists' preference was the difference between the physicists' notion of equilibrium and that of the physical chemists. Even though many aspects of the proposed theory may have been very similar to the respective physical theory, Lewis's aim was to articulate not so much *the* theory of physical chemistry, but rather the theory *of* physical chemistry by emphasizing the significance of the unambiguously measured quantities for the chemist. Lewis's work repeatedly attempted to formulate thermodynamics on what he considered to be an axiomatic basis where the emphasis was on defining parameters and procedures that would appear *convenient* to the chemists. Lewis became one of the very first, together with van't Hoff, to convince chemists of the importance of *theories* in chemistry and that chemical thermodynamics provided such a possibility. More significantly, Lewis tried to convince chemists of the usefulness, even the indispensability, of *mathematical* theories.

Like every form and expression of appropriation, opinions differed among the members of the chemistry community as to the use of mathematics. The chemist Edward Frankland predicted that the future of chemistry was to lay in its alliance with mathematics. The chemist Paul Schützenberger believed that mathematics would become an instrument as useful to the chemist as the balance (Coulson 1974, 10). Van't Hoff could not have been more mathematical in his systematic study of chemical thermodynamics. Ostwald's extensive use of mathematics would have been much more influential had it not been undermined by his insistence on energetics. Lewis was not less skilled in mathematics. Even Joseph Larmor and Joseph John Thomson before him tried to propose a mathematical framework for dealing with chemical problems. But there was also strong resistance against such programs.

As early as 1884, Henry E. Roscoe, one of the pillars of the British chemistry establishment and a person who was very sympathetic to the physicists' meddling into the chemists' affairs, was still not sure how successful mathematics would be for chemistry. He noted the importance of the physicists' research concerning the structure of the atom, but he held serious reservations as to the effectiveness of mathematics in chemistry: "How far this mathematical expression of chemical theory may prove consistent with the facts remains to be seen" (British Association for the Advancement of Science 1884, 342).

Arthur Smithells, the forceful spokesman of British chemistry, at the 1907 meeting of the British Association for the Advancement of Science expressed his excitement about the state of chemistry but also his worry about "the invasion of chemistry by mathematics," and the feeling of being "submerged and perishing in the great tide of physical chemistry, which was rolling up into our laboratories" (British Association for the Advancement of Science 1907, 477, 478). And Henry Armstrong was noting that "now that physical inquiry is largely chemical, now that physicists are regular excursionists into our territory, it is essential that our methods and our criteria be understood by them . . . It is a serious matter that chemistry should be so neglected by physicists" (British Association for the Advancement of Science 1907, 394).

This uneasy relationship between chemists and mathematics can also be traced during the emergence of quantum chemistry. All those who were directly involved in the development of quantum mechanics were confronted with the evaluation of the relations of chemistry to physics and by extension to mathematics. Longuet-Higgins, one of Coulson's students, went further in assessing the complex relation of quantum chemistry to mathematics. He turned the whole argument upside down. He did not consider that there was a danger that quantum chemistry might be subsumed under mathematics and boasted that the time had come for chemists to teach mathematics to the mathematicians. He introduced the paper "An Application of Chemistry to Mathematics" with the bold statement:

I imagine that the title of this paper will shock many of the readers of this Journal. It is generally taken for granted, at least by mathematicians, that in the hierarchy of the exact sciences mathematics holds first place, with physics second and chemistry an insignificant third. Organic chemistry is considered at best a practical necessity and at worst a rather noisome branch of cookery. In this paper I hope to show that pure mathematics is occasionally enriched not only by the fruits of physics, but also by those of chemistry, and to establish this thesis by proving a mathematical theorem of some intrinsic interest which was, in fact, suggested by an empirical generalization in organic chemistry. (Longuet-Higgins 1953, 99)

He concluded by pointing out that the discovery of many other theorems, with an intrinsic interest from the purely mathematical point of view, was prompted by chemical laws. He hoped that "the more trained mathematicians will come to recognize theoretical chemistry as a subject not altogether unworthy of their professional attention" (Longuet-Higgins 1953, 106).

In discussing the ways quantum chemists went about constructing their theories, it becomes necessary to discuss not only the problems that arise in their appropriation of physics, but also the resistances expressed in having *overtly* mathematized theories. It appears that since the last quarter of the 19th century, chemists were expressing their views about the elusive meaning of the term "overtly."

In one of his early papers, Pauling acknowledged his debt to Lewis and showed how his theory came to explain Lewis's schema of the shared electron-pair bond.

A comprehensive theory of the chemical bond based on the concept of resonance emerged out of the "Nature of the Chemical Bond" series, which was completed by 1933. In fact, Pauling believed that the task of the chemist should be "to attempt to make every new discovery into a general chemical theory."[1] The concept of resonance had played a fundamental role in the discovery of the hybridization of bond orbitals, the one-electron and the three-electron bond, and the discussion of the partial ionic character of covalent bonds in heteropolar molecules. Furthermore, the idea of resonance among several hypothetical bond structures explained in "an almost magical way" the many puzzles that had plagued organic chemistry.[2] Resonance established the connecting link between Pauling's new valence theory and the classical structural theory of the organic chemist, which Pauling classified as "the greatest of all theoretical constructs." Resonance—originally a physical concept—now became absolutely crucial in the formulation of a chemical theory.

> The theory as developed between 1852 and 1916 retains its validity. It has been sharpened, rendered more powerful, by the modern understanding of the electronic structure of atoms, molecules and crystals; but its *character* has not been greatly changed by the addition of bond orbitals, the theory of resonance, partial ionic character of bonds in relation to electronegativity, and so on. It remains a *chemical* theory, based on the tens of thousands of chemical facts, the observed properties of substances, their structure, their reactions. It has been developed almost entirely by induction (with, in recent years, some help from the ideas of quantum mechanics developed by the physicists). It is not going to be overthrown. (Pauling 1970, 998, emphasis ours)

It was as succinct a statement about the historical role of the newly emerging valence theory as there could be. Pauling was not willing to break ranks with the chemists. He argued that his was not a new theory, but a way of modernizing the very framework of chemists, which he viewed as being determined by structural theory. His was not a new theory as such, but part of a well-entrenched theoretical tradition of chemistry. Structural theory was a solid chemical theory, and developments in the form of resonance theory did not alter its character—despite "some help from the ideas of quantum mechanics developed by the physicists." Pauling spoke as a chemist to fellow chemists. His was an effort for ideological hegemony among the chemists. And he was perfectly suited for this role not having been tricked by the Sirens' song of the physicists' quantum mechanics. *His* use of quantum mechanics did not shadow the chemists' tradition as expressed by structural theory. It further augmented it.

Well into the 1970s, well into the period when it became clear that computers were bringing dramatic changes to quantum chemistry, E. Bright Wilson, the co-author with Pauling of *Introduction to Quantum Mechanics with Applications to Chemistry*, wrote a paper examining the impact of quantum mechanics on chemistry. He posed the following questions: Is quantum mechanics correct? Is ordinary quantum mechanics good enough for chemistry? Why should we believe that quantum mechanics is in principle accurate, even for the lighter atoms? Can quantum mechanical calculations replace experiments? Has quantum mechanics been important for chemistry? Can many-particle wave functions be replaced by simpler quantities? Based on the ways in which computers were being used in quantum chemistry, and worried about the lack of new ideas during the past 20 years, Wilson speculated on the possibility that the "computer age will lead to the partial substitution of computing for thinking." But he hoped for "new and better schemes," and he still believed that qualitative considerations would continue to dominate the applications of quantum chemistry. This was, after all, because of the special methodology of chemistry:

Chemistry has a method of making progress which is uniquely its own and which is not understood or appreciated by non-chemists. Our concepts are often ill-defined, our rules and principles full of exceptions, and our reasoning frequently perilously near being circular. Nevertheless, combining every theoretical argument available, however shaky, with experiments of many kinds, chemists have built up one of the great intellectual domains of mankind and have acquired great power over nature, for good or ill. (Wilson 1976, 47)

Wilson was encapsulating the development of quantum chemistry in an amazingly succinct, yet shocking, way. There was no attempt to polish the narrative or to turn the protagonists into heroes. Nor was there any attempt to be humble. And the message was clear: the history was messy, the result unique. From the very beginning, among the chemists, there was an ambivalent attitude toward any new proposal of "*how* to do quantum chemistry" or, rather, "*what* to do with quantum mechanics when doing quantum chemistry." To many physicists, the chemists' pragmatism appeared flippant. To some chemists or chemically oriented physicists, the physicists' mania to do everything from first principles appeared as unnecessarily cumbersome and tortuous. Disagreement over technical issues, more often than not, had its origins in differences of methodological and ontological commitments. Different cultural affinities brought about further murkiness, yet more and more new results. And throughout these developments, many chemists were attempting to convince chemists that quantum chemistry was a different ball game altogether: one needed to be convinced that chemistry will have different theoretical schemata, and that this state of affairs would be the constitutive aspect of the subdiscipline.

In this respect, Longuet-Higgins's view is of interest. He talked of three kinds of chemistry: experimental, theoretical, and computational. He asserted that even though most chemists tend to think of molecular computations as belonging to theoretical chemistry, it could be argued that such computations were really experiments. Conventional experiments are carried out on real atoms and molecules, "computational experiments are performed on more or less 'modest' and unreliable models of the real thing." So the chemist who does computations is obliged to have a convincing explanation why the numbers come out as they do. If not, there may be doubt as to whether they "may not be artefacts of his basic approximations." This, he considered, was the substance of most objections to heavy computations of molecular properties by ab initio methods. If such methods have been well attested for a given class of problems, then it is not unreasonable to "attach weight to the computational solution of a further problem in that particular class. Unfortunately, the most interesting problems are usually those with some element of novelty" (Longuet-Higgins 1977, 348).

Let us remember Coulson (1960, 174), again: "Chemistry itself operates at a particular level of depth. At that depth certain concepts have significance and—if the word may be allowed—reality. To go deeper than this is to be led to physics and elaborate calculation. To go less deep is to be in a field akin to biology." Coulson did his utmost to convince chemists—and, perhaps, physicists—that in quantum chemistry, the *role* of theory was not something static, and it had a lot to do, among other things, with the demands of the community, of its decisions concerning the "appropriate depth" at which quantum chemistry will operate.

Notwithstanding these persistent uncertainties, it was certainly an achievement of quantum chemists to have been able to reassess the role of theory in chemistry, to foster a reappraisal of the meaning of experiments, to rethink the role of visual representations, and to accommodate diverse modes of explanation. These, in fact, may have been the reasons behind Löwdin's choice of the title "Quantum Chemistry— A Scientific Melting Pot" for the meeting organized in 1977 to celebrate both the 500th anniversary of the University of Uppsala and the 50th anniversary of quantum chemistry (Löwdin et al. 1978).

But it was not only that quantum chemistry reassessed the role of theory in chemistry. The role of experiment was, also, redefined. Post–Second World War developments included a number of institutional initiatives. Since the early days of the war, the Mathematical Laboratory took shape at the University of Cambridge. In fall 1950, Slater established the Solid-State and Molecular Theory Group at MIT, which was initially housed in the premises of the new Research Laboratory of Electronics. Mulliken's Laboratory of Molecular Structure and Spectra was created in 1952. Coulson founded the Mathematical Institute within the School of Mathematics in 1952, with special premises for people to meet and discuss, and a decade later he was a member of a committee that started the first University Computing Laboratory (Altmann and Bowen 1974, 88–89). In 1958, the laboratory of Löwdin's Quantum Theory Group was inaugurated. As exemplified by all the cases listed above, many opted to associate their new groups with sites they chose deliberately to call *laboratories*. They were not, of course, experimental laboratories. But much like them, they were churning out numbers. Much like the experimental laboratories, these laboratories had a hierarchical structure, they were populated by scientists with different expertise, they included technicians, and they could accommodate distinctive practices and characteristic cultures. The new laboratories became the sites where successive generations of computers were adapted to the needs of quantum chemistry. Built, tested, used, and superseded by more powerful ones, they were often supported by contracts with military agencies eager to profit from them in the upcoming era of Big Science.[3]

Developments in computers forced quantum chemists to rethink the status of experimental practices and to reconceptualize the notion of experiment, not so much within the more traditional framework involving instrumentation and laboratories, but, in this case, almost exclusively within the framework of mathematics. Soon afterwards, the idea of a mathematical laboratory materialized and was explored successfully by quantum chemistry groups.

Quantum chemists were not only apt users of the new instruments but also played a role themselves both in developing hardware and software and in producing special codes for the numerical calculations of molecular quantities. They had previously enrolled expert "human computers" for their calculations; now they became themselves computer wizards. Computers and these laboratories emerged simultaneously and reshaped the culture of quantum chemistry.

And, of course, as is always the case and despite the celebrated autonomy of experiments, theory and experiment do have various ties between them, even in this new framework. The symposium on *Aspects de la Chimie Quantique Contemporaine* was held during 1970 in Menton, France, and was organized by the CNRS. Roald Hoffmann (1971, 133), then at Cornell University and future Nobel laureate for 1981, offered an analysis of the "meager achievements" of quantum chemistry in the field of the chemical reactivity of molecules in their excited states and outlined the ways to circumvent it. He sharply distinguished between two types of theoretical chemistry. He summoned "interpretative theoretical chemistry" to the promising search for the "theoretical framework used to relate the experimental measurement of some physical observable to a microscopic parameter of a molecule." Opposing this type of theoretical chemistry were the "electronic structure calculators," deemed to be not very successful, and prone to many extremes. He expressed a worry about a trend whereby chemists are encouraged not to do laboratory experiments but their substitutes through computer calculations, something to be surely avoided.

> If we consider a calculation on a molecule as a numerical experiment and focus on the observables that are measured (predicted) by such a numerical experiment on a small molecule of the size of butadiene, then I would bet that the experimentalist will be able to predict (on the basis of his experience, reasoning by analogy) more correctly the outcome of his theoretical colleague's numerical experiment than the theoretician could predict his experimental friend's laboratory observation. (Hoffmann 1971, 134)[4]

Hoffmann had no doubts that in the methodological approach of "interpretative theoretical chemistry" lay the future success of quantum chemistry. When properly applied it produced results of far more lasting value, "the hard facts of true molecular parameters," than those of the ephemeral approximate calculations

The Theoretical Particularity of Chemistry

The detour about the role of theory in chemistry and the subsequent efforts to clarify issues that pertain to the theoretical framework of chemistry provides sufficient material to discuss another issue, which is the *theoretical particularity* of chemistry—the character of its theories, and the differences from what are considered as theories in physics. If anything is clear, it is that chemical theories are not incomplete physical theories, they are not pre-theoretical schemata that will reach maturity when the physicists will (decide to) deal with them properly. Theories in chemistry—and as we hope to have shown, theories in quantum chemistry—have an autonomy of their own, they continuously adapt to the chemists' culture, re-forming it in the process. These theories may have been the result of intricate processes of appropriation and reappropriation of physical theories, but, at the end of the day, they "became" chemical theories. Of course, the specific role of mathematics in physics makes a number of philosophical problems to be unambiguously formulated. There are no intrinsic limitations as to how deep physics can probe. Whether it studies the planets, billiard balls, atoms, nuclei, electrons, quarks, or superstrings, it is still physics. Both chemistry and biology are particularly sensitive to such changes of scale, and this intrinsic characteristic reflects itself in the character(istics) of their theories. It appears that the history of (quantum) chemistry is also a history of the attempts of chemists to establish the autonomy of its theories with respect to the "analogous" physical theories. In a way, chemists had been obliged to do it. Otherwise, chemists would be continuously living in an identity crisis and would never be sure whether chemistry should be doing the describing and physics the explaining. Quantum chemists have passionately debated these issues, and the myth of the reflective physicist and the more pragmatic chemist is, if anything, historically untenable.

Throughout the history of quantum chemistry, it appears that in almost all the cases, the reasons for proposing new concepts or engaging in discussions about the validity of the various approaches were

1. To circumvent the impossibility to do analytical calculations.
2. To create a discourse with which chemists would have an affinity.
3. To make compatible two languages, the language of classical structure theory and that of quantum mechanics.

Perhaps it may be argued that the involvement in such discussions of almost all those who did pioneering work in quantum chemistry (and, certainly, of everyone whose work we analyze in this book)—either in their published papers or in their correspondence—had to do with *legitimizing the epistemological status of various concepts in order to be able to articulate the characteristic discourse of quantum chemistry*. Of course, the process of legitimization is not only related to the clarification of the content of the proposed concepts and the correctness of certain approaches. The process itself is a rigorously "social" process, involving rhetorical strategies, professional alliances, institutional affirmations, presence in key journals and at conferences, and so forth.[5] Nevertheless, any of the philosophical repercussions appear to have been the unintended implications of such a strategy. The relations between the epistemological status of the proposed concepts in the discourse being formed and the philosophical aspects of these concepts is no trivial matter, and many times the validity of the former cannot be assessed without recourse to the latter—even if such a recourse has been done by the quantum chemists in a philosophically naïve manner. It was the successes of quantum mechanics in chemistry that induced some chemists and some philosophers to bring to the fore a number of philosophical issues about chemistry or to discuss problems other philosophers of science had been discussing, but now within the context of chemistry. Reductionism turned out to be one of the pivotal issues.

Perhaps, Dirac's claim should be considered as nothing more than a physicists' projection of what physics can do for chemistry, yet the question still remains as to how the chemists' practice had come to terms with reductionism—not whether theories are reducible, but whether the ontology is reducible. At a trivial level there is much in favor of reductionism: Both physics and chemistry deal with atoms and electrons. They comprise the ontological stratum of all the phenomena involved in these disciplines. Again, in a trivial manner, there is a serious difficulty with emergence: It is almost impossible to "build" the phenomena related to both disciplines starting from the building blocks. Hence, such an asymmetry brings in serious complications in the discussion of the philosophical problem. R. Bishop (2005) insists upon a different point. Much of what is associated with reductionism is the claim that physics is the only science offering the possibility of a complete description of the physical world. If that is so, then reductionism will eventually dominate. But is this epistemic claim about physics historically tenable? Might it be the case that reductionism is a historically (not even epistemologically) contingent claim? If despite these objections, one insists on introducing the problem of reductionism, is it not the case that the ultimate statement of reductionism is that all chemistry is explainable in terms of spin—a purely quantum mechanical notion? It may just be the case that reductionism cannot be satisfactorily discussed independent of the character of theory in chemistry.

Dirac's 1929 pronouncement encapsulated what was already part of the physicists' culture for many decades. And, with Dirac's specific contributions to the development of quantum mechanics, it became possible to articulate this reductionist program: Chemistry after the Heitler–London paper could be perceived as being the different manifestations of spin, and spin, after all, was under the jurisdiction of the physicists. And though physicists believed that the new quantum mechanics had also taken care of chemistry, the chemists themselves did not appear to have been under any panic that their identity was being transformed and they were being turned into physicists.

Nor did they believe that their very existence was being threatened, as it appeared that what they have been doing could be done much better by the physicists. The appropriation of quantum mechanics, the attempts to overcome cultural resistances within the chemical community on how to appropriate quantum mechanics, and the different views on how to form the appropriate discourse became issues related to the problematic of reductionism (Simões 1993; Gavroglu and Simões 1994; Gavroglu 1995).

Let us now raise a different but correlated question: whether reductionism may be a misplaced category if one wants to discuss a number of questions for chemistry. Perhaps reductionism is a physicist's analytical tool and not a chemist's. Might it be the case that the whole notion of reductionism expresses a trend that is dear to the physicists' own culture rather than that of the chemists? Though physicists took for granted the reduction of chemistry to physics and did little about it, the chemists did not have the luxury of waiting for history to fulfill such an agenda. The benign neglect by chemists of their doomsday so clearly planned by the physicists is, certainly, worth taking note. For reductionism may have been a program, but it was nearly impossible to realize it because as it became evident right at the beginning, one could not deal analytically with any of the other elements except hydrogen and helium, even in grossly approximate terms.

Are there any other dimensions to reductionism, whose discussion may be considered more fruitful in addressing the same set of problems? Perhaps discussing the (uneasy) relationship of chemists with mathematics may provide additional insights. We provided some instances to show that the chemists' relationship with the appropriation of mathematics into their culture was far more complex and difficult than their appropriation of physics. And though the two cannot be considered as totally independent of each other, it can, in fact, be argued that, historically, chemists were more resistant in accepting the use of mathematics rather than the physical concepts and the physical techniques.

We have insisted throughout the book that what we want to articulate is not the philosophical considerations of reductionism, the discussion of which in the case of quantum chemistry, and thus quantum mechanics, has been greatly enriched by contributions of Hans Primas (1983, 1988), Jeff Ramsey (1997), Eric Scerri (2007), and J. van Brakel (2000), among others, but rather the ways it has marked the culture of quantum chemists, the way the awareness of such a problem by the community of (quantum) chemists—in naïve philosophical terms—permeated their practices. Though a number of them had expressed their "worries," reductionism *in any of its variants* was certainly not a paralyzing factor. Perhaps, one of the intriguing aspects of reductionism is that much of the discussion depends on the theoretical framework with respect to which such a discussion is realized. But, how has this problem appeared in the context of another theory in chemistry, that of chemical thermodynamics? What can one say about reductionism in this case? Is the claim of reductionism valid in this case? What does the unquestionable validity of thermodynamics, and its applications in chemistry, tell us about the relationship between physics and chemistry? Much like the quantum mechanical case, we have a "similar" ontology in physics and chemistry when viewed through chemical thermodynamics. But how can we formulate the problem of reductionism within the framework of chemical thermodynamics by taking into consideration its descriptive and explanatory strengths and weaknesses at the time when chemical thermodynamics was being projected as *the* theory for chemistry, much before the all embracing role of quantum mechanics came to the fore? It appears that the chemists at the time were not so much disturbed by the fact that a theory of physics was being "translated" to cater to their needs but by the generalized use of mathematics in chemistry—and this appears to be independent of whether one subscribed to Ostwald's energetics or to British atomism. As we have already noted, chemical thermodynamics, though sharing many common features with thermodynamics, differed from it in various respects. The differences were not only because chemical thermodynamics included new and, at times, arbitrary parameters, but because *chemical theories were formulated by chemists with fundamentally different cultural outlooks compared with those of the physicists.*

Compared with physicists, these chemists expressed a different culture when it came to formulate a theory and to impose their demands on such a theory—such as the constitutive and regulatory role of empirical data in theory building. Can one, then, pose the question whether reductionism may not be independent of the subculture of chemists and physicists? Is it the case that the question up to now has been formulated almost exclusively in terms of the physicists' culture? Different scientific communities impose different explanatory demands on their theories, and this constitutes an important cultural characteristic of each community. It may, perhaps, be the case that the formulation of many of the philosophical issues in the different scientific (sub)disciplines is dependent on the specific cultural characteristics of the particular communities that comprise the practitioners of each of these (sub)disciplines. And, it may thus be the case that the discussion of these issues may be greatly enriched through such a perspective when the history of the (sub)disciplines does not only come to the fore in order to clarify theoretical or technical issues, but also in order to articulate practices and cultural strands.

Thus, let us formulate a more general question: What was *at stake* every time such philosophical/theoretical issues were raised during the history of quantum chemistry? When quantum chemists were trying to articulate their views about, say "visualizability," or when they were disagreeing on how a "proper" theory is to be built, what was the character of the ensuing discussions? In what respects did such discussions bring about changes to entrenched mentalities and start to articulate the strands of the emerging subcultures of the emerging (sub)disciplines? Was there, in other words, a philosophical agenda on the part of some of the quantum chemists, or might it have been the case that these issues and discussions were not necessarily aiming at the clarification of their philosophical implications? One cannot claim, of course, that quantum chemists were unaware of the philosophical character of many of the concepts they were introducing or of the philosophical implications of the discussions they were having. But, at the same time, one cannot claim that it was philosophy that they had in mind when they were proposing the concepts or when they were pursuing their discussions.[6]

In this realm, the metaphor of the artificial fiber proposed by Coulson (1953, 37) is quite suggestive. To underline "how much the validity of the scientist's account depends on the degree of interlocking between its elements," Coulson called attention to the fact that "the strength of an artificial fiber depends on the degree of cross-linking between the different chains of individual atoms." In the same manner, one might argue that the explanatory success of quantum chemistry throughout successive developmental stages rested on the degree of interlocking among constitutive elements—chemical concepts, mathematical notions, numerical methods, pictorial representations, experimental measurements, virtual experiments—to such an extent that it was not the relative contribution of each component that mattered, but the way in which the whole was reinforced by the *cross-linking* and *cross-fertilization* of all elements. Furthermore, its success depended not only on epistemological but also on social aspects of this *cross-fertilization*. It involved the establishment and permanent negotiation of alliances among members of a progressively more international community of practitioners, intense networking, and adjustments and readjustments within the community at the individual, institutional, and educational levels—in short it involved a huge rearrangement in the material culture of quantum chemistry. Going a step further, we call attention to Coulson's advocacy of diversity of styles as a core characteristic of quantum chemistry and, we might add, as a fundamental component of "in-between" disciplines.

Therefore, if we regard the beginnings of "in-between" disciplines as processes where (technical) success has to be accompanied by an intricate strategy of legitimization, then the different outlooks, be they different methodological priorities or different philosophical viewpoints or different ontological commitments, coalesce into articulating the different cultures of how to actually practice quantum chemistry. The "globality/homogeneity" of a particular (sub)discipline when it has reached a mature stage may, perhaps, be understood better in terms of the cultures of (sub)disciplines. Of course, there were issues related to personal antagonisms, to priorities, to genuine disagreements, to different interpretations—but do not all these inscribe/define each separate culture for doing quantum chemistry? But the consensus was achieved by eclecticism, by members of the "second generation" for whom neither philosophical sophistication nor historical consciousness is a necessary condition for continuing to practice what they inherited: As the first generation moved away from center stage, the cultural wars ended not because someone scored a clear victory, but because at the beginning there were diverging trends that through eclecticism eventually gave way to a process of confluence. What started as a patchwork evolved into a seamless whole. A theoretical framework that accommodated a number of epistemologically and, at times, socially induced partitions started progressively to look smoother because the reasons for continuing to sustain the partitions—be they technical, institutional, or personal—began to wane. And they began to wane as there were more and more technical successes, as the discipline started to be institutionally visible and assertive, and as professional successes helped deflate the egos of the protagonists. In other words, legitimization, theoretical homogeneity, and cultural homogeneity were interdependent processes.

In fact, by 1970, members of the first generation of quantum chemists were in their sixties and seventies. Some had already passed away: Hellmann was executed in 1938, London and Lennard-Jones both died in 1954, and Hartree died 4 years later, in 1958. Heitler, Hückel, Hund, and Van Vleck were no longer contributors to the discipline. Pauling had been estranged from the discipline he founded and planned to dominate. Already by wartime, his attention was drifting away to problems that shaped molecular biology. In fact, still active were just Mulliken, who was awarded the Nobel Prize in 1966, Slater, and Coulson. But contrary to the persistent and domineering participation of Mulliken and Coulson, Slater re-entered the field after his reconciliation with the turn of events in the aftermath of the Second World War. Their groups nurtured many of the members of the new generation of quantum chemists. They started defining the agenda of the discipline: Raymond Daudel, Bernard and Alberte Pullman, Kotani and Löwdin, Parr, Pariser, and Pople, Crawford, Shull, Platt, Roothaan, Scherr, Ransil, Barnett, Boys, Clementi, McWeeny, Hall, Appel, Calais, Lindenberg, Fröman, and many more.

Circulation, networking, exchange programs, textbooks, international meetings, and summer schools became the constitutive elements of the training of this whole new generation of practitioners. The concern for bigger molecules extended the field of application of quantum chemistry to inorganic chemistry and solid-state physics (metallic complexes, crystals), as well as to biology, medicine, and pharmacology. The change of scale from very small molecules to big molecules and macromolecules introduced new constraints into the discussion (role of the environment in inducing properties in molecules). And this trend helped the emergence of quantum biochemistry, quantum biology (and to a lesser extent quantum pharmacology), as well as computational chemistry, molecular engineering, and materials science and engineering (Bensaude-Vincent 2001). In a sense, with quantum chemistry's forays in biology, medicine, and pharmacy, the centuries' old relations of the discipline with the precursors to these specialties resurfaced again, in the context of a sustained relation with physics and mathematics. Even the emergence of philosophy of chemistry has been closely associated with quantum chemistry.[7]

The genesis and development of quantum chemistry as an autonomous subdiscipline owed much to those scientists who were able to realize that "what had started as an extra bit of physics was going to become a central part of chemistry." Those who managed to escape successfully from the "thought forms of the physicist" (Coulson 1970, 259) by implicitly or explicitly addressing issues such as the role of theory in chemistry, the methodological status of empirical observations and virtual experiments, helped to create a new space for chemists to go about practicing their discipline. The ability to "cross boundaries" between disciplines was perhaps the most striking and permanent characteristic of those who consistently contributed to the development of quantum chemistry. Moving at ease between physics, chemistry, mathematics, and later biology became a prerequisite to be successful in borrowing techniques, appropriating concepts, devising new calculational methods, and developing legitimizing strategies. With the era of computers and the development of computer science, quantum chemists were among the first scientists to explore the potentialities of the new instrument and even to collaborate in its development. In this way, they also became participants in what many dubbed as the Second Instrumental Revolution in chemistry (Morris 2002; Reinhardt 2006). The discussion over changing practices and their implications for the evolving identity of quantum chemistry shows how the history of quantum chemistry illustrates one of the trends that more forcefully characterized "in-between" disciplines emerging throughout the 20th century—the exploration of frontiers and the crossing of disciplinary boundaries reinforced by the mediation of many new instruments.

Notes

The following abbreviations are used in the notes:

- AHQP Archives for the History of Quantum Physics, Niels Bohr Library, American Institute of Physics
- AIP American Institute of Physics
- BC R. T. Birge Correspondence, The Brancroft Library, University of California at Berkeley
- BL Bancroft Library, Lewis Correspondence, Berkeley
- CP Coulson Papers, Bodleian Library, Modern Manuscripts Collection, Oxford
- HP Hartree Papers, Christ's College, Cambridge, UK
- LA Fritz London Archives, Duke University
- L-J P Lennard-Jones Papers, Churchill College, Cambridge, UK
- LP London Papers, Duke University Library
- MacP MacInnes Papers, The Rockefeller University Archives
- MP Mulliken Papers, University of Chicago Library
- PP Ava Helen and Linus Pauling Papers, Special Collections, Oregon State University
- SBPK Papers of Erich Hückel, Staatsbibliothek zu Berlin–Preußischer Kulturbesitz
- Sl P Slater Papers, American Philosophical Society, Philadelphia
- SP Sidgwick Papers, Lincoln College, Oxford

Chapter 1

1. On the chemical aspects of Bohr's 1913 theory, see Kragh (1977).
2. In the water molecule, Bohr considered the oxygen nucleus positioned in the middle of the line joining the two hydrogen nuclei. A ring of four electrons encircled the oxygen nucleus whereas two larger rings, having three electrons each, rotated in a plane parallel to the oxygen ring and perpendicularly to the line joining all three nuclei at mid-distance from the oxygen to each hydrogen nucleus.
3. Letter Niels Bohr to Carl W. Oseen, December 20, 1915, quoted in Assmus (1992, 223).
4. Many of the original papers referred to in this chapter were translated into English in Hettema (2000).
5. AHQP, Interview with Pauling.
6. AHQP, Interview with Heitler.

7. AHQP, Interview with Heitler, emphasis added.
8. One of the main drawbacks of their approach was the absence of the contribution of the ionic terms. Though the inclusion of these terms does not lead to any appreciable differences in the case of the hydrogen molecule, the effects of their a priori absence in similar kinds of calculations has been systematically examined in Van Vleck and Sherman (1935). For a recent discussion of Heitler and London's results, see Park (2009), whom we thank for pointing out an oversight on our part concerning the paper by Heitler and London.
9. Reminiscences by Prof. Sir Brian Pippard to Kostas Gavroglu.
10. AHQP, Interview with Heitler.
11. The hypothesis of "electron spin" and the anomalous magnetism to which it is related was first put forward by Goudsmit and Uhlenbeck (1925, 1926) and then employed in a semiclassical fashion with great success in the description of atomic spectra. The linking of this hypothesis with the formulations of quantum mechanics occurred later in the writings of Pauli (1927), Darwin (1927), and Dirac (1928).
12. AHQP, Interview with Heitler.
13. Rodebush, who was a student of Lewis, does not mention Heitler and London by name when he comes to the quantum mechanical treatment of the hydrogen molecule.
14. Mulliken (1929) was also addressed specifically to chemists. Although dealing mainly with band spectra, Mulliken discussed briefly Heitler and London's paper.
15. A mathematical group consists of a set of elements that are related to each other according to certain rules. The kind of elements relevant to the symmetries of molecules are symmetry elements. With each symmetry element, there is an associated symmetry operation.
16. LA, Heitler to London, September (?) 1927.
17. LA, Heitler to London, September (?) 1927.
18. AHQP, Interview with E. Wigner.
19. AHQP, Interview with Heitler.
20. LA, Heitler to London, December 7, 1927.
21. AHQP, Interview with Heitler.
22. AHQP, Interview with Heitler.
23. AHQP, iInterview with Heitler.
24. Hückel expressed this point explicitly in later papers (1937, 1937a).
25. W. Hückel (1931) included quantum interpretations and was very influential when eventually translated into English (Berson 1996a; Kragh 1996; Karachalios 1997).
26. In this respect, it is interesting to quote what Otto Schmidt—who had worked as a chemist and became the director of BASF and continued his research at the main laboratory of I.G. Farbenindustrie AG in Ludwigshafen until 1938—said during the discussions of the board meeting: "The problem of valence is something that the organic chemist definitely needs as a tool of research. In the dye laboratories of I.G. [Farben], the old Kekulé formula is still used as a basis for their entire research activities. One would like to have information on the labile electrons and the positions particularly susceptible to photochemical action. Furthermore, technicians would like to receive orientation on modern concepts of valence. There is such an abundance of noteworthy new conceptions in this area that organic chemists could very much profit by" (Karachalios 2003, 121).

27. Biographical information on Hund can be found in AHQP, Interview with Hund; Kuhn et al. (1976); and in Kragh (1996a). On the Göttingen group, see Kuhn (1972); Hermann (1978); Parr (1977). Hund's view on the development of quantum physics is found in Hund (1974).
28. In an adiabatic transition, one assumes that the transition process is so slow that the electrons are able to adjust continuously to the changing internuclear separation.
29. Mulliken received his Ph.D. in 1921, became assistant professor in New York University in 1926, associate professor in the University of Chicago in 1928 and full professor in 1931.
30. LA, Hund to London, July 13, 1928.

Chapter 2

1. This volume contains a complete list of Mulliken's publications. An assessment of Mulliken's contributions to quantum chemistry is Löwdin and Pullman (1964). Mulliken's own assessment is Mulliken (1989).
2. BC, Box 21, Folder 1922–1925, Letter Mulliken to Birge, June 1925.
3. BC, Boxes 21, 31, and 32.
4. In MP, Box 81, Folder 16, "Molecular Models" and "Dynamics of Molecular Electrons," there appear several sketches of molecular structures that Mulliken might have drawn in the process leading to the suggestion of the above postulates.
5. BC, Box 21, Letter Mulliken to Birge, October 17, 1926.
6. In Kemble et al. (1926), in the concluding chapter, Kemble showed how the recent developments in quantum theory could be incorporated into the theory of band spectra and introduced Hund's 1926 paper to the American physicists and chemists.
7. MP, Box 72, Folder 14, Talk February 24, 1928.
8. AHQP, Interview with Mulliken.
9. MP, Box 72, Folder 14, Talk February 24, 1928; AHQP, Reel 49, Van Vleck Correspondence, Letter Mulliken to Van Vleck, April 11, 1928.
10. AHQP, Reel 49, Van Vleck Correspondence, Letter Mulliken to Van Vleck, May 21, 1928: "Have you looked at the MSS on electron quantum numbers that I sent to Tate a while ago? I am anxious to get someone to find fault with it." The answer followed soon: "I am afraid that I am unable to supply any "severe criticism" such as you requested, although I read the paper through twice. The subject is indeed an interesting one, and a contribution from you as a man thoroughly conversant with the experimental data are a valuable supplement or rather complement to the rather abstract work of Hund." Letter from Van Vleck to Mulliken, June 23, 1928.
11. In the old quantum theory, such a drastic change could only be justified when the molecule was formed by violent means such as collision or light absorption.
12. MP, Box 80, Folder 17.
13. AHQP, Interview with Mulliken.
14. MP, Box 107, Folder 2, Letter Gale to Mulliken, March 6, 1928; Ideal Situation; Draft of a letter from Mulliken to Gale; Letter Gale to Mulliken, March 22, 1928; Draft of a letter from Mulliken to Gale, March 24, 1928; Letter Gale to Mulliken, April 6, 1928.
15. MP, Box 82, Folders 8, 9, 10, 11.

16. See also Pauling (1928a). PP, Box 58, G. N. Lewis 1913–1947, Reprints and correspondence, letter Pauling to Lewis, March 7, 1928.

17. Slater wrote an unpublished autobiography titled *A Physicist of the Lucky Generation*. The manuscript is held at the Slater Papers, American Philosophical Society. Van Vleck (1964).

18. As far as we know he is a singular case, in that it was a (physical) chemist, not a physicist, a mathematician, or an astronomer, who responded to relativity ahead of members of supposedly more receptive scientific communities. In one paper, he derived the mass–energy relation from his former ideas on light pressure, without the help of the relativity principle; in another he derived Einstein's equations from conservation laws together with the principle of relativity. His imprint was left forever in quantum theory by christening light corpuscles as photons. (Lewis 1908; Lewis and Wilson 1912; Lewis 1926, 1926a). The role of Lewis in the reception of special relativity in the United States is addressed in Goldberg (1968, 1984).

19. Here we follow Simões (2007). However, Lewis's ideas on the chemical bond have been discussed in the history of science literature since the 1970s. To Kohler (1971, 1975) we owe a perceptive analysis of the development of Lewis's ideas in the context of other chemical atomic models and its appropriation by Irving Langmuir. Servos (1990) subsequently discussed them in the context of the development of physical chemistry in the United States and Gavroglu and Simões (1994) in the context of the emergence of quantum chemistry in the United States. Arabatzis and Gavroglu (1997), Gavroglu (2001), and Arabatzis (2006) further discussed this topic in the framework of the two contrasting models of atomic structure—models that came to be known as the chemical and the physical atom.

20. Lewis returned to quantum theory and relativity (Lewis 1926, 1926a). His imprint was left forever in quantum theory by christening light corpuscles as photons. He also delved into the field of American prehistory, geology (he wrote a paper on the thermodynamics of glaciations), and economics (he wrote two papers on price stabilization).

21. Translated in the framework of Lewis's theory, normal valence meant the number of electrons that occupied the outer cube, whereas contra-valence meant the number of places left vacant in the outer cube.

22. Earlier, Friedrich A. Kekulé had also assumed an oscillation of atoms in benzene around equilibrium positions.

23. Evidence that Arsem's model might have influenced Lewis is provided in Kohler (1971, 362–374).

24. It is possible that van't Hoff stereochemistry might have influenced Lewis's search for a three-dimensional atomic model. Both van't Hoff and Lewis wished their models to be "visualizable."

25. The exceptions, molecules such as NO, NO_2, or ClO_2, were called "odd molecules," and in a way their existence confirmed the "rule of two," for they all were unstable molecules that tended to react to form compounds by pairing. In contrast, new spectroscopic data resulting from Henry Moseley's X-ray spectrum analysis showed helium to include just two electrons, contrary to Lewis's former assumption.

26. Isomerism is the phenomenon in which two or more substances have the same composition yet show different properties owing to a different arrangement of their atoms.

27. For the role of symbolic representations in theory change, see Darden (1991).

28. For biographical details see Paradowski (1991), Hager (1995, 1998), and for a discussion of Pauling's structural chemistry see Paradowski (1972). A complete bibliography of Pauling's papers from 1920 to 1990 compiled by Z. S. Hermann and D. B. Munro is contained in Mead (1991). See also Rich and Davidson (1968).

29. In a seminar talk given in 1925, Pauling discussed the idea of binuclear orbits (Pauling Papers, Box 242, Popular Scientific Lectures 1925–1935, "Modern Theories of Valence and Chemical Combination").

30. PP, Box 71, Noyes A. A. Correspondence 1921–1938, Letter Pauling to Noyes, November 22, 1926.

31. PP, Box 71, Noyes A. A. Correspondence 1921–1938, Letter Pauling to Noyes, May 22, 1926.

32. PP, Box 71, Noyes A. A. Correspondence 1921–1938, Letter Pauling to Noyes, April 25, 1926.

33. PP, Box 71, Noyes A. A. Correspondence 1921–1938, Letters Pauling to Noyes, April 25, May 22, 1926.

34. AHQP, Interview with Pauling.

35. PP, Box 71, Noyes A. A. Correspondence 1921–1938, Letter Pauling to Noyes, May 22, 1926.

36. PP, Box 71, Noyes A. A. Correspondence 1921–1938, Letter Pauling to Noyes, July 12, 1926, emphasis not in original. The work was presented in a meeting on magnetism organized by Debye in Zürich and was published soon thereafter (Pauling 1927).

37. PP, Box 71, Noyes A. A. Correspondence 1921–1938, Letter Pauling to Noyes, July 12, November 22, December 17, 1926.

38. The letter written by Sommerfeld is in PP, Box 97, Sommerfeld A., Reprints and Correspondence.

39. AHQP, Interview with Pauling.

40. PP, Box 113, Van Vleck Correspondence, Letter Pauling to Van Vleck, August 9, 1927.

41. PP, Box 210, LP Calculations and Manuscripts, vol. III, 1926–1927.

42. There is no relation between Pauling's earlier work on molecular orbitals and Mulliken's research on the same topic.

43. AHQP and PP, Box 210, LP Notes and Calculations, vol. III, 1926–1927, "Work on Molecular Orbitals."

44. AHQP, Interview with Pauling.

45. AHQP, Interview with Pauling.

46. AHQP, Interview with Pauling. PP, Box 113, Van Vleck Correspondence, Letter Pauling to Van Vleck, August 9, 1927.

47. PP, Box 113, Van Vleck Correspondence, Letter Pauling to Van Vleck, August 9, 1927.

48. AHQP and PP, Box 210, LP Notes and Calculations, vol. III, 1926–1927, "Work on Molecular Orbitals," 1926. AHQP and PP, Box 210, LP Calculations and Manuscripts, vol. III, 1926–1927, "He Spectrum, or H2. Two Helium Atoms," Zürich 1927 June, July.

49. AHQP and PP, Box 210, LP Calculations and Manuscripts, vol. III, 1926–1927, "He Spectrum, or H2. Two Helium Atoms," Zürich 1927 June, July.

50. AHQP, Interview with Pauling.

51. The Bancroft Library, Birge Correspondence, Box 33, Folder Jan–Mar 1931, Letter Birge to Mulliken, April 18, 1931.

52. PP, Box 168, American Chemical Society: Correspondence 1925–1944, Letter from Pauling to Lamb, February 11, 1931. In AHQP, in the interview with Pauling, Pauling referred to the frequent delays in publication in the JACS, but especially to the problems he previously experienced with editor and referees, for his were largely theoretical papers. In 1929, Pauling published a paper on "The Principles Determining the Structure of Complex Ionic Crystals," which he had to reduce to half to comply with the editor's request.
53. Sl P, Folder on "Wave Mechanics in the Classical Decade 1923-1932."
54. Sl P, Folder *A Physicist of the Lucky Generation. An Autobiography*, Chapter "Harvard 1926-1928 – Newlyweds in Cambridge."
55. Sl P, Box on Pauling, Folder 1, Letter Pauling to Slater, July 6, 1930.
56. Sl P, Box on Van Vleck, Folder 2, Postcard Van Vleck to Slater, sometime 1930.
57. Sl P, "Scientific Notes: Molecules, Directed Valence," October 10, 1930.
58. Sl P, Folder on "Wave Mechanics in the Classical Decade 1923-1932."
59. Sl P, Box on Pauling, Letter Slater to Pauling, April 11, 1931. On the contexts of simultaneous discovery, see Park (2000).
60. Sl P, Folder No. 1 on Pauling, Letter Pauling to Slater, October 23, 1931.
61. PP, Box 212, LP Berkeley Lectures: Quantum Mechanics 1929–1933, First Lecture on the "Nature of the Chemical Bond," March 25, 1931.
62. Private communication to Ana Simões.
63. PP, Box 214, LP Calculations and Manuscripts 1930–1936, "Eigenfunctions for Chemical Bonds," which are the calculations for the first paper of the series on "The Nature of the Chemical Bond." Pauling later added a note to this manuscript to the effect that he wrote almost everything in one evening.
64. PP, Box 212, LP Berkeley Lectures: Quantum Mechanics 1929–1933, introduction to first lecture on "The Nature of the Chemical Bond." Later on he dedicated his textbook *The Nature of the Chemical Bond* (1939) to Lewis.
65. PP, Box 209, LP Notes and Calculations, vol. II 1923–1929, "1928-London's paper. General Ideas on Bonds." How Pauling came to opt for hybridization may give us an extra instance of the interaction between physical and biological modes of thought in his thought processes. See Nye (2000).
66. PP, Box 212, LP Berkeley Lectures: Quantum Mechanics 1929–1933, conclusion to the lectures on "The Nature of the Chemical Bond." Pauling stated that "the rules have worked out so well that even if there were no such justification they could be accepted on an empirical basis."
67. In paramagnetic substances the magnetic moment induced by an applied field is in the direction of the field. Paramagnetism in atoms is electronic in origin and is generally associated with an odd number of electrons.
68. Besides the results on the electron-pair bond, Pauling hinted at two results that would be the subject of future papers. He suggested that the transition between the electron-pair bond and the ionic bond might be analyzed in terms of the mixing of the wave functions for M:X and M^+X^-, where MX molecules are molecules such as HF or NaCl, and that the compound HF had probably an ionic structure, a conclusion that was supported by the formation of hydrogen bonds with fluorine. He further suggested that although the electron-pair bond was the most common

type of bond, there might be situations that would give rise to a new type of bond, which he named the three-electron bond, and which might explain the electronic structure of the O_2 molecule and account for its paramagnetism.

69. Sl P, Box on Pauling, Letter Pauling to Slater, October 23, 1931.

70. AHQP, Interview with Pauling.

71. AHQP, Interview with Pauling. In a report on work in progress prepared in 1930, Mulliken stressed the importance of molecular spectra for both physicists and chemists: "[Band Spectra's] methods of analysis are those of physics both in the experimental and theoretical steps of the process; the application of results is of special interest to chemistry. In other words this field of molecular spectroscopy is of interest to both physicists and chemists, although it is pure science from the standpoint of the physicist, but more nearly applied science from that of the chemist" in MP, Box 84, Folder 10, Draft of Information to the President, March 15, 1930, emphasis not in original.

72. BC, Box 33, Letter Birge to Mulliken, February 1, 1930.

73. BC, Box 33, Letter Birge to Mulliken, February 1, 1930, emphasis not in original. Probably Birge did not understand Mulliken's ideas on valence theory. Compare with Birge's comment on Pauling's 1931 lectures.

74. BC, Box 21, Folder 1929–1948, Letter Mulliken to Birge, March 26, 1931, emphasis not in original.

75. BC, Box 32, Folder Jan–Sep 1928, Letter Birge to Mulliken, March 17, 1928.

76. BC, Box 21, Folder 1929–48, Letter Mulliken to Birge, Leipzig [1930].

77. PP, Box 168, American Chemical Society: Correspondence, 1925–1944, Letters Pauling to Lamb, February 11, 1931, June 4, 1931, February 21, 1933, Letter Lamb to Pauling, February 28, 1933.

78. PP, Box 168, American Chemical Society: Correspondence, 1925–1944, Letters Pauling to Lamb, February 21, 1933.

79. PP, Box 212, LP Berkeley Lectures: Quantum Mechanics 1929–1933, "The Three-Electron Bond," April 12, 1933.

80. PP, Box 210, Work on Molecular Orbitals, 1926.

81. PP, Box 212, LP Berkeley Lectures: Quantum Mechanics 1929–1933, "The Three-Electron Bond," April 12, 1933.

82. PP, Box 212, LP Berkeley Lectures: Quantum Mechanics 1929–1933, "The Nature of the Chemical Bond," lecture on "The Three-Electron Bond," April 17, 1931.

83. Pauling then concluded that the normal state of alkali halide molecules such as LiF, NaCl, and CsF was essentially ionic; HF was largely ionic; and that the normal state of hydrogen halide molecules such HCl, HBr, HI was largely covalent. This was a confirmation of Pauling's reasoning for the ionicness of HF developed already in Pauling (1928a).

84. BL, Lewis Correspondence, CU-30, Box 2, Folder on K. T. Compton, Letter Compton to Lewis, August 6, 1932.

85. PP, Linus Pauling Lectures: Quantum Mechanics 1929–1933, Box 212, Fortieth Lecture, April 26, 1929, "The Shared Electron Pair Bond."

86. As we saw in chapter 1, aromatic compounds are organic compounds such as benzene containing benzene rings in their structure and possessing a planar ring of atoms linked by alternate

single and double bonds; conjugated systems are compounds containing alternating single and double bonds in their structures.

87. With such a distinction, one could derive a set of essentially constant bond energies as long as one restricted its application to those molecules represented by one single valence bond structure. This set was then used to compute the energy values associated with several valence bond structures. For compounds of the second class, the difference between the energies of formation given by thermochemical data and the values calculated for each of various hypothetical structures had to be always positive (or zero), and this energy difference should be interpreted as the resonance energy of the molecule. This method enabled Pauling and Sherman to calculate resonance energies for a vast amount of compounds.

88. SBPK, Box 6, Folder 5. 213, Letter Hückel to Pauling, Stuttgart (Bad Cannstatt), December 28, 1933.

89. SBPK, Box 6, Folder 5. 115, Letter Pauling to Hückel, California Institute of Technology, Pasadena, April 17, 1934 [original English]. Andreas Karachalios (2003, 141) remarks that "It is evident from Pauling's response that he had difficulty recognizing that "method II," notwithstanding its "great simplifications," was in any way more suitable for benzene. Pauling was mistaken that "method II," with its orthogonal orbitals, does not fully obey the Pauli exclusion principle. On the contrary it does so in an extremely elegant and simple way. It is difficult to understand how Pauling could arrive at this false theoretical verdict."

90. AHQP, Reel 49, Van Vleck Correspondence, Letter Mulliken to Van Vleck, February 28, 1932; March 28, 1932.

91. AIP, Van Vleck Papers, Letter Mulliken to Van Vleck, Box 38, Folder 47, April 16, 1935.

92. The scientific trajectories of Hund and Mulliken ran parallel in the decade 1927–1937 when they both contributed to the creation and development of molecular orbitals. They split afterwards. Hund, who loved to embrace new areas and always feared to become a narrow specialist, moved away from molecular orbital theory, whereas Mulliken consistently built his career around molecules and the theory of molecular orbitals. This fact was noted by Mulliken in a letter to his longtime friend: "You speak of having been too lazy to continue long with one field, but it seems to me that you have done most of the important things very quickly and shown great speed and activity in doing this and going to something new—just the opposite of being lazy. Whereas I have been lazy in not going very much into new areas. In any event, you have been very much hindered by outside events" in MP, Box 24, Folder 14, Letter from Mulliken to Hund, February 14, 1965.

93. AHQP, Reel 49, Van Vleck Correspondence, Letter from Van Vleck to Mulliken, February 22, 1927.

94. The same is stressed in a letter to Van Vleck in AHQP, Van Vleck Correspondence, Reel 49, July 1, 1935.

95. MP, Box 74, Folder 2, "Electron States and the Structure of Molecules."

96. This is an excerpt from the Lewis's award speech (1960).

97. AHQP, Interview with Mulliken.

98. AHQP, Interview with Mulliken.

99. AHQP, Interview with Wigner.

100. Letter Pauling to Reymond Holmon, March 1987, as quoted in Holmen (1990).

Notes

101. AHQP, Reel 49, Van Vleck Correspondence, Letter Van Vleck to Slater, November 12, 1928, Letter Slater to Van Vleck, November 17, 1928.
102. AIP, Van Vleck Correspondence, Box 39, Folder 60, Letter Van Vleck to Slater, March 19, 1974, concerning the "geriatric session" commemorating the 75th anniversary of the American Physical Society.
103. Sl P, Box on Van Vleck, Folder 3, Letter Van Vleck to Editor of Wiley-Interscience, February 27, 1975, on Slater's autobiography.
104. Sl P, Box on Van Vleck, Folder 2, Letter Slater to Van Vleck, December 7, 1971.
105. Sl P, Folder *A Physicist of the Lucky Generation. An Autobiography*, Chapter "Harvard 1928-1929."
106. Sl P, Folder *A Physicist of the Lucky Generation. An Autobiography*, Chapter "Harvard 1928-1929."
107. Sl P, Folder *A Physicist of the Lucky Generation. An Autobiography*, Chapter "Harvard 1928-1929."
108. AIP, Van Vleck Correspondence, Box 39, Folder 60, "1920-1930. The First Ten Years of John Slater's Scientific Career."
109. AIP, Van Vleck Correspondence, Box 39, Folder 60, Letter Slater to Van Vleck, March 26, 1974.
110. Sl P, Folder *A Physicist of the Lucky Generation. An Autobiography*, Chapter "Harvard 1928-1929."
111. AIP, Van Vleck Correspondence, Box 39, Folder 60, "1920-1930. The First Ten Years of John Slater's Scientific Career."
112. Sl P, Folder on "Wave Mechanics in the Classical Decade 1923-1932."
113. Sl P, Folder *A Physicist of the Lucky Generation. An Autobiography*, Chapter "Leipzig 1929."
114. AHQP, Interview with Heitler.
115. Heitler considered that "it was really in this paper that we then could see what the chemists mean by a chemical formula, when they write down a chemical formula. And we could associate a chemical bond immediately with an electron pair with antiparallel spin in the wave function describing the molecules" (AHQP, Interview with Heitler).
116. Sl P, Box on Born, Letter Slater to Born, October 16, 1930.
117. AHQP, Interview with Slater.
118. AHQP, Interview with Slater.
119. AIP, Van Vleck Correspondence, Box 51, Folder 115, "Reminiscences of my Scientific Rapport with R. S. Mulliken."
120. AHQP, Van Vleck Correspondence, Letter Van Vleck to Mulliken, January 17, 1927.
121. AIP, Van Vleck Correspondence, Box 51, Folder 115, "Reminiscences of my Scientific Rapport with R. S. Mulliken."
122. AHQP, Van Vleck Correspondence, Letter Mulliken to Van Vleck, March 28, 1928.
123. In the 1970s Mulliken and Van Vleck helped organize the Symposium on the Fifty Years of Valence Theory. There were historical talks by Van Vleck, Mulliken, Coulson, and Daudel. Van Vleck and Mulliken aimed at placing Lewis and Langmuir in proper perspective. AIP, Van Vleck Correspondence, Box 38, Folder 47, Letter Mulliken to Van Vleck, July 28, 1969. In the same letter Mulliken confided that he used "to think Pauling was unfair to Slater about hybridization

and directional valence, but a perusal of Pauling's short preliminary paper suggests that he already had sp hybridization in mind then, at least roughly, although he didn't come out with the details until his JACS 1931 article."

124. AHQP, Interview with Van Vleck conducted by Thomas Kuhn.
125. LA, Heitler to London, November 4, 1935.
126. LA, Heitler to London, November 12, 1935.
127. LA, Heitler to London, November 12, 1935.
128. LA, Heitler to London, November 12, 1935.
129. LA, London to Heitler, October or November 1935.
130. LA, London to Heitler, November 17, 1935.
131. LA, Heitler to London, November 22, 1935.
132. LA, Heitler to London, November 28, 1935.
133. LA, London to Heitler, October or November 1935.
134. LA, Heitler to London, beginning of December 1935.
135. LA, Heitler to London, December 13, 1935, emphasis in original.
136. LA, Heitler to London, December 13, 1935, emphasis in original.
137. LA, Heitler to London, February 6, 1936, emphasis in original.
138. LA, Heitler to London, October 7, 1936.
139. LA, London to Heitler, November (?) 1936.
140. LA, London to Born, October 1, 1936.
141. LA, Born to London, October 10, 1936.
142. LA, Born? to London, October 7, 1936.
143. LA, Heitler to London, February 18, 1951.
144. Among his papers are the notebooks containing his philosophical essays. SP, III-Notebooks. Item 24. Philosophical essays 1895–1896. Item 25. Philosophical essays 1896–1897.
145. SP. V-Correspondence. Item 61. Letter Debye to Sidgwick, November 16, 1933.
146. SP. Item 49, Lecture on "Atomic Cohesion," Edgar Fahs Smith Lecture, Philadelphia, May 22, 1931.
147. SP. V-Correspondence. Item 71. Rutherford. Letter Rutherford to Sidgwick, July 26, 1937.
148. As already discussed in this chapter (see section "The Development of Pauling's Program: The Nature of the Chemical Bond"), tautomerism undermined the idea that a substance has one unique formula. It corresponds with the situation in which a substance can exist in two different forms, although it has proved impossible to separate the isomers.
149. SP. V-Correspondence. Item 68. Letter London to Sidgwick, May 10, 1934.
150. SP. V-Correspondence. Item 79. Sutton 1934–1951. Letter Sutton to Hamp (abbreviation of Hampson), March 11, 1934.
151. SP. Item 52. "Atomic Cohesion," Edgar Fahs Smith Lecture, Philadelphia, May 22, 1931. Item 53: "Some problems in molecular structure," Glasgow University Alchemists' Club, February 13, 1935. Item 55: "Multiple links," Birbeck College, February 14, 1938.
152. SP. Item 53: "Some problems in molecular structure," Glasgow University Alchemists' Club, February 13, 1935.
153. PP. Box 96, Sidgwick 1935–1951. Correspondence. Letter Pauling to Sidgwick, May 25, 1936.

154. Sl P, Box on K. T. Compton, Folder 3, Application to the Rockefeller Foundation for Assistance in Securing a Science Research Fund for the Massachusetts Institute of Technology.

155. Sl P, Box on Compton, Folder 1, Letter Slater to Compton, May 23, 1930; Letter Compton to Slater, May 28, 1930.

156. Sl P, Box on Compton, Folder 4, Research Program in Physics prepared by Slater.

157. Sl P, Box on Compton, Folder 4, Research Program in Physics prepared by Slater.

158. Sl P, Folder on *A Physicist of the Lucky Generation. An Autobiography*, letter transcribed in the chapter "Harvard 1928-1929."

159. Sl P, Box on Pauling, Folder 1, Letter Slater to Pauling, January 21, 1931.

160. Sl P, Box on Pauling, Folder 1, Letter Slater to Pauling, January 21, 1931.

161. Sl P, Box on Pauling, Folder 1, Letter Pauling to Slater, February 9, 1931.

162. Sl P, Box on Pauling, Folder 1, Letter Slater to Pauling, July 13, 1931.

163. Sl P, Folder of First Draft of Talk on "Philosophy and Physics," Franklin Institute, Philadelphia, December 9, 1937. See also Schweber (1990).

164. SlP, Folder on J. E. Mayer, Letter Slater to Mayer, September 6, 1950.

165. PP, Box 243, Popular Scientific Lectures 1956–1965, "The Theory of Valence and of the Structure of Atoms, Molecules, and Crystals," 1957.

166. According to Pauling, the textbook *Introduction to Quantum Mechanics with Applications to Chemistry* holds the record of continuous publication without modifications by McGraw-Hill (1935–1985). After 1985 it was published as a Dover edition. It is the second most cited of all Pauling's publications, *The Nature of the Chemical Bond* being the first. Private communication to Ana Simões. For an analysis of the strategies used by Pauling to popularize resonance theory among chemists, see Park (1999).

167. PP, Box 58, Lewis G. N.: 1913–1947. Reprints and Correspondence. Letter Lewis to Pauling, August 25, 1939.

168. PP, Box 299, The Nature of the Chemical Bond 1932–1959. Letter Mayer to Pauling, July 27, 1939.

169. PP, Box 372, Chapters for Book on Quantum Mechanics of Organic Molecules, Chapter and Section Topics.

170. PP, Box 115, W Correspondence, File on Wheland, Letter Pauling to Wheland, October 30, 1936, March 13, 1937, March 30, 1937, July 28, 1937, Letter Wheland to Pauling, November 29, 1936.

171. PP, Box 115, W Correspondence, File on Wheland, Letter Pauling to Wheland, December 1, 1944, June 27, 1945, Letter Wheland to Pauling, December 6, 1944.

172. Hückel cited in his article the following papers by Ingold, Arndt, and Eistert: Ingold (1934, especially 250–252); Arndt and Eistert (1936).

173. Years later, Hückel (1957, 873) criticized Pauling by emphasizing the contradictory meaning of the term "structure": "Here in one and the same sentence Pauling uses two meanings for the word "structure": once with the meaning of a diagram that is assigned to a specific function that can (but does not have to!) be used as one of the initial functions for the perturbation calculation; the other time, to designate the real state of the molecule; for such functions are not changed anymore. Only specific linear combinations are sought, in particular those that lead to the lowest energy in the perturbation calculation."

174. It was not, of course, the case that such sentiments were shared by all the chemists of the community. Characteristic of the differences is the editorial note to the first article where it is stressed that any particular way of dealing with chemical phenomena should not be excluded on a priori grounds, but should first be closely studied. In the same article, Ya. Syrkin and M. Dyatkina were also attacked. They were the authors of the excellent book *The Chemical Bond and the Structure of Molecules* and had translated Pauling's book into Russian.

175. PP, Box 261. Letter M. V. King to Pauling, January 23, 1953; letter Coulson to Pauling, October 7, 1953; letter Coulson to King, January 18, 1954; letter King to Pauling, February 9, 1954.

176. PP, Box 115, Letter Wheland to Pauling, January 20, 1956.

177. PP, Box 115, Letter Wheland to Pauling, January 20, 1956.

178. PP, Box 115, Letters Pauling to Wheland, January 26 and February 8, 1956.

179. PP, Box 115, Letters Pauling to Wheland, January 26 and February 8, 1956.

180. In order to follow Mulliken's evolution from a position closer to Pauling to the position advocated in the Nobel lecture, let us recall his assessment of the state of quantum chemistry right after the war. In 1947, in the paper titled "Quantum-Mechanical Methods and the Electronic Spectra and Structure of Molecules," Mulliken reviewed past accomplishments and the current state in order to delineate a strategy to push quantum chemistry forward after the "vacuum" created by the war. Recalling Dirac's famous statement expressed in 1929, Mulliken pointed to the failure of theoretical physicists to contribute to quantum chemistry and briefly reviewed available methods that he classified into four categories, having in mind their "relative quantitativeness and their relative content of pure theory and empirical data": "qualitative method," "semiempirical method," "approximate theoretical method," and "accurate theoretical method." He claimed that he was inclined to favor "the semiempirical method" in which certain theoretically defined integrals were substituted by empirical parameters. But his methodological allegiance was soon to change (Mulliken 1947, 203–204). For these developments, see chapter 4.

Chapter 3

1. For the British context and specifically the London–Manchester school of theoretical organic chemistry, see Nye (1993).

2. Named after the mathematician John Couch Adams, the prize is awarded each year by the Faculty of Mathematics of the University of Cambridge and St. John's College to outstanding mathematical work by a young scholar working in the United Kingdom. The Adams Prize essay was revised and extended into Fowler (1929).

3. AHQP, Notes taken by L. H. Thomas on "The Quantum Theory of Spectra," Lent Term 1923–1924, Cavendish, and "Recent Developments of Quantum Theory," May 1925, Trinity College. There are also lecture notes on these same courses taken by P. A. M. Dirac. In the first course the following comment concerning the application of quantum theory to chemistry was made: "Nothing definite has so far come from this theory as to chemical combination. . . . The general octet idea has, however, been of value, e.g. in the hands of Lewis and Kossel (and Langmuir). The octet seems to play a very fundamental part and the idea appears to be good that in molecules octets act fully or partially by sharing e.g. 4 electrons, 2 electrons. This, however, though not contrary to Bohr's theory, has not been deduced from it."

4. AHQP, Letter Fowler to Kronig, May 25, 1925.
5. AHQP, Letter Fowler to Kramers, December 1, 1925.
6. AHQP, Letter Kramers to Fowler, December 9, 1925.
7. Also quoted in Buckingham (1987, 113).
8. AHQP, Interview with N. Mott.
9. AHQP, Letters Fowler to Kramers, October 13, 1925, Fowler to Kramers, October 27, 1925, Fowler to Kramers, November 17, 1925, referred to the translation of some of Bohr's papers. In 1928, Fowler was inviting Kronig to stay in Cambridge for around 15 days to give a set of 6 lectures on "The Structure of Molecules and their Spectra," Letter Fowler to Kronig, October 30, 1928. Later on, in 1933, Heisenberg was invited to give the Scott lectures at the Cavendish aimed at students of "Physics and Theoretical Physics" and the more formal and public Rouse Ball Lecture in which the speaker was expected "to discuss in a general way some aspects of Mathematics or Mathematical Physics." Letter Fowler to Heisenberg, November 30, 1933.
10. AHQP. In Van Vleck to Fowler, May 21, 1932, Van Vleck suggested that, if he was looking for volumes to be published by Oxford Press, Condon and Shortley's manuscript on atomic spectra might be of interest and contribute to try to overcome the "deficit of books on quantum mechanics as applied sufficiently in detail to problems of atomic structure, emphasising the methods of Slater and the like."
11. L-J P. II. Scientific and personal correspondence. 14. Letters Chapman to Jones, May 11, June 5, June 22, 1924; letter H. A. Myers to Jones, May 30, 1924; Letters Jones to Chapman, June 19, July 6, 1924.
12. L-J P. II. Scientific and personal correspondence. 15. Letters A. M. Tyndall to Jones, January 30, February 2, 6, 9, 1925; letter Registrar to Jones, March 3, 1925.
13. L-J P. II.5. Report of work carried out during tenure of the 1851 Exhibition senior research studentship, 1922–1924; Investigations proposed for a third year. The report comprised the following sections: 1. Free paths in a nonuniform rarefied gas with an application to the escape of molecules from isothermal atmospheres; 2. The equation of state of a gas; 3. On the determination of molecular fields. From the variation of the viscosity of a gas with temperature; 4. On the determination of molecular fields. From the equation of state of a gas; 5. On the determination of molecular fields. From crystal measurements and kinetic theory data (incomplete); 6. The mobilities of ions in an electric field (incomplete); 7. The escape of calcium atoms from the sun. The first five sections correspond with five different papers bearing similar titles.
14. Among the papers that preceded his dissertation was Hartree (1923a).
15. See Park (2009) for Hartree's role on the development of computational methods in the history of quantum chemistry.
16. Sl P, Box on Hartree, Folder 1, Letter Hartree to Slater, December 18, 1928.
17. LP, Letter Hartree to London, September 16, 1928. [In England, 'Physics' usually means 'Experimental Physics;' until the last few years 'Theoretical Physics' has hardly been recognized as a subject like it is here, and, I understand, in your country. In Cambridge particularly the bias has been very much toward the experimental side, and most people now doing research on theoretical physics studied mathematics, not physics.]
18. Sl P. B Sl 2p. Folder 1. Letter Slater to Hartree, November 25, 1931.
19. Sl P. B Sl 2p. Folder 1. Letter Hartree to Slater, December 21, 1931.
20. HP, Recollections of his daughter Margaret Hartree Booth written in 1986.

21. A similar statement was expressed on Garner and Lennard-Jones (1929a, 611).
22. See Hall (1991) for an assessment of this paper.
23. This approximation came to be known as the linear combination of atomic orbitals (LCAO) approximation.
24. AHQP, Interview with Mulliken.
25. Two of Lennard-Jones's papers (1931b, 1931c) gave a detailed analysis in simple but exact terms of the mathematical foundation of the new mechanics.
26. The same wording was used in Lennard-Jones (1931a, 478).
27. The same sort of criticism was made by Wigner to Heitler. AHQP. Interview with E. P. Wigner.
28. It was held at the Department of Physical Chemistry of the University of Cambridge in September 28–30, 1933, and was, once again, chaired by Lowry. Most of the 150 participants were British or German (more than 20 Germans were present). Other national groups represented in considerably less numbers were those coming from the United States, Denmark, and New Zealand.
29. Slater contributed to the discussion after their papers, stated his agreement with their conclusions, and made some predictions as to future developments in the theory of solids. In *International Conference on Physics. Paper and Discussions* (1935, 53).
30. They were A. J. Berry, R. E. Felson, F. G. Mann, B. I. O. Masson, R. W. G. Norriss, H. S. Patterson, P. L. Robinson, W. Wardlaw, R. Whyllaw-Esay, P. F. Spencer, and P. O. Finch. University of Cambridge Archives, O XIV 54.
31. *Cambridge Historical Register Supplement* (1942, 12). The professorship was assigned to the Faculty of Physics and Chemistry.
32. Collection of Lecture notes deposited by C. A. Coulson in Cambridge University Library, Shelfmark: Add. 7807. This is not accurate: London in 1929 started teaching a course titled "Quantum Mechanics and Chemistry" at the University of Berlin as a privatdozent.
33. L-J P. IV. 85. Lecture Notes for Courses from 1938 to 1952. Syllabus of Lectures in *Theoretical Chemistry*, introductory course of 16 lectures given in 1949. Notes deposited at the Lennard-Jones Archives by professor G. G. Hall of the University of Nottingham.
34. LJ Papers, II.6. Report of the Faculty Board of Mathematics on the need for a Computing Laboratory sent by Lennard-Jones, probably written prior to December 2, 1936.
35. LJ Papers, II.6. Note on naming the laboratory "computing" and not "calculating."
36. Lennard-Jones resumed his academic activities in 1946, after the end of the war during which he held the position of chief superintendent of armament research in the Ministry of Supply.
37. In realizing his program, Lennard-Jones obtained equations for the molecular orbitals to which electrons can be assigned in both normal and excited states of a molecule. He had also shown that for a molecule of the type XY_2, these equations could be transformed in such a way as to be satisfied by a distribution about one of the bonds XY and a similar distribution about the other XY bond. Such distributions were given special significance by Lennard-Jones and they were called equivalent orbitals, as each was the mirror image of the other in the plane of symmetry of the molecule. The next step was to generalize the approach by applying it to molecules of the type XY_n, where n atoms are symmetrically distributed about the central atom X (Lennard-Jones 1949a). An additional conclusion was that both the extensive and localized properties of molecules could be dealt with by the same mathematical formulation (Lennard-Jones 1950).

38. AHQP, Interview with Mrs. Hartree.
39. AHQP, Interview with Mrs. Hartree.
40. HP, Box 154 iv, Letter Hartree to Lindsay, October 20, 1935. Lindsay's interest in foundational issues in physics materialized in some books such as *Foundations of Physics* written with H. Margenau in 1936, reprinted in 1957, and in 1981 as a Dover publication. Some of his publications in the history and philosophy of physics are Lindsay (1968, 1970a, 1973).
41. HP, Box 154 iv, Letter Hartree to Lindsay, February 18, 1937.
42. HP, Box 154 iv, Letter Hartree to Lindsay, April 27, 1937.
43. HP, Box 154 iv, Letter Hartree to Lindsay, October 20, 1935.
44. These are problems where the first derivative of x with respect to t depended on the value taken by x, or some other quantity depending on x, at a previous instant $t - a$, besides depending on the value taken by x at time t.
45. HP, Box 154 iv, Letter Hartree to Lindsay, October 11, 1936.
46. Sl P. B Sl 2p. Folder 1. Letter Hartree to Slater, September 1, 1939.
47. Sl P. B Sl 2p. Folder 1. Letter Hartree to Slater, September 1, 1939.
48. Sl P. B Sl 2p. Folder 1. Letter Slater to Hartree, September 18, 1939.
49. Sl P. B Sl 2p. Folder 1. Letter Slater to Hartree, September 18, 1939.
50. Sl P. B Sl 2p. Folder 1. Letter Hartree to Slater, November 18, 1939.
51. Sl P. B Sl 2p. Folder 1. Letter Slater to Hartree, December 14, 1939.
52. Sl P. B Sl 2p. Folder 1. Letter Hartree to Slater, February 1, 1940.
53. Brodetsky, who was professor of mathematics at the University of Leeds, later became Minister of Education in the first Israeli Government.
54. CP, "What makes a beautiful proof?" B.10.2, 1952–1960, pp. 5–6; published as Coulson (1969a).
55. Collection of Lecture notes taken 1928–1932 deposited by C. A. Coulson in Cambridge University Library, Shelfmark: Add. 7807.
56. On the religious views of Raven and Eddington, see Bowler (2001), Stanley (2007).
57. "Graduates engaged in post graduate work are reminded that their supervisor is a university officer and when visiting him officially in that capacity they should dress as they would in visiting any officer of the university or of their own college (e.g. a tutor). Gowns, however, need not be worn in the chemical laboratory." (CP, Ms. Coulson 150, Box G.11, Correspondence L-LE, G.11.3, G.11.3-J.E. Lennard-Jones, Note Lennard-Jones to Coulson, June 14, 1933.)
58. In the final period of Ph.D. work, Coulson applied to the Appointments Committee of the Mathematics Faculty. Lennard-Jones wrote him in a critical mood: "I would like to know what is the present position with regard to your work. I have seen so little of you lately that I feel out of touch with what you are doing. I should have preferred that you come to see me from time to time so that I could judge what progress you are making." (CP, Ms. Coulson 150, Box G.11, Correspondence L-LE, G.11.3, G.11.3-J.E. Lennard-Jones, Letter Lennard-Jones to Coulson, April 22, 1936.)
59. CP, Ms. Coulson 48, Box B.8, B.28.3, Letter Coulson to R. H. Fowler, March 13, 1934. "Report on Work done and doing since June 1933."
60. CP, Ms. Coulson 150, Box G.11, Correspondence L-LE, G.11.3, G.11.3-J.E. Lennard-Jones, Letter Lennard-Jones to Coulson, January 4, 1935.

61. CP, Ms. Coulson 40, Box B.20, B.20.12, "The Electronic Structure of Molecules from the Standpoint of the Theory of Molecular Orbitals," 1934. The thesis was never published. Its sections were the following: Preface; I-General theory of valence; II-Comparison of the electron-pair theory and the molecular-orbital theory; III-Energy levels of homonuclear diatomic molecules; IV-Energy levels of heteronuclear diatomic molecules; V-Energy levels of equilateral triatomic molecules; VI-Energy levels of some special polyatomic molecules; VII-Energy level and structure of methane; VIII-Discussion of integrals.

62. CP, Ms. Coulson 51, Box B.31., B.31.6. One page of topics for "programme of work on aromatic molecules," written on December 1, 1936. Also Ms Coulson 50, Box B.30., B.30.14; "Comparison of resonance energies of molecules by the orbital and the pair-bond theories" written on December 9, 1936, and "Some thoughts on the aromatic molecules" written on December 25, 1936.

63. CP, Ms. Coulson 154, Box G.15, G.15.5, Letter Coulson to Mulliken, February 20, 1942, italics ours.

64. CP, Ms. Coulson 154, Box G.15, G.15.5, Letter Mulliken to Coulson, August 8, 1942.

65. CP, Ms. Coulson 150, Box G.11, Correspondence L-LE, G.11.3, G.11.3-J.E. Lennard-Jones, Letter Coulson to Lennard-Jones, January 24, 1939.

66. CP, Ms. Coulson 64, Box C.1., C.1.11. Correspondence with Max Born. Letter of February 3, 1972. Coulson writes to a certain Peter and refers to his relationship with Born, trying to excuse Born for his tone in the Born–Einstein correspondence.

67. CP, Ms. Coulson 23, Box B.3, B.3.11, Valence Lectures, Dundee 1939–43, Course of sixteen lectures. Ms. Coulson 24, Box B.4, Lectures delivered at Dundee University.

68. Also CP, Ms. Coulson 33, Box B.3., B.33.3. Correspondence concerning paper "The lengths of the links of unsaturated hydrocarbon molecules." Letter Coulson to Urey, May 12, 1939.

69. CP, Ms. Coulson 33, Box B.3., B.33.3. Letter Urey to Coulson, June 29, 1939.

70. CP, Ms. Coulson 33, Box B.3., B.33.3. Letter Coulson to Urey, July 20, 1939.

71. On this topic, see also Park (2005).

72. Sutton Papers, Bodleian Library, Modern Manuscripts Collections contain information on the strategies put forth to create a molecular research group at Oxford. For more information on Sutton, see Whiffen (1994) and Gavroglu and Simões (2002).

73. CP, Ms. Coulson 151, G.12.6. Correspondence Coulson-Longuet-Higgins 1946–49. Letter L-H to Coulson, February 19, 1948.

74. CP, Ms. Coulson 151, G.12.6. Correspondence Coulson-Longuet-Higgins 1946–49. Letter Coulson to Longuet-Higgins, May 9, 1948.

75. CP, Ms. Coulson 150, Box G.11, Correspondence L-LE, G.11.3, G.11.3-J.E. Lennard-Jones, Letter Coulson to Lennard-Jones, March 2, 1948. After Coulson's inaugural lecture where he had mentioned his relations and debts to Lennard-Jones, the latter confided in a letter of February 26, 1948: "It is very good of you to think of referring to me in your inaugural lecture. May I say in return that it was a pleasure to have you as a research student and that I found you an interesting and stimulating colleague. I am reminded now of those early days in Cambridge as I try to start up from scratch again on my return to academic work." See also letter Lennard-Jones to Coulson, March 4, 1948, and May 11, 1948.

76. Longuet-Higgins was a postdoctoral fellow at the University of Chicago during 1948–1949 (Mulliken 1989, 129). CP, Ms. Coulson 151, Box G.12, G.12.7. Correspondence with Longuet-Higgins 1950–1968, Letter Longuet-Higgins to Coulson, February 6, 1952, Department of Chemistry of the University of Manchester.
77. CP, Ms. Coulson 91, Box C.28, C.28.7. Correspondence with Erich Hückel. Letter Hückel to Coulson, April 18, 1948.
78. CP, Ms. Coulson 151, G.12.7. Correspondence Coulson-Longuet-Higgins 1950–68. Letter Coulson to Longuet-Higgins, November 21, 1950.
79. CP, Ms. Coulson 91, Box C.28, C.28.3. Letter Coulson to Heisenberg, February 12, 1948.
80. Among the British students were Juliane Jacobs, M. P. Barnett, Norman March, Peter Davies, George Lester, Tim Greenwood, Alan Lidiard, Peter Higgs, Eric Pitts, and Robert Taylor. From Sweden came Inga Fischer, from Holland Joop de Heer, John van der Waals, and A. Buseman, from the United States Carl Moser, from Australia Ron Brown, from France Roland Lefebvre, and from Argentina S. L. Altmann. The Australians David Craig and Allan MacColl from University College collaborated also with Coulson.
81. On Coulson's ideas on science and religion, see Simões (2004).
82. CP, Ms. Coulson 150, Box G.11, Correspondence L-LE, G.11.3, G.11.3-J.E. Lennard-Jones, Letter Lennard-Jones to Coulson, March 25, 1948. Ms. Coulson 150, Box G.11, Correspondence L-LE, G.11.3, G.11.3-J.E. Lennard-Jones, Letter Coulson to Lennard-Jones, March 29, 1948.
83. CP, Ms. Coulson 150, Box G.11, Correspondence L-LE, G.11.3, G.11.3-J.E. Lennard-Jones, Letter Coulson to Lennard-Jones, March 10, 1950.
84. CP, Ms. Coulson 91, Box C.28, C.28.5., Letter Hinshelwood to Coulson, March 6, 1950.
85. Mulliken and Coulson both agreed that configuration interaction may become very important in the more complicated molecules when one starts with molecular orbital configurations. "Of course, this is all very familiar, but I think it is interesting to see how configuration interaction becomes important in different places in the AO and the MO methods." (CP, Ms. Coulson 154, Box G.15, G.15.5. Letter Mulliken to Coulson, January 24, 1949.)
86. CP, Ms. Coulson 150, Box G.11., G.11.3., Correspondence L-LE, Letter Lennard-Jones to Coulson, October 9, 1950.
87. About the same time, in late 1950, the expression "ab initio" was introduced in the literature in Parr, Craig, and Ross (1950) (see chapter 4).
88. CP, Ms. Coulson 115, Box D.8, D.8.8, "Chemistry by Computation."
89. CP, Ms. Coulson 30, B.10.6, 1965–1973, "Making Models," Open University Program, on 1.
90. Before this talk he published three papers on the topic (Coulson 1941b, 1947a, 1947b).
91. The second edition appeared in 1961 and the third edition in 1979 as *Coulson's Valence*, edited by Roy McWeeny (Oxford University Press). We shall refer to the first edition except when we state otherwise.
92. CP, Ms Coulson 155, G.16.3, Letter Coulson to Pauling, August 1, 1951. Nevertheless, on May 12, 1952, Pauling wrote to Coulson that he was still eagerly waiting to receive a copy of the book.
93. PP. Pauling's copy of *Valence*. Marginalia. Handwritten comments in the blank end-page of the book.

94. The book was also reviewed by Roberts (1952).
95. CP, Ms Coulson 155, G.16.3, Letter Coulson to Pauling, September 8, 1952.
96. CP, Ms Coulson 155, G.16.3, Letter Pauling to Coulson, September 25, 1952.
97. CP, Ms Coulson 155, G.16.3, Letter Pauling to Coulson, September 25, 1952.
98. CP, Ms Coulson 155, G.16.3, Letter Pauling to Coulson, November 4, 1952.
99. CP, Ms Coulson 155, G.16.3, Letter Coulson to Pauling, July 14, 1953.
100. CP, Ms Coulson 155, G.16.3, Letter Coulson to Pauling, September 12, 1953.
101. CP, Ms. Coulson 36, B.16.4, on 2.
102. A historian of science could even write, Coulson says in a critical mood: "the truth is that Chemistry has no place in the strict scientific scheme . . . the ultimate generalizations of chemistry are all derivable, and indeed must inevitably have been reached sooner or later from the development of physics itself . . . the part played by chemistry in the growth of science has been a pragmatic, a heuristic one . . . chemistry rightly figures prominently in the history of science: in the philosophy of science it should figure not at all." CP, Ms. Coulson 36, B.16.4, on 3.
103. Also in CP, Ms. Coulson 36, B.16.4, on 9.

Chapter 4

1. Nye (1993) addresses the contributions of the Paris School of Theoretical Organic Chemistry in the period 1880–1930.
2. Interview with Alberte Pullman by Udo Anders, October 28, 1997, p. 2. Available at: <http://www.quantum-chemistry-history.com/Pull1>.
3. Interview with Alberte Pullman by Udo Anders, October 28, 1997, p. 2. Available at: <http://www.quantum-chemistry-history.com/Pull1>.
4. Reference to Schmidt's papers in Interview with Alberte Pullman by Udo Anders, October 28, 1997, p. 2. Available at: <http://www.quantum-chemistry-history.com/Pull1>.
5. Interview with Alberte Pullman by Udo Anders, October 28, 1997, p. 2. Available at: <http://www.quantum-chemistry-history.com/Pull1>.
6. Interview with Alberte Pullman led by Udo Anders, October 28, 1997, p. 2. Available at: <http://www.quantum-chemistry-history.com/Pull1>.
7. Interview with Alberte Pullman by Udo Anders, October 28, 1997, p. 9, p. 11. Available at: <http://www.quantum-chemistry-history.com/Pull1>. According to her reminiscences: "but all the publications of Coulson, Longuet-Higgins came later, some of them in 1939, and in these years we were not aware of them. We discovered the whole method of molecular orbital development by Coulson and others after the end of the war, after 1945." And later on: "the developments in America were different. There was a big contribution by Mulliken, which we were not really aware of. Oh, well, perhaps I heard him mentioned. You know, between Pauling and Mulliken there was always a kind of antagonism. Mulliken is very little quoted in Slater and Pauling." Gaston Berthier recalled that it was after the 1948 Paris Conference that French scientists became familiar with Mulliken's work. See Interview with Gaston Berthier by Udo Anders, June 2, 1997. Available at: <http://www.quantum-chemistry-history.com/Berth1.htm>. 4. In the Interview with Alberte Pullman led by Udo Anders, October 28, 1997, p. 9, p. 11 (<http://www.quantum-chemistry-history.com/Pull1>), she recalled keeping the microfilm copy of Pauling's textbook

throughout the years as a memory of those early days, and that due to Raymond Daudel's inability to read German, she translated most of the literature for him.

8. Interview with Alberte Pullman by Udo Anders, October 28, 1997, p. 7. Available at: <http://www.quantum-chemistry-history.com/Pull1>.

9. Interview with Alberte Pullman by Udo Anders, October 28, 1997, p. 4. Available at: <http://www.quantum-chemistry-history.com/Pull1>.

10. Formerly an analytical chemist, Gaston Berthier (1923–) got tired of working in the automobile industry and joined the Pullmans' group. Interview with Gaston Berthier conducted by Udo Anders, June 2, 1997, p. 1. Available at: <http://www.quantum-chemistry-history.com/Berth1.htm>.

11. In fact, this meeting was preceded by another one on *Échanges isotopiques et structure moléculaire*, which took place at the Institut du Radium and was attended by many of the participants of the meeting on the chemical bond, which started just 2 days after its end.

12. All papers delivered at the conference were published in the *Journal de Chimie Physique* 45 (1948): 141–250.

13. AHQP. Interview with Mulliken.

14. Interview with Alberte Pullman by Udo Anders, October 28, 1997, p. 5. Available at: <http://www.quantum-chemistry-history.com/Pull1>.

15. Furthermore, in Paris, at the École Normale Supérieure, a quantum chemistry group was founded under the supervision of Josiane Serres.

16. Interview with Gaston Berthier by Udo Anders, June 2, 1997, p. 5. Available at: <http://www.quantum-chemistry-history.com/Berth1>.

17. During the 1950s, this and other groups in France were using desk calculators, such as Peerless and Fridens, to perform most calculations. The first ones in use were confiscated from the Germans by French troops after the end of the war. Julg calculated the self-consistent field of the π-system of azulene (10 electrons), involving 4500 molecular integrals, before making use of the SCF iterations, which showed strong divergence that resisted the standard convergence methods of the time. The problem was solved by finding a very efficient graphical method that enabled Julg to arrive at the solution before his competitors who were already using a computer. He estimated that the whole process took him more than 4000 hours of hard work. Interview with Alberte Pullman, October 28, 1997, and Gaston Berthier, June 2, 1997, both by Udo Anders. Available at: <http://www.quantum-chemistry-history.com>. See also A. Julg, "Une page d'histoire: La méthode LCAO améliorée" at <http://www.quantum-chemistry-history.com/Julg_af>.

18. Interview with Gaston Berthier by Udo Anders, June 2, 1997, p. 2. Available at: <http://www.quantum-chemistry-history.com/Berth1>.

19. Barriol (1971, original publication 1966). This textbook originated as a course given at the Faculty of Sciences in Nancy. The English translator notes interestingly that "in view of the significant contributions made to quantum chemistry by various European groups, *it seems appropriate to have available in English an example of their approach in preparing their students*. This text is a particularly good example of this approach" (Barriol 1971, v, italics ours). Barriol's interest for the teaching of the mathematical background of quantum mechanics emerged while he was held prisoner during the Second World War. His prison mates were given lectures on Dirac's *Principles of Quantum Mechanics*, for instance. See Blondel-Mégrelis (2001).

20. A follow-up to this textbook, a sort of largely transformed second edition, appeared under the following names and title: Daudel et al. (1983). An earlier textbook was Daudel (1956).
21. Szent-Györgyi had a short-lived correspondence with Fritz London attempting to get the latter involved in the investigation of the behavior of biomolecules within the framework of macroscopic quantum phenomena proposed by London as a result of his work in low-temperature physics.
22. Citation from Coulson (1960, 177).
23. MacP, Letter Mulliken to MacInnes, September 18, 1950.
24. MacP, Letter MacInnes to Mulliken, September 22, 1950.
25. MacP, Letter Mulliken to MacInnes, September 29, 1950. Also Letter Mulliken to MacInnes, October 23, 1950.
26. MacP, Letter MacInnes to Mulliken, October 2, 1950.
27. MacP, Letter Mulliken to MacInnes, November 10, 1950.
28. MacP, Letter MacInnes to Mulliken, December 4, 1950.
29. MacP, Proposal for task order under contract N6ori-20 to the O. N. R. from the University of Chicago, prepared by Mulliken, December 7, 1950. The title was the original one.
30. CP, Ms. Coulson 154, Box G.15, G.15.5. Letter Mulliken to Coulson, October 31, 1950. Also Ms. Coulson 154, Box G.15, G.15.5. Letter Coulson to Mulliken, November 11, 1950.
31. MacP, Letter Mulliken to participants, July 13, 1951.
32. CP, Ms. Coulson 154, Box G.15, G.15.5. Letter Mulliken to Coulson, June 16, 1949.
33. CP, Ms. Coulson 154, Box G.15, G.15.5. Letter Coulson to Mulliken, September 2, 1949.
34. CP, Ms. Coulson 154, Box G.15, G.15.5. Letters Mulliken to Coulson, January 17, 1950, January 23, 1950. Mulliken talked about Parr's paper on butadiene using Roothaan's LCAO self-consistent-field method, to which Coulson replied that he believed Parr's work on butadiene to be very similar to what his group had been doing.
35. In the early and mid 1950s tabulating and publishing lists with the numerical values of integrals appearing in molecular calculations became an undertaking expected to give specific results. One, among many, of the outcomes of such frantic activity was the publication in 1965 of the rather impressive *Dictionary of π-electron Calculations* by Coulson and A. Streitwieser, among others, from the University of California. This book had an interesting history. In 1954, Coulson together with Raymond Daudel in Paris had prepared a publication that they called *Dictionary of Values of Molecular Constants*. It was privately (cyclostyled) circulated and included the results of numerical calculations for a wide variety of simple π-electron systems, both hydrocarbons and heteromolecules. A new edition came out in 1959, and in 1961 Coulson decided that there should be a proper commercial publication. In the meantime, Streitwieser was preparing an analogous volume at Berkeley that would have included a greater variety of hydrocarbon molecules than those envisaged by Coulson. They decided to join forces and prepare two volumes. One would include data from the relatively simple molecules and the other data from complicated ones. Only one of the volumes eventually came out. The volume included an introduction to molecular orbital theory and proceeded to derive the formulas for calculating energy levels, bond orders, free valences, and polarizabilities. They considered 150 linear and cyclic polyacenes, and the data they included had been recomputed at the Mercury Computer at the Oxford University Computing Laboratory, and the printed output had been photographed directly to avoid errors in trans-

ferring data. In 1965, Streitwieser and J. I. Brauman published the *Supplemental Tables of Molecular Orbital Calculations* in two volumes. The second volume included the whole of the dictionary, and thus the whole enterprise must have had Coulson's consent. The volumes included various quantities for 460 organic compounds, the calculations being done with the IBM 704 at the University of California.

36. See papers included in Parr and Crawford (1952).
37. MacP, Letter B. L. Crawford to MacInnes, October 4, 1951.
38. CP, Ms. Coulson 154, Box G.15, G.15.5. Letter Mulliken to Coulson, October 17, 1951.
39. MacP, Letter Edwin Wilson to Mulliken, February 28, 1952.
40. MacP, Letter Mulliken to Parr and Crawford, March 3, 1952.
41. MacP. Draft of the final report by R. G. Parr and B. L. Crawford.
42. "Minutes of informal meetings on molecular integral problems. First session and Second Session," in *Symposium on Molecular Physics* (1954, 109–119).
43. Tributes to Löwdin and assessments of his contributions are included in Calais, Goscinski, Lindenberg, and Öhrn (1976) and Brandas and Kryachko (2003). Lindenberg (2002) used the phrase "quantum chemistry as a lifestyle," which we borrowed for our title, to characterize Löwdin's scientific style and personality.
44. Table with suggested budget is included in Fröman and Lindenberg (2007, 10).
45. They were recorded and summarized by Hall. A rather complete version of the proceedings came out in the Technical Note 16 issued by the Quantum Chemistry Group and known unofficially as Acta Valadalensia. A selection of the discussions was published for Löwdin's *Festschrift* (Fröman and Goscinski 1976). See also Hall (1959).
46. The Sanibel Symposia were supported by the U. S. Air Force Office of Scientific Research, and, later on, those devoted to computational quantum chemistry were supported by IBM.
47. Since their beginning, one day of a Sanibel Symposium was dedicated to Quantum Biology. After 1974, special symposia on quantum biology were organized, and in 1979 the odd-year meeting honored Bernard Pullman. In 1976, the Sanibel Symposia were renamed International Symposia on "Atomic, Molecular and Solid-State Theory, Collision Phenomena, Quantum Statistics and Computational Methods."
48. Subsequent Sanibel meetings used often the new journal as an outlet for their communications. Later, Löwdin also published another journal titled *Advances in Quantum Chemistry*.
49. Löwdin enrolled in theoretical reflections on the state, aims, and methods of quantum chemistry in many of his publications. Among those, see Löwdin (1957, 1977, 1986, 1986a).
50. After the declaration of martial law, Roothaan decided to return to the family's home in Nijmegen. He and his brother John were made prisoners when the German security police and the collaborationist Dutch police came after his brother Victor, who had been involved in underground activities and had already fled from home thereby escaping imprisonment.
51. For the impact of computers, see Park (1999, 2003). Participants also offered their reflections. Examples include Clementi (1967), Ohno (1978), Ostlund (1979).
52. For an extended discussion of the role of computations in the early years of quantum chemistry, see Park (2009).
53. In the paper "Broken Bottlenecks and the Future of Molecular Quantum Mechanics," coauthored with Roothaan, they voiced: "we think it is no exaggeration to say that the workers in

this field are standing on the threshold of a new era" (Mulliken and Roothaan 1959, 398). In his autobiography, he spoke about the "dawn of a new era" (Mulliken 1989, 159).

54. Mulliken was in the audience of the Symposium on Molecular Structure and Spectroscopy, held at Ohio State University on June 9, 1952, when the result was first presented and expressed his opinion about how significant he considered this development to be (Pariser 1990, 321).

55. When Longuet-Higgins succeeded Lennard-Jones as professor of theoretical chemistry in 1954, Boys remained a lecturer. Their scientific styles were very different, and there were often conflicts between the two, "sometimes quite painful to watch" (Handy, Pople, and Shavitt 1996, 6014).

56. For more information on computational chemistry in the United Kingdom see Smith and Sutcliffe (1997).

57. Sessions dealt with atoms and small molecules, the many-body problem, density matrices, methods to deal with atoms in molecules, complex molecules, nature of the chemical bond, problems in structure and spectra, spectroscopy methods, reaction rates, and intermolecular forces.

58. By 1990, Martin Karplus (born 1930) suggested replacing the two-dimensional Pople diagram by a three-dimensional one including as an extra dimension the estimated accuracy of calculation for the system under consideration. At the same time, he changed the linear scale of the axis in Pople's diagram representing the size of the molecule (which covered 1–100 electrons) by a logarithmic scale going up to 106 electrons. This change highlighted the possibility of conducting ab initio computations at a satisfactory accuracy for reasonably complex molecules and their reactions. Furthermore, Karplus recognized that density functional methods appeared to violate the "hyperbola of quantum chemistry" in the sense that they fall within the range of accuracy and sophistication of Hartree–Fock type calculations but handle molecules with a larger number of electrons within available computer time (Karplus 1990).

59. This project was suggested in 1962 by Harrison Shull and initiated in 1963 by quantum chemists Keith Hall and Frank Presser. Since 1966, it was running essentially as a collection and distribution agency, without a quantum chemist in charge (the staff consisted of one secretary, three work-study students, and one quantum chemist who spent one fourth of his time preparing the newsletter). Programs submitted were not critically evaluated but were tested to ensure that they performed as stated by the submitter. There was no interaction with users requesting programs; there was no one in QCPE with whom to consult for advice (National Academy of Sciences 1971, 14).

Chapter 5

1. PP, Box 242, Popular Scientific Lectures 1925–1955, "Recent Work on the Structure of Molecules," Talk given to the Southern Section of the American Chemical Society, 1936.

2. PP, Box 242, Popular and Scientific Lectures 1925–1955, "Resonance and Organic Chemistry," 1941.

3. In fact, human computers, often females or students, gave way to computing machines, increasingly fast and potent. From the ENIAC, the acronym for Electronic Numerical Integrator and Computer, built in 1945, to the EDVAC, the acronym for Electronic Discrete Variable Arith-

metic Computer, built in 1952, and the UNIVAC, the acronym for Universal Automatic Computer, generations of computers replaced former ones at an amazing pace.

4. The eminently experimental side of all chemistry, quantum chemistry included, was reiterated in another contribution to the conference (Zahradnik 1971), which also stressed the minor role the quantum theory of the chemical bond had played to date in the domain of chemical reactions.

5. This is very clear, for instance, when one contrasts the impact of both Hückel and Hellmann in popularizing their ideas among chemists vis-à-vis the efforts by, let us say, Pauling.

6. This situation slowly changed after the 1970s when some participants began discussions about issues involving philosophical aspects and later even began contributing to philosophy of chemistry journals.

7. Note that in the 1977 symposium "Quantum Chemistry—A Scientific Melting Point" there was a section especially devoted to philosophical issues in the quantum sciences, and R.G. Woolley was one of the participants. See, for example, Löwdin, Calais, and Goscinski (1978). Primas and Josep Del Re are also instances of quantum chemists turned philosophers of science, and Eric Scerri was invited to participate in the volumes *Fundamental World of Quantum Chemistry. A Tribute to the Memory of P-O Löwdin*. See Brandas and Kryachko (2003). And many other chemists (the Nobel laureate Roald Hoffmann and Pierre Lazlo, to name just a few) have been active players in the emergence and development of philosophy of chemistry. So there seems to be a strong interaction between the two disciplines.

Bibliography

Abe, Y. 1981. Pauling's revolutionary role in the development of quantum chemistry. *Historia Scientiarum* 20:107–124.

Altmann, S. L., and E. J. Bowen. 1974. Charles Alfred Coulson 1910–1974. *Biographical Memoirs of Fellows of the Royal Society. Royal Society (Great Britain)* 20:75–134.

Anderson, P. W. 1972. More is different. *Science* 177:393–396.

Arabatzis, T. 2006. How the electrons spend their leisure time: the chemists' perspective. In *Representing electrons. A biographical approach to theoretical entities*, 175–199. Chicago: University of Chicago Press.

Arabatzis, T., and K. Gavroglu. 1997. The chemists' electron. *European Journal of Physics* 18:150–163.

Armit, J. W., and R. Robinson. 1925. Polynuclear heterocyclic aromatic types. Part II. Some anhydronium bases. *Journal of the Chemical Society (London)* 127:1604–1618.

Arndt, F., and B. Eistert. 1936. Über den Chemismus der Claisen-Kondensation. *Berichte der Deutschen Chemischen Gesellschaft* 69:2381–2398.

Arsem, W. C. 1914. A Theory of valency and molecular structure. *Journal of the American Chemical Society* 36:1655–1675.

Assmus, A. 1991. *Molecular structure and the genesis of the American quantum physics community, 1916–1926.* Ph.D. dissertation, Harvard University.

Assmus, A. 1992. The molecular tradition in early quantum theory. *Historical Studies in the Physical Sciences* 22:209–231.

Assmus, A. 1993. The Americanization of molecular physics. *Historical Studies in the Physical Sciences* 23:1–33.

Baird, D., E. R. Scerri, and L. C. McIntyre, eds. 2006. *Philosophy of chemistry. Synthesis of a new discipline*. Dordrecht: Kluwer Academic Publishers.

Bantz, D. A. 1980. The structure of discovery: evolution of structural accounts of chemical bonding. In *Scientific discovery: case studies*, ed. T. Nickles, 291–329. Dordrecht: Reidel Publishing Company.

Barkan, D. K. 1999. *Walther Nernst and the transition to modern physical science*. Cambridge, UK: Cambridge University Press.

Barriol, J. 1966. *Elements de méchanique quantique*. Paris: Masson et Cie.

Barriol, J. 1971. *Elements of quantum mechanics with chemical applications*. New York: Barnes & Nobles.

Bauer, E. 1933. *Introduction à la théorie des groupes et à ses applications à la mécanique quantique*. Paris: Presses universitaires de France.

Bensaude-Vincent, B. 2001. The construction of a scientific discipline: materials science in the United States. *Historical Studies in the Physical Sciences* 31:223–248.

Bernal, M. J. M., and S. F. Boys. 1952. Electronic wave functions. VII. Methods of fundamental coefficients for the expansion of vector-coupled Schrödinger integrals and some values of these. *Philosophical Transactions of the Royal Society of London* 245:116–138.

Bernal, M. J. M., and S. F. Boys. 1952a. Electronic wave functions. VIII. A calculation for the ground states Na⁺, Ne, F⁻. *Philosophical Transactions of the Royal Society of London* 245:139–154.

Berson, J. A. 1996. Erich Hückel – Pionier der organischen Quantenchemie: Leben, Wirken und späte Anerkennung. *Angewandte Chemie* 108:2923–2937.

Berson, J. A. 1996a. Erich Hückel, pioneer of organic quantum chemistry: reflections on theory and experiment. *Angewandte Chemie International Edition in English* 35:2750–2764.

Berson, J. A. 1999. *Chemical creativity. Ideas from the work of Woodward, Hückel, Meerwein, and others*. Weinheim: Wiley-VCH.

Berthier, G. 1953. Structure électronique du fulvène: étude par la méthode du champ moléculaire self-consistant. *Journal de Chimie Physique* 50:344–351.

Berthier, G., and J. Mich. 1993. Foreword. *Journal of Molecular Structure THEOCHEM* 85:1–2.

Berthier, G., C. A. Coulson, H. H. Greenwood, and A. Pullman. 1948. Structure électronique des hydrocarbures aromatiques à 4 noyaux benzéniques accolés. Etude par la méthode des orbitales moléculaires. *Comptes Rendus Hebdomadaires des Séances de l'Académie des Sciences* 226:1906–1908.

Bethe, H. 1929. Termaufspaltung in Kristallen. *Annalen der Physik* 3:133–208.

Bethe, H. 1930. Termaufspaltung in Kristallen. *Zeitschrift fur Physik* 60:248.

Birge, R. T. 1926. The energy levels of the carbon monoxide molecule. *Nature* 117:229–230.

Birge, R. T. 1926a. The band spectra of carbon monoxide. *Physical Review* 28:1157–1181.

Birge, R. T. 1926b. The structure of molecules. *Nature* 117:300–302.

Bishop, R. 2005. Patching physics and chemistry together. *Philosophy of Science* 72:710–722.

Blondel-Mégrelis, M. 2001. Between disciplines: Jean Barriol and the Theoretical Chemistry Laboratory in Nancy. In *Chemical sciences in the 20th century. Bridging boundaries*, ed. C. Reinhardt, 105–118. New York: Wiley-VCH.

Bogaard, P. A. 1978. The limitations of physics as a chemical reducing agent. *Philosophy of Science Association* 2:345–356.

Bogaard, P. A. 2003. How G.N. Lewis reset the terms of the dialogue between chemistry and physics. *Annals of the New York Academy of Sciences* 988:307–312.

Bohr, N. 1913. On the constitution of atoms and molecules. *Philosophical Magazine* 26:857–875.

Bohr, N. 1924. *Theory of spectra and atomic constitution*. Cambridge, UK: Cambridge University Press.

Bolcer, D., and R. B. Herman. 1994. Development of computational chemistry in the USA. *Reviews in Computational Chemistry* 5:1–63.

Born, M. 1930. Zur Quantentheorie der chemischen Kräften. *Zeitschrift fur Physik* 64:729–740.

Born, M. 1930a. Berichtung. *Zeitschrift fur Physik* 65:718.

Born, M. 1960. *The mechanics of the atom* [transl. from German by J.W. Fischer and revised by D.R. Hartree, 1st German edition 1924]. New York: F. Ungar Pub. Co.

Born, M., and E. Hückel. 1923. Zur Quantentheorie mehratomiger Molekeln. *Zeitschrift fur Physik* 24:1–12.

Bowler, P. J. 2001. *Reconciling science and religion. The debate in early-twentieth-century Britain*. Chicago: University of Chicago Press.

Boys, S. F. 1950. Electronic wave functions. I. A general method of calculation for stationary states of any molecular system. *Proceedings of the Royal Society of London. Series A* 200:542–554.

Boys, S. F. 1950a. Electronic wave functions. II. A calculation for the ground state of the beryllium atom. *Proceedings of the Royal Society of London. Series A* 201:125–137.

Boys, S. F. 1951. Electronic wave functions. III. Some theorems on integrals of antisymmetric functions of equivalent orbital form. *Proceedings of the Royal Society of London. Series A* 206:125–137.

Boys, S. F. 1951a. Electronic wave functions. IV. Some general theorems for the calculation of Schrödinger integrals between complicated vector-coupled functions for many-electron atoms. *Proceedings of the Royal Society of London. Series A* 207:181–197.

Boys, S. F. 1951b. Electronic wave functions. V. Systematic reduction methods for all Schrödinger integrals of conventional systems of antisymmetric vector-coupled functions. *Proceedings of the Royal Society of London. Series A* 207:197–215.

Boys, S. F. 1952. Electronic wave functions. VI. Some theorems facilitating the evaluation of Schrödinger integrals of vector-coupled functions. *Philosophical Transactions of the Royal Society of London* 245:95–111.

Boys, S. F. 1952a. Electronic wave functions. IX. Calculations for the three lowest states of the beryllium atom. *Proceedings of the Royal Society of London. Series A* 217:136–150.

Boys, S. F. 1952b. Electronic wave functions. X. A calculation of eight variational poly-detor wavefunctions for boron and carbon. *Proceedings of the Royal Society of London. Series A* 217:235–251.

Boys, S. F. 1960. Mathematical problems in the complete quantum prediction of chemical phenomena. *Reviews of Modern Physics* 32:285–295.

Boys, S. F., and V. E. Price. 1954. Electronic wave functions. XI. A calculation of eight variational wavefunctions for Cl, Cl-, S and S-. *Philosophical Transactions of the Royal Society of London* 246:451–462.

Boys, S. F., and R. C. Sahni. 1954. Electronic wave functions. XII. The evaluation of general vector-coupling coefficients by automatic computation. *Philosophical Transactions of the Royal Society of London* 246:463–479.

Boys, S. F., G. B. Cook, C. M. Reeves, and I. Shavitt. 1956. Automatic fundamental calculations of molecular structure. *Nature* 178:1207–1209.

Bragg, W. H. 1935. Opening survey. In *International conference on physics. Paper and discussions. Volume II. The solid state of matter*, 1–6. London: Physical Society.

Brandas, E. J., and E. S. Kryachko, eds. 2003–2004. *Fundamental world of quantum chemistry. A tribute to the memory of Per-Olov Löwdin, 3 vols.* Dordrecht: Kluwer Academic Publishers.

Brion, H., R. Daudel, and S. Odiot. 1954. Théorie de la localisabilité des corpuscules. II. Niveaux élécroniques des átomes et énérgies associées. *Journal de Chimie Physique* 51:358–360.

British Association for the Advancement of Science. 1884. *Proceedings*. London: John Murray.

British Association for the Advancement of Science. 1907. *Proceedings*. London: John Murray.

Brock, W. H. 1992. *The Norton history of chemistry*. New York: W. W. Norton & Company.

Brush, S. G. 1970. Lennard-Jones, John Edward. *Dictionary of Scientific Biography* 8:185–187.

Brush, S. G. 1999. Dynamics of theory change in chemistry, part 1: the benzene problem 1865–1945. *Studies in History and Philosophy of Science* 30:21–79.

Brush, S. G. 1999a. Dynamics of theory change in chemistry, part 2: benzene and molecular orbitals 1945–1980. *Studies in History and Philosophy of Science* 30:263–302.

Buchwald, J. Z., and A. Warwick, eds. 2001. *Histories of the electron. The birth of microphysics.* Cambridge, MA: MIT Press.

Buckingham, A. D. 1987. Quantum chemistry. In *Schrödinger. Centenary celebration of a polymath*, ed. C. W. Kilmister, 112–118. Cambridge, UK: Cambridge University Press.

Burrau, O. 1927. Berechnung des Energiewertes des Wasserstoffmolekel-Ions (H_2^+). *Videnskabernes Selskab Matematisk-Fysiske Meddelelser* 7:1–18.

Busham, N., and S. M. Rosenfeld, eds. 2000. *Of minds and molecules. New philosophical perspectives on chemistry.* Oxford: Oxford University Press.

Butler, L. J. 1994. Robert S. Mulliken and the politics of science and scientists. *Historical Studies in the Physical Sciences* 25:25–35.

Calais, J.-L., O. Goscinski, J. Linderberg, and Y. Öhrn, eds. 1976. *Quantum science. Methods and structure. A tribute to Per-Olov Löwdin.* New York: Plenum Press.

Cambridge Historical Register Supplement 1931–1940. Cambridge, UK: Cambridge University Press.

Cartwright, N. 1987. Philosophical problems of the quantum theory: the response of American physicists. In *The probabilistic revolution.* Vol. 1. ed. L. Krüger and G. Gigerenzer, 417–435. Cambridge, MA: MIT Press.

Carson, C. 1996. The peculiar notion of exchange forces – I, origins in quantum mechanics, 1926–1928. *Studies in the History and Philosophy of Modern Physics* 27:23–45.

Cassebaum, H., and G. B. Kauffman. 1971. The periodic system of chemical elements: the search for its discoverers. *Isis* 62:314–327.

Ceruzzi, P. E. 1998. *A history of modern computing.* Cambridge, MA: MIT Press.

Chamberlin, T. C. 1897. The method of multiple working hypotheses. *Journal of Geology* 5:837–848.

Charpentier-Morize, M. 1997. *Jean Perrin (1870–1942). Savant et homme politique.* Paris: Belin.

Clark, G. L. 1928. A symposium on atomic structure and valence. An introduction. *Chemical Reviews* 5:361–364.

Clementi, E. 1967. Chemistry and computers. *International Journal of Quantum Chemistry* 1S:307–312.

Clementi, E. 1972. Computation of large molecules with the Hartree-Fock model. *Proceedings of the National Academy of Sciences of the United States of America* 69:2942–2944.

Compton, K. T. 1927. Specialization and cooperation in scientific research. *Science* 66:435–442.

Condon, E. U. 1927. Wave mechanics and the normal state of the hydrogen molecule. *Proceedings of the National Academy of Sciences of the United States of America* 13:466–470.

Condon, E. U., and G. H. Shortley. 1935. *The theory of atomic spectra.* Cambridge, UK: Cambridge University Press.

Coulson, C. A. 1935. The electronic structure of H_3^+. *Proceedings of the Cambridge Philosophical Society* 31:244–259.

Coulson, C. A. 1937. The evaluation of certain integrals occurring in studies of molecular structure. *Proceedings of the Cambridge Philosophical Society* 33:104–110.

Coulson, C. A. 1937a. A note on the criterion of maximum overlapping of wave functions. *Proceedings of the Cambridge Philosophical Society* 33:111–114.

Coulson, C. A. 1937b. The electronic structure of methane. *Transactions of the Faraday Society* 33:388–398.

Coulson, C. A. 1937c. The energy and screening constants of the hydrogen molecule. *Transactions of the Faraday Society* 33:1479–1492.

Coulson, C. A. 1938. The electronic structure of some polyenes and aromatic molecules. IV. The nature of the links of certain free radicals. *Proceedings of the Royal Society of London. Series A* 164:383–396.

Coulson, C. A. 1938a. Self-consistent field for molecular hydrogen. *Proceedings of the Cambridge Philosophical Society* 34:204–212.

Coulson, C. A. 1939. The electronic structure of some polyenes and aromatic molecules. VII. Bonds of fractional order by the molecular orbital method. *Proceedings of the Royal Society of London. Series A* 169:413–428.

Coulson, C. A. 1939a. The lengths of the links of unsaturated hydrocarbon molecules. *Journal of Chemical Physics* 7:1069–1071.

Coulson, C. A. 1940. On the calculation of the energy in unsaturated hydrocarbon molecules. *Proceedings of the Cambridge Philosophical Society* 36:201–203.

Coulson, C. A. 1941. Momentum distribution in molecular systems. Part I. The single bond. *Proceedings of the Cambridge Philosophical Society* 37:55–66.

Coulson, C. A. 1941a. Momentum distribution in molecular systems. Part III. Bonds of higher order. *Proceedings of the Cambridge Philosophical Society* 37:74–81.

Coulson, C. A. 1941b. Quantum theory of the chemical bond. *Proceedings of the Royal Society of Edinburgh* A61:115–139.

Coulson, C. A. 1941c. *Waves*. Edinburgh and London: University Mathematical Texts.

Coulson, C. A. 1942. Two-center integrals occurring in the theory of molecular structure. *Proceedings of the Cambridge Philosophical Society* 38:210–223.

Coulson, C. A. 1946. Two communications in General Discussion on bond number and peroxides. *Transactions of the Faraday Society* 42:265.

Coulson, C. A. 1947. Introductory survey – the theory of free radicals. The theory of the structure of free radicals. *Discussions of the Faraday Society* 2:7–8, 9–18.

Coulson, C. A. 1947a. The meaning of resonance. *Endeavour* 6:42–47.

Coulson, C. A. 1947b. Representation of simple molecules by molecular orbitals. *Quarterly Review of the Royal Society of London* 1:144–178.

Coulson, C. A. 1948. Wave mechanics in physics, chemistry and biology. *Science Progress* 36:436–449.

Coulson, C. A. 1948a. *Electricity*. Edinburgh and London: University Mathematical Texts.

Coulson, C. A. 1949. Free valence in organic reactions. *Journal of Chemical Physics* 45:243–248.

Coulson, C. A. 1950. The Christian religion and contemporary science. *Modern Churchman* 40(September):205–215.

Coulson, C. A. 1950a. Significant progress and present problems. *Discussions of the Faraday Society* 9:1–4.

Coulson, C. A. 1951. Critical survey of the method of ionic homopolar resonance. *Proceedings of the Royal Society of London. Series A* 207:63–73.

Coulson, C. A. 1952. *Valence*. 1st ed. Oxford: Clarendon Press.

Coulson, C. A. 1952a. What is a chemical bond? *Scientific Journal of the Royal College of Science* 21:11–29.

Coulson, C. A. 1953. *Christianity in an age of science*. London: Oxford University Press.

Coulson, C. A. 1953a. *The spirit of applied mathematics*. Oxford: Clarendon Press.

Coulson, C. A. 1954. Present tendencies in the theory of large molecules. In *Symposium on molecular physics. Held at Nikko on the occasion of the International Conference on Theoretical Physics*, 23–33. Tokyo: Maruzen Co., Ltd.

Coulson, C. A. 1955. *Science and Christian belief*. London: Oxford University Press.

Coulson, C. A. 1955a. The contributions of wave mechanics to chemistry. (The Tilden Lecture.) *Journal of the Chemical Society* 2069–2084.

Coulson, C. A. 1958. *Science and the idea of God*. Cambridge, UK: Cambridge University Press.

Coulson, C. A. 1958–1959. Mathematics and the real world. In *Golden Jubilee Commemoration, Part II*, 261–269. Calcutta: Calcutta Mathematical Society.

Coulson, C. A. 1960. "Present state of molecular structure calculations." Conference on molecular quantum mechanics, University of Colorado at Boulder, June 21–27. *Reviews of Modern Physics* 32:170–177.

Coulson, C. A. 1961. *Valence*. 2nd ed. Oxford: Oxford University Press.

Coulson, C. A. 1964. R.S. Mulliken – his work and influence on quantum chemistry. In *Molecular orbitals in chemistry, physics and biology – a tribute to R.S. Mulliken*, ed. P.-O. Löwdin and B. Pullman, 1–15. New York: Academic Press.

Coulson, C. A. 1969. "On liking mathematics." Presidential Address to the Mathematical Association, Cambridge, April 9, 1969. *Mathematical Gazette* LII:227–239.

Coulson, C. A. 1969a. What makes a beautiful mathematical proof? *Bulletin of the Mathematical Association of India* 1(3):39–44.

Coulson, C. A. 1970. Recent developments in valence theory. Symposium "Fifty years of valence theory." *Pure and Applied Chemistry* 24:257–287.

Coulson, C. A. 1973. *The shape and structure of molecules*. Oxford: Oxford Chemistry Series.

Coulson, C. A. 1973a. Samuel Francis Boys 1911–1972. *Biographical Memoirs of Fellows of the Royal Society. Royal Society (Great Britain)* 19:94–115.

Coulson, C. A. 1974. *Theoretical chemistry past and future* (S. L. Altmann, ed). Oxford: Clarendon Press.

Coulson, C. A., and M. P. Barnett. 1951. The evaluation of integrals occurring in the theory of molecular structure. I and II. *Philosophical Transactions of the Royal Society of London. Series A* 243:221–249.

Coulson, C. A., and R. Daudel, eds. 1954. *Dictionary of values of molecular constants: calculated theoretically by wave mechanical methods (molecular orbital method)*. Paris: Centre de Chimie Théorique de France.

Coulson, C. A., and W. E. Duncanson. 1941. Momentum distribution in molecular systems. Part II. Carbon and the C-H bond. *Proceedings of the Cambridge Philosophical Society* 37:67–73.

Coulson, C. A., and I. Fischer. 1949. Notes on the molecular orbital treatment of the hydrogen molecule. *Philosophical Magazine* 40:386–393.

Coulson, C. A., and J. Jacobs. 1951. Electronic levels in simple conjugated systems. II. Butadiene. *Proceedings of the Royal Society of London. Series A* 206:287–296.

Coulson, C. A., and H. C. Longuet-Higgins. 1947. The electronic structure of conjugated systems. I. General theory. *Proceedings of the Royal Society of London. Series A* 191:39–60.

Coulson, C. A., and H. C. Longuet-Higgins. 1947a. The electronic structure of conjugated systems. II. Unsaturated hydrocarbons and their hetero-derivatives. *Proceedings of the Royal Society of London. Series A* 192:16–32.

Coulson, C. A., and H. C. Longuet-Higgins. 1948. The electronic structure of conjugated systems. III. Bond orders in unsaturated molecules. *Proceedings of the Royal Society of London. Series A* 193:447–456.

Coulson, C. A., and H. C. Longuet-Higgins. 1948a. The electronic structure of conjugated systems. IV. Force constants and interaction constants in unsaturated hydrocarbons. *Proceedings of the Royal Society of London. Series A* 193:456–464.

Coulson, C. A., and H. C. Longuet-Higgins. 1948b. The electronic structure of conjugated systems. V. The interaction of two conjugated systems. *Proceedings of the Royal Society of London. Series A* 195:188–197.

Coulson, C. A., and W. E. Moffitt. 1949. The properties of certain strained hydrocarbons. *Philosophical Magazine* 40:1–35.

Coulson, C. A., and G. S. Rushbrooke. 1940. Note on the method of molecular orbitals. *Proceedings of the Cambridge Philosophical Society* 36:193–200.

Coulson, C. A., D. P. Craig, and J. Jacobs. 1951. Electronic levels in simple conjugated systems. III. The significance of configuration interaction. *Proceedings of the Royal Society of London. Series A* 206:297–308.

Coulson, C. A., B. O'Leary, and R. B. Mallion. 1978. *Hückel theory for organic chemists*. London: Academic Press.

Coulson, C. A., A. Streitwieser, M. D. Poole, and J. I. Brauman. 1965. *Dictionary of π-electron calculations*. San Francisco: Freeman and Co.

Craig, D. P. 1950. Electronic levels in simple conjugated systems. I. Configuration interaction in cyclobutadiene. *Proceedings of the Royal Society of London. Series A* 202:498–506.

Davenport, D. A. 1980. Linus Pauling—chemical educator. *Journal of Chemical Education* 57:35–37.

Darden, L. 1991. *Theory change in science. Strategies from Mendelian genetics*. Oxford: Oxford University Press.

Darwin, C. G. 1927. The electron as a vector wave. *Proceedings of the Royal Society of London* 116:227–253.

Darwin, C. G. 1958. Douglas Rainer Hartree. *Biographical Memoirs of the Fellows of the Royal Society of London* 4:103–116.

Darwin, C. G., and R. H. Fowler. 1922. The partition of energy. *Philosophical Magazine* 44:450–479.

Darwin, C. G., and R. H. Fowler. 1922a. The partition of energy. II. *Philosophical Magazine* 44:823–842.

Darwin, C. G., and R. H. Fowler. 1922b. Fluctuations in an assembly in statistical equilibrium. *Proceedings of the Cambridge Philosophical Society* 21:391–404.

Darwin, C. G., and R. H. Fowler. 1923. Some refinements of the theory of dissociation equilibrium. *Proceedings of the Cambridge Philosophical Society* 21:730–745.

Daudel, P. 1946. Sur la méthode des diagrammes moléculaires. *Comptes Rendus Hebdomadaires des Séances de l'Académie des Sciences* 223:947–948.

Daudel, R. 1952. Remarque sur le role de l'indiscernabilité des électrons en chimie théorique. *Comptes Rendus Hebdomadaires des Séances de l'Académie des Sciences* 235:886–888.

Daudel, R. 1956. *Les Fondements de la chimie théorique*. Paris: Gauthiers Villars. [Translated into Daudel, R. 1968. *The fundamentals of theoretical chemistry. Wave mechanics applied to the study of atoms and molecules*. New York: Pergamon Press.]

Daudel, R. 1973. The localization of electrons and the concept of the chemical bond. In *Wave mechanics. The first fifty years*, ed. W. C. Price, S. S. Chissick, and T. Ravensdale, 61–71. London: Butterworth Group.

Daudel, R. 1992. About the nature of the chemical bond. *Journal of Molecular Structure THEOCHEM* 261:113–114.

Daudel, R. 1992a. Quantum chemistry. *Encyclopedia of Physical Science and Technology* (vol 13, pp 629–640). New York: Academic Press.

Daudel, R., and A. Pullman. 1945. Sur le calcul de la répartition du nuage électronique dans les molécules aromatiques. *Comptes Rendus Hebdomadaires des Séances de l'Académie des Sciences* 220:888–889.

Daudel, R., and A. Pullman. 1946. Sur une méthode génerale de construction de diagrammes moléculaires. *Comptes Rendus Hebdomadaires des Séances de l'Académie des Sciences* 222:663–664.

Daudel, R., and A. Pullman. 1946a. L'étude des molécules par la méthode de la Mésomérie. *Journal of Physics* 7:59–64, 74–83, 105–111.

Daudel, R., and P. Salzedo [the future P. Daudel]. 1944. A propos de la variation de la covalaffinité d'un element avec sa valence. *Comptes Rendus Hebdomadaires des Séances de l'Académie des Sciences* 28:972–973.

Daudel, R., H. Brion, and S. Odiot. 1955. Localisability of electrons in atoms and molecules. Application to the study of the notion of shell and of the nature of chemical bonds. *Journal de Chimie Physique* 23:2080–2083.

Daudel, R., A. Bucher, and H. Moreu. 1944. Une nouvelle méthode d'étude des valences dirigées. Son application à la détermination de la structure des pentahalogénures de phosphore. *Comptes Rendus Hebdomadaires des Séances de l'Académie des Sciences* 218:917–918.

Daudel, R., R. Lefebvre, and C. Moser. 1959. *Quantum chemistry. Methods and applications*. New York: Interscience Publishers.

Daudel, R., G. Leroy, D. Peeters, and M. Sana. 1983. *Quantum chemistry*. New York: John Wiley & Sons.

Daudel, R., S. Odiot, and H. Brion. 1954. Théorie de la localisabilité des corpuscules. I. La notion de loge et la signification géométrique de la notion de conche dans le cortège électronique des atomes. *Journal de Chimie Physique* 51:74–77.

Davidson, E. R. 1984. General introduction: computational quantum chemistry – 1984. *Faraday Symposium Chemical Society* 19:2–15.

Debye, P., and L. Pauling. 1925. The inter-ionic attraction of ionized solutes. IV. The influence of variation of dielectric constant on the limiting law for small concentrations. *Journal of the American Chemical Society* 47:2129–2134.

Devonshire, A. F. 1936. The interaction of atoms and molecules with solid surfaces III. The condensation and evaporation of atoms and molecules. *Proceedings of the Royal Society of London. Series A* 156:6–28.

Devonshire, A. F. 1936a. The interaction of atoms and molecules with solid surfaces V. The diffraction and reflection of molecular rays. *Proceedings of the Royal Society of London. Series A* 156:37–44.

Dewar, M. J. S. 1969. *The molecular orbital theory of organic molecules*. New York: McGraw-Hill.

Dickinson, R. G., and L. Pauling. 1923. The crystal structure of molybdenite. *Journal of the American Chemical Society* 45:1466–1471.

Dirac, P. A. M. 1926. On the theory of quantum mechanics. *Proceedings of the Royal Society of London. Series A* 112:661–677.

Dirac, P. A. M. 1928. Quantum theory of the electron. *Proceedings of the Royal Society of London. Series A* 117:610–624.

Dirac, P. A. M. 1929. Quantum mechanics of many-electron systems. *Proceedings of the Royal Society of London. Series A* 123:714–733.

Early, J. E., ed. 2003. Chemical explanation: characteristics, development, autonomy. *Annals of the New York Academy of Sciences* 988.

Everitt, C. W. F., and W. M. Fairbank. 1973. Fritz London. *Dictionary of Scientific Biography* 8:473–479.

Eyring, H., J. Walter, and G. Kimball. 1944. *Quantum chemistry*. New York: Wiley Publishers.

Fellows, F. H. 1985. *J.H. Van Vleck: the early life and work of a mathematical physicist*. Ph.D. dissertation, University of Minnesota.

Fischer, C. F. 2004. *Douglas Rayner Hartree: his life in science and computing*. London: World Scientific Publishing.

Fock, V. 1930. Näherungsmethode zur Lösung des quantemechanischen Mehrkörperproblemes. *Zeitschrift fur Physik* 62:126–148.

Ford, B., and G. G. Hall. 1970. Model building – an educational philosophy for applied mathematics. *International Journal of Mathematical Education in Science and Technology* 1:77–83.

Fowkes, F. M. 1972. Harkins, William Draper. *Dictionary of Scientific Biography* 6:117–119.

Fowler, R. H. 1921. A simple extension of Fourier's integral theorem and some physical applications, in particular to the theory of quanta. *Proceedings of the Royal Society of London. Series A* 22:462–471.

Fowler, R. H. 1923. Bohr's atom in relation to the problem of covalency. *Transactions of the Faraday Society* 19:459–468.

Fowler, R. H. 1926. On dense matter. *Monthly Notices of the Royal Astronomical Society* 87:114.

Fowler, R. H. 1926a. General forms of statistical mechanics with special reference to the requirements of the new quantum mechanics. *Proceedings of the Royal Society of London. Series A* 113:432–449.

Fowler, R. H. 1927. Matrix and wave mechanics. *Nature* 119:239–241.

Fowler, R. H. 1929. *Statistical mechanics. The theory of the properties of matter in equilibrium based on an essay awarded the Adams Prize in the University of Cambridge 1923–1924*. Cambridge, UK: The University Press.

Fowler, R. H. 1932. A report on homopolar valency and its mechanical interpretation. In *Chemistry at the Centenary Meeting of the British Association for the Advancement of Science*, 226–246. Cambridge, UK: W. Heffer and Sons Ltd.

Freed, K. 1955. Building a bridge between ab initio and semi-empirical theories of molecular electronic structure. In *Structure and dynamics of atoms and molecules: Conceptual trends*, ed. J. L. Calais and E. S. Kryachko, 25–67. Dordrecht: Kluwer Academic Publishers.

Fröman, A., and O. Goscinski. 1976. Acta Valadalensia revisited: Per-Olov Löwdin in scientific discussion. In *Quantum science. Methods and structure. A tribute to Per-Olov Löwdin*, ed. J.-L. Calais, O. Goscinski, J. Linderberg, and Y. Öhrn, 33–51. New York: Plenum Press.

Fröman, A., and J. Lindenberg. 2007. *Inception of quantum chemistry at Uppsala*. Uppsala: Uppsala University.

Fry, H. S. 1928. A pragmatic system of notation for electronic valence conceptions in chemical formulas. *Chemical Reviews* 5:557–568.

Garner, W. E., and J. E. Lennard Jones. 1929. Molecular spectra and molecular structure. *Nature* 124:584–588.

Garner, W. E., and J. E. Lennard Jones. 1929a. Introduction – molecular spectra and molecular structure. A general discussion. *Transactions of the Faraday Society* 25:611–627.

Garner, W. E., and J. E. Lennard Jones. 1929b. Summary. *Transactions of the Faraday Society* 25:942–949.

Gavroglu, K. 1995. *Fritz London—a scientific biography*. Cambridge, UK: Cambridge University Press.

Gavroglu, K. 1996. Can theories of chemistry provide an argument against realism? In *Realism and anti-realism in the philosophy of science*, ed. R. S. Cohen, R. Hilpinen, and Q. Renzong, 149–170. Dordrecht: Kluwer Academic Publishers.

Gavroglu, K. 1997. Philosophical issues in the history of chemistry. *Synthese* 111:283–304.

Gavroglu, K., ed. 2000. Special issue. Theoretical chemistry in the making: appropriating concepts and legitimizing techniques. *Studies in the History and Philosophy of Science* 31B(4).

Bibliography

Gavroglu, K. 2000a. Controversies and the becoming of physical chemistry. In *Scientific controversies: philosophical and historical perspectives*, ed. P. K. Machamer, M. Pera, and A. Baltas, 177–198. Oxford: Oxford University Press.

Gavroglu, K. 2001. The physicists' electron and its appropriation by the chemists. In *Histories of the electron. The birth of microphysics*, ed. J. Buchwald and A. Warwick, 363–400. Cambridge, MA: MIT Press.

Gavroglu, K., and Y. Goudaroulis. 1991. *Through measurement to knowledge: the selected papers of Heike Kamerlingh Onnes 1853–1926*. Dordrecht: Kluwer Academic Publishers.

Gavroglu, K., and A. Simões. 1994. The Americans, the Germans and the beginnings of quantum chemistry: the confluence of diverging traditions. *Historical Studies in the Physical Sciences* 25:47–110.

Gavroglu, K., and A. Simões. 2000. One face or many? The role of textbooks in building the new discipline of quantum chemistry. In *Communicating chemistry. Textbooks and their audiences, 1789–1939*, ed. A. Lundgren and B. Bensaude-Vincent, 415–449. Canton, MA: Science History Publications.

Gavroglu, K., and A. Simões. 2002. Preparing the ground for quantum chemistry to appear in Great Britain: the contributions of the physicist R.H. Fowler and the chemist N.V. Sidgwick. *British Journal for the History of Science* 35:187–212.

Gold, A. 1961. Quantum chemistry summer institute. *Physics Today* 14(1):40–42.

Goldberg, S. 1968. *Early responses to Einstein's theory of relativity: a case study in national differences, 1905–1911*. Ph.D. dissertation, Harvard University.

Goldberg, S. 1984. *Understanding relativity: origins and impact of a scientific revolution*. Boston: Birkhäuser.

Goodstein, J. R. 1984. Atoms, molecules, and Linus Pauling. *Social Research* 51:691–708.

Goudsmit, S., and G. E. Uhlenbeck. 1925. Ersetzung der Hypothese von unmechanischen Zwang durch eine Forderung bezüglich des inneren Verhaltens jedes einzelnen Elektrons. *Naturwissenschaften* 13:953–954.

Goudsmit, S., and G. E. Uhlenbeck. 1926. Spinning electrons and the structure of spectra. *Nature* 107:264–265.

Graham, L. R. 1964. A Soviet Marxist view of structural chemistry: the theory of resonance controversy. *Isis* 55:20–31.

Guéron, J., and M. Magat. 1971. A history of physical chemistry in France. *Annual Review of Physical Chemistry* 22:1–23.

Haas, A. 1930. *Quantum chemistry. A short introduction in four non-mathematical lectures*. London: Constable & Company Ltd.

Hacking, I. 1985. Styles of scientific reasoning. In *Post-analytic philosophy*, ed. J. Rajchman and C. West, 145–165. New York: Columbia University Press.

Hacking, I. 1992. Styles for historians and philosophers. *Studies in History and Philosophy of Science* 23:1–20.

Hager, T. 1995. *Linus Pauling and the chemistry of life*. New York: Simon & Schuster.

Hager, T. 1998. *Force of nature: the life of Linus Pauling*. Oxford: Oxford University Press.

Hale, G. E. 1919. The national importance of scientific and industrial research. The purpose of the National Research Council. *National Research Council, Bulletin* 1:1–7.

Hall, G. G. 1950. The molecular orbital theory of chemical valency. VI. Properties of equivalent orbitals. *Proceedings of the Royal Society of London. Series A* 202:336–344.

Hall, G. G. 1951. The molecular orbital theory of chemical valency. VIII. A method of calculating ionization potentials. *Proceedings of the Royal Society of London. Series A* 205:541–552.

Hall, G. G. 1954. The electronic structure of *trans*-butadiene calculated by the standard excited state method. *Transactions of the Faraday Society* 50:773–779.

Hall, G. G. 1955. The bond orders of alternant hydrocarbon molecules. *Proceedings of the Royal Society of London. Series A* 229:251–259.

Hall, G. G. 1957. The bond orders of some conjugated hydrocarbon molecules. *Transactions of the Faraday Society* 53:573–581.

Hall, G. G. 1959. Chemistry and quantum chemistry. *Nature* 183:158–159.

Hall, G. G. 1972. Modelling—a philosophy for applied mathematicians. *Bulletin—Institute of Mathematics and its Applications* 8:226–228.

Hall, P. J. 1986. The Pauli exclusion principle and the foundation of chemistry. *Synthese* 69:267–272.

Hall, G. G. 1989. Some reminiscences of a Japanese civil servant. *Bulletin—Institute of Mathematics and its Applications* 25:162–164.

Hall, G. G. 1991. The Lennard-Jones paper of 1929 and the foundations of molecular orbital theory. *Advances in Quantum Chemistry* 22:1–6.

Hall, G. G. 1991a. Computational quantum chemistry—then and now. *Journal of Molecular Structure THEOCHEM* 234:13–18.

Hall, G. G. 1996. Invited Paper: Samuel Francis Boys 1911–1972. *Molecular Physics* 88:309–314.

Hall, G. G. 1999. Recollections and reflections. *International Journal of Quantum Chemistry* 74:439–454.

Hall, G. G., and J. Lennard-Jones. 1950. The molecular orbital theory of chemical valency. III. Properties of molecular orbitals. *Proceedings of the Royal Society of London. Series A* 202:155–165.

Hall, G. G., and J. Lennard-Jones. 1951. The molecular orbital theory of chemical valency. VII. Molecular structure in terms of equivalent orbitals. *Proceedings of the Royal Society of London. Series A* 205:357–374.

Handy, N. C., J. A. Pople, and I. Shavitt. 1996. Samuel Frances Boys. *Journal of Physical Chemistry* 100:6007–6016.

Harris, M. 2007. *The physico-chemical nature of the chemical bond: valence bonding and the path of physico-chemical emergence*. Ph.D. dissertation, University of Toronto.

Harris, M. 2008. Chemical reductionism revisited: Lewis, Pauling, and the physico-chemical nature of the chemical bond. *Studies in History and Philosophy of Science* 39:78–90.

Hartmann, H. 1954. *Theorie der chemischen Bindung auf quantentheoretischer Grundlage*. Berlin: Springer Verlag.

Hartmann, H., and H. C. Longuet-Higgins. 1982. Erich Hückel. 9 August 1896–16 February 1980. *Biographical Memoirs of Fellows of the Royal Society. Royal Society (Great Britain)* 28:153–162.

Hartree, D. R. 1923. On the propagation of certain types of electromagnetic waves. *Philosophical Magazine* 46:454–460.

Hartree, D. R. 1923a. On some approximate numerical applications of Bohr's theory of spectra. *Proceedings of the Cambridge Philosophical Society* 21:625–641.

Hartree, D. R. 1928. The wave mechanics of an atom with a non-Coulomb central field. Part I. Theory and methods. *Proceedings of the Cambridge Philosophical Society* 24:89–110.

Hartree, D. R. 1928a. The wave mechanics of an atom with a non-Coulomb central field. Part II. Some results and discussion. *Proceedings of the Cambridge Philosophical Society* 24:111–132.

Hartree, D. R. 1928b. The wave mechanics of an atom with a non-Coulomb central field. Part III. *Proceedings of the Cambridge Philosophical Society* 24:426–437.

Hartree, D. R. 1933. Results of calculations of atomic wave functions. I. Survey and self-consistent fields for Cl^- and Cu^+. *Proceedings of the Royal Society of London. Series A* 143:282–301.

Hartree, D. R. 1933a. Results of calculations of atomic wave functions. II. Results for K^+ and Cs^+. *Proceedings of the Royal Society of London. Series A* 143:506–517.

Hartree, D. R. 1935. The differential analyser. *Nature* 135:940–943.

Hartree, D. R. 1935a. Results of calculations of atomic wave functions. III. Results for *Be, Ca* and *Hg*. *Proceedings of the Royal Society of London. Series A* 149:210–231.

Hartree, D. R. 1936. Theory of complex atoms. *Nature* 138:1080–1082.

Hartree, D. R. 1937. The effect of configuration interaction on the low terms of the spectra of oxygen. *Proceedings of the Cambridge Philosophical Society* 33:240–249.

Hartree, D. R. 1947. *Calculating machines. Recent and prospective developments and their impact on mathematical physics*. Cambridge, UK: Cambridge University Press.

Hartree, D. R. 1948. A discussion on computing machines. A historical survey of digital computing machines. *Proceedings of the Royal Society of London. Series A* 195:265–271.

Hartree, D. R. 1948a. The calculation of atomic structures. *Reports on Progress in Physics* 11:113–143.

Hartree, D. R. 1957. *The calculations of atomic structures*. New York: Wiley.

Hartree, D. R., and W. Hartree. 1935. Self-consistent field with exchange, for beryllium. *Proceedings of the Royal Society of London. Series A* 150:9–33.

Heilbron, J. L. 1983. The origins of the exclusion principle. *Historical Studies in the Physical Sciences* 13:261–310.

Heisenberg, W. 1926. Mehrkörperproblem und Resonanz in der Quantenmechanik. *Zeitschrift fur Physik* 38:411–426.

Heisenberg, W. 1926a. Über die Spektra von Atomsystemen mit zwei Elektronen. *Zeitschrift fur Physik* 39:499–518.

Heisenberg, W. 1927. Mehrkorperprobleme und Resonanz in der Quantenmechanik. II. *Zeitschrift fur Physik* 41:239–267.

Heisenberg, W. 1932. Contribution to the discussion on the structure of simple molecules. In *Chemistry at the Centenary Meeting of the British Association for the Advancement of Science*, 247–248. Cambridge, UK: W. Heffer & Sons.

Heitler, W. 1928. Störungsenergie und Austausch beim Mehrkörperproblem. *Zeitschrift fur Physik* 46:47–72.

Heitler, W. 1928a. Zur Gruppentheorie der homöopolaren Chemischen Bindung. *Zeitschrift fur Physik* 47:835–858.

Heitler, W. 1936. *Quantum theory of radiation*. Oxford: Oxford University Press.

Heitler, W. 1955. The theory of chemical bond. *Arkiv für Fhysik* 10:145–156.

Heitler, W. 1956 [1st ed. 1945]. *Elementary wave mechanics with applications to chemistry*. Oxford: Oxford University Press.

Heitler, W. 1967. Quantum chemistry: the early period. *International Journal of Quantum Chemistry* 1:13–36.

Heitler, W., and F. London. 1927. Wechselwirkung neutraler Atome und homöopolare Bindung nach der Quantenmechanik. *Zeitschrift fur Physik* 44:455–472.

Heitler, W., and G. Poschl. 1934. Ground states of Cs and O_2 and the theory of valency. *Nature* 137:833–834.

Heitler, W., and G. Rumer. 1931. Quantentheorie der Chemischen Bindung in mehratomige Molecule. *Zeitschrift fur Physik* 68:12–31.

Hellmann, H. 1937. *Einfuhrung in die Quantenchemie*. Leipzig and Vienna: Franz Deuticke.

Hellmann, H., and W. Jost. 1934. Zum Verständnis der "chemischen Kräfte" nach der Quantenmechanik. *Z. Elektrochem. Angew. Phys. Chem.* 40:806–814.

Hendry, R. F. n/d. The metaphysics of chemistry. Available at: <http://www.dur.ac.uk/r.f.hendry>. Accessed January 8, 2008.

Hendry, J. 1981. Bohr-Kramers-Slater: a virtual theory of virtual oscillators and its role in the history of quantum mechanics. *Centaurus* 25:189–221.

Hendry, J. 1984. *The creation of quantum mechanics and the Bohr-Pauli dialogue*. Dordrecht: Reidel.

Hendry, R. F. 2001. Mathematics, representation and molecular structure. In *Tools and modes of representation in the laboratory sciences*, ed. U. Klein, 221–236. Dordrecht: Kluwer Academic Publishers.

Hendry, R. F. 2003. Autonomy, explanation, and theoretical values. Physicists and chemists on molecular quantum mechanics. *Annals of the New York Academy of Sciences* 988:44–58.

Hendry, R. F. 2004. The physicists, the chemists and the pragmatics of explanation. *Philosophy of Science* 71:1048–1059.

Hermann, A. 1978. Born, Max. *Dictionary of Scientific Biography* 15(1):39–44.

Hettema, H. 2000. *Quantum chemistry. Classic scientific papers*. London: World Scientific.

Hicks, B. 1937. The shape of the Compton line for helium and molecular hydrogen. *Physical Review* 52:436–442.

Hiebert, E. N. 1983. Walter Nernst and the application of physics to chemistry. In *Springs of scientific creativity: essays on founders of modern science*, ed. R. Aris, 203–231. Minneapolis: University of Minnesota Press.

Hildebrand, J. H. 1958. Gilbert Newton Lewis. *Biographical Memoirs. National Academy of Sciences (U. S.)* 31:209–235.

Hill, A. V. 1943. Note on W. Hartree, The ethical dilemma of science. *Nature* 152:154–156.

Hoffmann, R. 1971. Chemical reactivity of molecules in their excited states. In *Colloque International sur les Aspects de la Chimie Quantique Contemporaine*, 8–13 July 1970, Menton, France, org. R. Daudel and A. Pullman, 133–153. Paris: Éditions du Centre National de la Recherche Scientifique.

Holmen, R. 1990. Kasimir Fajans. *Bulletin for the History of Chemistry* 6:7–15.

Holton, G. 1988. On the hesitant rise of quantum physics research in the United States. In *Thematic Origins of Scientific Thought: Kepler to Einstein,* Rev. ed., 147–187. Cambridge, MA: Harvard University Press.

Hückel, E. 1925. Zur Theorie konzentrierter wässeriger Lösungen starker Elektrolyte. *Zeitschrift für Physik* 26:93–147.

Hückel, E. 1928. Adsorption und Kapillarkondensation. In *Kolloidforschung in Einzeldarstellungen*, Part 7. Leipzig: Akademische Verlagsgesellschaft.

Hückel, E. 1930. Zur Quantentheorie der Doppelbindung. *Zeitschrift für Physik* 60:423–456.

Hückel, W. 1931. *Theoretische Grundlagenden der organischen Chemie*. Leipzig: Akademische Verlagsgesellschaft.

Hückel, E. 1931. Quantentheoretische Beiträge zum Benzolproblem. I. Die Elektronenkonfiguration des Benzols und verwandter Verbindungen. *Zeitschrift für Physik* 70:204–286.

Hückel, E. 1931a. Quantentheoretische Beiträge zum Benzolproblem. II. Quantentheorie der induzierten Polaritäten. *Zeitschrift für Physik* 72:310–337.

Hückel, E. 1932. Quantentheoretische Beiträge zum Problem der aromatischen und ungesättigten Verbindungen. III. *Zeitschrift für Physik* 76:628–648.

Hückel, E. 1934. Theory of free radicals of organic chemistry. *Transactions of the Faraday Society* 30:40–52.

Hückel, E. 1935. Aromatic and unsaturated molecules: contributions to the problem of their constitution and properties. In *International conference on physics. Paper and discussions. Volume II. The solid state of matter*, 9–35. London: Physical Society.

Hückel, E. 1937. Grundzüge der Theorie ungesättigter und aromatischer Verbindungen. *Zeitschrift für Elektrochemie* 43:752–788.

Hückel, E. 1937a. Grundzüge der Theorie ungesättigter und aromatischer Verbindungen. *Zeitschrift für Elektrochemie* 43:827–849.

Hückel, E. 1957. Zur modernen Theorie ungesättigter und aromatischer Verbindungen. *Zeitschrift für Elektrochemie* 61:866–890.

Hund, F. 1926. Zur Deutung einiger Erscheinungen in den Molekelspektren. *Zeitschrift für Physik* 36:657–674.

Hund, F. 1927. *Linienspektren und periodisches System der Elemente*. Berlin: J. Springer.

Hund, F. 1927a. Zur Deutung der Molekelspektren. I. *Zeitschrift für Physik* 40:742–764.

Hund, F. 1927b. Zur Deutung der Molekelspektren. II. *Zeitschrift für Physik* 42:93–120.

Hund, F. 1929. Chemical binding. *Transactions of the Faraday Society* 25:646–648.

Hund, F. 1931. Zur Frage der chemischen Bindung. *Zeitschrift für Physik* 73:1–30.

Hund, F. 1933. Allgemeine Quantenmechanik des Atom – und Molekelbaues. In *Handbuch der Physik* (zweite Auflage), Vol. 24, 1st Part: *Quantentheorie*, 561–693. Berlin: Verlag von Julius Springer.

Hund, F. 1935. Description of the binding forces in molecules and crystal lattices on quantum theory. In *International Conference on Physics. Paper and discussions. Volume II. The solid state of matter*, 36–45. London: Physical Society.

Hund, F. 1974. *The history of quantum theory*. London: Harrap.

Hunsberger, I. M. 1954. Theoretical chemistry in Russia. *Journal of Chemical Education* 32:504–514.

Hylleraas, E. A. 1928. Über den Grundzustand des Heliumatoms. *Zeitschrift fur Physik* 48:469–494.

Hylleraas, E. A. 1929. Neue Berechnung der Energie des Heliums im Grundzustande, sowie des tiefsten Terms von Ortho-Helium. *Zeitschrift fur Physik* 54:347–366.

Hylleraas, E. A. 1963. Reminiscences from early quantum mechanics of two-electron atoms. *Reviews of Modern Physics* 35:421–431.

Ihde, A. J. 1964. *The development of modern chemistry*. New York: Harper & Row.

Ingold, C. K. 1934. Principles of an electronic theory of organic reactions. *Chemical Reviews* 15:225–274.

Institut International de Chimie Solvay. 1925. *Rapports et discussions sur cinq questions d'actualité: Premier Conseil de Chimie, 21 au 27 Avril 1922*. Paris: Gauthiers Villars.

Institut International de Chimie Solvay. 1926. *Structure et activité chimique. Rapports et discussions: Deuxieme Conseil de Chimie, 16 au 24 Avril 1925*. Paris: Gauthiers Villars.

Institut International de Chimie Solvay. 1928. *Rapports et discussions sur des questions d'actualité: Troisieme Conseil de Chimie, 12 au 18 Avril 1928*. Paris: Gauthiers Villars.

International Conference on Physics. Paper and discussions. 1935. Volume II. The solid state of matter. London: Physical Society.

James, J. 2008. *Naturalizing the chemical bond: discipline and creativity in the Pauling program 1927–1942*. Ph.D. dissertation, Harvard University.

James, H. M., and A. S. Coolidge. 1933. The ground state of the hydrogen molecule. *Journal of Chemical Physics* 1:825–835.

Jammer, M. 1966. *The conceptual development of quantum mechanics*. New York: McGraw-Hill.

Janich, P., and N. Psarros, eds. 1998. *The autonomy of chemistry*. Würzburg: Königshausen & Neumann.

Jansen, L. 1977. P.-O. Löwdin's scientific and other activities. *International Journal of Quantum Chemistry* 11:897–905.

Jeffreys, B. S. 1987. Douglas Rayner Hartree (1897–1958). *Comments Atomic and Molecular Physics* 20:189–198.

Julg, A. 1988. Is hybridization just an artifact or does it reflect some physical reality. *Journal of Molecular Structure THEOCHEM* 169:125–136.

Julg, A. 1988a. From atom to cosmos. *Journal of Molecular Structure THEOCHEM* 166:33–38.

Julg, A., and P. Julg. 1978. Vers une nouvelle interpretation de la liaison chimique. *International Journal of Quantum Chemistry* 13:483–497.

Julg, A., and A. Pullman. 1953. Structure éléctronique du fulvène. Introduction de l'interaction configurationnelle. *Journal de Chimie Physique* 50:459–467.

Jost, W. 1935. Zum Verständnis der "chemischen Kräfte" nach der Quantenmechanik. II. *Z. Elektrochem. Angew. Phys. Chem.* 41:667–674.

Karachalios, A. 1997. Die Entstehung und Entwicklung der Quantenchemie in Deutschland. *Mitt. Ges. Deut. Chem. Fachgr. Gesch. Chem.* 13:163–179.

Karachalios, A. 2000. On the making of quantum chemistry in Germany. *Studies in the History and Philosophy of Science* 31B(4):493–510.

Karachalios, A. 2001. Giovanni Battista Bonino and the making of quantum chemistry in Italy. In *Chemical sciences in the 20th century. Bridging boundaries*, ed. C. Reinhardt, 75–104. New York: Wiley-VCH.

Karachalios, A. 2003. *Erich Hückel (1896–1980): from physics to quantum chemistry* [English transl.]. Ph.D. dissertation, Johannes Gutenberg-Universität, Mainz.

Karachalios, A. 2010. *Erich Hückel (1896–1980). From physics to quantum chemistry, translated into English by Ann M. Hentschel*. Berlin: Springer Publishers.

Kargon, R. H. 1977. Temple to science: cooperative research and the birth of the California Institute of Technology. *Historical Studies in the Physical Sciences* 8:3–31.

Karplus, M. 1960. Weak interactions in molecular quantum mechanics. *Reviews of Modern Physics* 32:455–460.

Karplus, M. 1990. Three-dimensional 'Pople diagram.' *Journal of Physical Chemistry* 94:5435–5436.

Kasha, M. 1962. Quantum chemistry in molecular biology. In *Horizons in biochemistry. Albert Szent-Györgyi dedicatory volume*, ed. M. Kasha and B. Pullman, 583–599. New York: Academic Press.

Kauffman, G. B. 1985. William Draper Harkins (1873–1951). A controversial and neglected American physical chemist. *Journal of Chemical Education* 62:758–761.

Kemble, E. C. 1929. General principles of quantum mechanics. Part I. *Physical Review, Supplement* 1:157–215.

Kemble, E. C. 1937. *The fundamental principles of quantum mechanics, with elementary applications*. New York: McGraw-Hill.

Kemble, E. C., and E. H. Hill. 1930. General principles of quantum mechanics. Part II. *Reviews of Modern Physics* 2:1–59.

Kemble, E. C., R. T. Birge, W. F. Colby, F. W. Loomis, and L. Page. 1926. Molecular spectra in gases. *National Research Council, Bulletin* 11:1–358.

Kevles, D. J. 1971. *The physicists. The history of a scientific community in modern America*. Cambridge, MA: Harvard University Press.

King, R. B. 2000. The role of mathematics in the experimental/theoretical/computational trichotomy of chemistry. *Foundations of Chemistry* 2:221–236.

Klein, M. J. 1970. The first phase of the Bohr-Einstein dialogue. *Historical Studies in the Physical Sciences* 2:1–39.

Klein, U. 1999. Do we need a philosophy of chemistry? In *Ars Mutandi: issues in philosophy and history of chemistry*, ed. N. Psarros and K. Gavroglu, 17–27. Leipzig: Leipziger Universitätsverlag.

Klein, U., ed. 2001. *Tools and modes of representation in the laboratory sciences*. Dordrecht: Kluwer Academic Publishers.

Kohler, R. E. 1971. The origin of G. N. Lewis's theory of the shared pair bond. *Historical Studies in the Physical Sciences* 3:343–376.

Kohler, R. E. 1973. Lewis, Gilbert Newton. *Dictionary of Scientific Biography* 8:289–294.

Kohler, R. E. 1974. Irving Langmuir and the 'octet' theory of valence. *Historical Studies in the Physical Sciences* 4:39–87.

Kohler, R. E. 1975. G. N. Lewis's views on bond theory 1900–1916. *British Journal for the History of Science* 8:233–239.

Kohler, R. E. 1975a. The Lewis-Langmuir theory of valence and the chemical community, 1920–28. *Historical Studies in the Physical Sciences* 6:431–468.

Konno, H. 1983. Slater's evidence for the genesis of the Bohr-Kramers-Slater theory. *Historia Scientiarum* 25:39–52.

Kotani, M. 1984. Scientific reminiscence – my pilgrimage through quantum molecular science. *International Journal of Quantum Chemistry: Quantum Chemistry Symposium* 18:11–20.

Kotani, M., Y. Kakiuchi, and G. Araki. 1954. Preface. In *Symposium on molecular physics. Held at Nikko on the occasion of the International Conference on Theoretical Physics*. Tokyo: Maruzen Co., Ltd.

Kotani, M., A. Amemiya, E. Ishiguro, and T. Kimura. 1955. *Table of molecular integrals*. Tokyo: Maruzen Co., Ltd.

Kragh, H. 1977. Chemical aspects of Bohr's 1913 theory. *Journal of Chemical Education* 54:208–210.

Kragh, H. 1990. *Dirac: A scientific biography*. Cambridge: Cambridge University Press.

Kragh, H. 1996. The young Erich Hückel. His scientific work until 1925. Invited paper at the Erich Hückel's FestKolloquium, 28 October 1996, University of Marburg.

Kragh, H. 1996a. Quantum interdisciplinarity: Friedrich Hund and early quantum chemistry. Invited paper at the symposium in Göttingen in honor of the 100th birthday of F. Hund, 6 February 1996, Georges-August Universität.

Kragh, H. 2001. Before quantum chemistry: Erich Hückel and the physics-chemistry interface. *Centaurus* 43:1–16.

Krishnamurthy, R. S., ed. 1996. *The Pauling Symposium. A discourse on the art of biography*. Corvallis, OR: Special Collections, Oregon State University Libraries.

Kronig, R. 1935. *The optical basis of valency*. Cambridge, UK: Cambridge University Press.

Kuhn, H. G. 1972. Franck, James. *Dictionary of Scientific Biography* 5:117–118.

Kuhn, T. S. 1978. *Black-body theory and the quantum discontinuity, 1894–1912*. New York: Oxford University Press.

Kuhn, T. S., J. L. Heilbron, P. Forman, and L. Allen. 1976. *Sources for the history of quantum physics. An inventory and report*. Philadelphia: The American Philosophical Society.

Kursanov, D. N., M. G. Gonikberg, B. Dubinin, M. I. Kabachnik, E. D. Kaverneva, E. N. Prilezhaeva, N. D. Sokolov, and R. Kh. Freidlina. 1952. The present state of the chemical structural theory. [Transl. into English by I.S. Bengelsdorf.] *Journal of Chemical Education* (January):2–13.

Lacassagne, A. 1955. Préface. In *Cancérisation par les substances chimiques et structure moléculaire*, ed. A. Pullman and B. Pullman, 5–8. Paris: Masson & Cie.

Lagowski, J. J. 1966. *The chemical bond (Classical researches in general chemistry)*. Boston: Houghton Mifflin.

Langmuir, I. 1912. The dissociation of hydrogen into atoms. *Journal of the American Chemical Society* 34:860–877.

Langmuir, I. 1919. The arrangement of electrons in atoms and molecules. *Journal of the American Chemical Society* 41:868–934.

Langmuir, I. 1919a. Isomorphism, isosterism and covalence. *Journal of the American Chemical Society* 41:1543–1559.

[Lennard-]Jones. J.E. 1924. On the determination of molecular fields I. From the variation of the viscosity of a gas with temperature. *Proceedings of the Royal Society (London)* 106:441–462.

[Lennard-]Jones. 1924a. On the determination of molecular fields II. From the equation of state of a gas. *Proceedings of the Royal Society (London)* 106:463–477.

[Lennard-]Jones. 1924b. On the determination of molecular fields III. From crystal measurements and kinetic theory data. *Proceedings of the Royal Society (London)* 106:709–718.

[Lennard-]Jones. 1925. On the atomic fields of helium and neon. *Proceedings of the Royal Society (London)* 107:157–170.

Lennard-Jones, J. E. 1925a. On the forces between atoms and ions. *Proceedings of the Royal Society (London)* 109:584–597.

Lennard-Jones, J. E. 1929. The electronic structure of some diatomic molecules. *Transactions of the Faraday Society* 25:668–686.

Lennard-Jones, J. E. 1931. The nature of cohesion. *Nature* 128:462–463.

Lennard-Jones, J. E. 1931a. Cohesion. *Proceedings of the Physical Society* 43:461–482.

Lennard-Jones, J. E. 1931b. Wave-functions of many-electron atoms. *Proceedings of the Cambridge Philosophical Society* 27:469–480.

Lennard-Jones, J. E. 1931c. The quantum mechanics of atoms and molecules. *Proceedings of the London Mathematical Society* 33:290–298.

Lennard-Jones, J. E. 1932. Evidence from molecular spectra. In *Chemistry at the Centenary Meeting of the British Association for the Advancement of Science*, 210–225. Cambridge, UK: W. Heffer and Sons Ltd.

Lennard-Jones, J. E. 1934. The electronic structure and the interaction of some simple radicals. *Transactions of the Faraday Society* 30:70–85.

Lennard-Jones, J. E. 1934a. Some aspects of the electronic theory of valency. *Journal of the Society of Chemical Industry* 53:223–225, 249–251.

Lennard-Jones, J. E. 1937. The electronic structure of some polyenes and aromatic molecules. I. The nature of the links by the method of molecular orbitals. *Proceedings of the Royal Society of London. Series A* 158:280–296.

Lennard-Jones, J. E. 1949. The molecular orbital theory of chemical valency. I. The determination of molecular orbitals. *Proceedings of the Royal Society of London. Series A* 198:1–13.

Lennard-Jones, J. E. 1949a. The molecular orbital theory of chemical valency. II. Equivalent orbitals in molecules of known symmetry. *Proceedings of the Royal Society of London. Series A* 198:14–25.

Lennard-Jones, J. E. 1950. The molecular orbital theory of chemical valency. III. Properties of molecular orbitals. *Proceedings of the Royal Society of London. Series A* 202:155–165.

Lennard-Jones, J. E. 1953. Quantum theory, theory of molecular structure and valence. *Annual Review of Physical Chemistry* 4:167–188.

Lennard-Jones, J. E. 1955. New ideas in chemistry. *Scientific Monthly* 80:175–184.

Lennard-Jones, J. E., and W. R. Cook. 1926. The molecular fields of hydrogen, nitrogen and neon. *Proceedings of the Royal Society (London)* 112:214–229.

Lennard-Jones, J. E., and W. R. Cook. 1927. The equation of state for a gaseous mixture. *Proceedings of the Royal Society (London)* 115:334–348.

Lennard-Jones, J. E., and A. F. Devonshire. 1936. The interaction of atoms and molecules with solid surfaces IV. The condensation and evaporation of atoms and molecules. *Proceedings of the Royal Society of London. Series A* 156:29–36.

Lennard-Jones, J. E., and A. F. Devonshire. 1937. The interaction of atoms and molecules with solid surfaces VI. The behaviour of adsorbed helium at low temperatures. *Proceedings of the Royal Society of London. Series A* 158:242–252.

Lennard-Jones, J. E., and G. G. Hall. 1951. A survey of the principles determining the structure and properties of molecules. Part II. The ionization potentials and resonance energies of hydrocarbons. *Discussions of the Faraday Society* 10:18–26.

Lennard-Jones, J. E., and A. E. Ingham. 1925. On the calculation of certain crystal potential constants and on the cubic crystal of least potential energy. *Proceedings of the Royal Society (London)* 107:636–653.

Lennard-Jones, J. E., and J. A. Pople. 1950. The molecular orbital theory of chemical valency. IV. The significance of equivalent orbitals. *Proceedings of the Royal Society of London. Series A* 202:166–180.

Lennard-Jones, J. E., and J. A. Pople. 1951. The molecular orbital theory of chemical valency. IX. The interaction of paired electrons in chemical bonds. *Proceedings of the Royal Society of London. Series A* 210:190–206.

Lennard-Jones, J. E., and J. A. Pople. 1951a. A survey of the principles determining the structure and properties of molecules. Part I. The factors responsible for molecular shape and bond energies. *Discussions of the Faraday Society* 10:9–18.

Lennard-Jones, J. E., and C. Strachan. 1935. The interaction of atoms and molecules with solid surfaces I. The activation of adsorbed atoms to higher vibrational states. *Proceedings of the Royal Society of London. Series A* 150:442–455.

Lennard-Jones, J. E., and P. A. Taylor. 1925. Some theoretical calculations of the physical properties of certain crystals. *Proceedings of the Royal Society (London)* 109:476–508.

Lennard-Jones, J. E., and J. Turkevich. 1937. The electronic structure of some polyenes and aromatic molecules. II. The nature of the links of some aromatic molecules. *Proceedings of the Royal Society of London. Series A* 158:297–306.

Lewis, G. N. 1907. Outline of a new system of thermodynamic chemistry. *Proceedings of the American Academy of Arts and Sciences* 43:259–293.

Lewis, G. N. 1908. A revision of the fundamental laws of matter and energy. *Philosophical Magazine* 16:705.

Lewis, G. N. 1913. Valence and tautomerism. *Journal of the American Chemical Society* 35:1448–1455.

Lewis, G. N. 1916. The atom and the molecule. *Journal of the American Chemical Society* 38:762–785.

Lewis, G. N. 1923a. Valence and the electron. *Transactions of the Faraday Society* 19:452–458.

Lewis, G. N. 1924. The magnetochemical theory. *Chemical Reviews* 2:231–248.

Lewis, G. N. 1926. The nature of light. *Proceedings of the National Academy of Sciences of the United States of America* 12:22–29.

Lewis, G. N. 1926a. Light waves and light corpuscles. *Nature* 117:236–239.

Lewis, G. N. 1926b. *The anatomy of science.* New Haven: Yale University Press.

Lewis, G. N. 1933. The chemical bond. *Journal of Chemical Physics* 1:17–28.

Lewis, G. N. 1966/1923. *Valence and the structure of atoms and molecules.* New York: The Chemical Catalog Company [reprinted by Dover Publications].

Lewis, E. 1998. *A biography of distinguished scientist Gilbert Newton Lewis.* New York: Edwin Mellen Press.

Lewis, G. N., and M. Randall. 1923. *Thermodynamics and the free energy of chemical substances.* New York: McGraw-Hill.

Lewis, G. N., and E. B. Wilson. 1912. The space-time manifold of relativity: the non-Euclidian geometry of mechanics and electromagnetism. *Proceedings of the American Academy* 48:389.

Linderberg, J. 2002. Quantum chemistry as a lifestyle. Per-Olov Löwdin, 1916–2000, Obituary notice, 13 February 2002. Department of Chemistry, Uppsala University.

Lindsay, R. B. 1968. *The nature of physics; a physicist's views on the history and philosophy of his science.* Providence: Brown University Press.

Lindsay, R. B. 1970. Hartree, Douglas Rayner. *Dictionary of Scientific Biography* 6:147–148.

Lindsay, R. B. 1970a. *Men of physics: Lord Rayleigh. The man and his work.* Oxford: Pergamon Press.

Lindsay, R. B. 1973. *Men of physics: Julius Robert Mayer. Prophet of energy.* Oxford: Pergamon Press.

Lochak, G. 1992. *Louis de Broglie.* Paris: Flammarion.

London, F. 1919. Die Bedeutung der Quantentheorie für die Chemie (Planck Festschrift). *Naturwissenschaften* 17:516–529.

London, F. 1928. Quantentheorie und chemische Bindung. In *Quantentheorie und chemie,* 59–84. Leipzig: Hirzel.

London, F. 1928a. Zur Quantentheorie der homoopolaren Valenzzahlen. *Zeitschrift für Physik* 46:455–477.

London, F. 1928b. Zur Quantenmechanik der homoopolaren Valenzchemie. *Zeitschrift für Physik* 50:24–51.

London, F. 1929. Die Bedeutung der Quantentheorie für die Chemie (Planck Festschrift). *Naturwissenschaften* 17:516–529.

London, F. 1954. *Superfluids*. New York: Wiley.

London, F., and E. Bauer. 1939. *La théorie de l'observation en mécanique quantique*. Paris: Hermann & Cie.

Longuet-Higgins, H. C. 1953. An application of chemistry to mathematics. *Scientific Journal of the Royal College of Science* 23:99–106.

Longuet-Higgins, H. C. 1977. Closing remarks. *Faraday Society Discussions* 62:347–348.

Longuet-Higgins, H. C. 1990. Robert Sanderson Mulliken. *Biographical Memoirs of Fellows of the Royal Society. Royal Society (Great Britain)* 35:329–354.

Longuet-Higgins, H. C., and L. Salem. 1959. The alternation of bond length in long conjugated chain molecules. *Proceedings of the Royal Society of London. Series A* 251:172–185.

Longuet-Higgins, H. C., and L. Salem. 1960. The alternation of bond length in long conjugated chain molecules. III. The cyclic polyenes $C_{18}H_{18}$, $C_{24}H_{24}$, $C_{30}H_{30}$. *Proceedings of the Royal Society of London. Series A* 257:445–456.

Löwdin, P.-O. 1950. The non-orthogonality problem connected with the use of atomic wave functions in the theory of molecules and crystals. *Journal of Chemical Physics* 18:365–375.

Löwdin, P.-O. 1954. A method of alternant molecular orbitals. In *Symposium on molecular physics. Held at Nikko on the occasion of the International Conference on Theoretical Physics*, 13–16. Tokyo: Maruzen Co., Ltd.

Löwdin, P.-O. 1955. Quantum theory of many-particle systems. I. Physical interpretation by means of density matrices, natural spin-orbitals, and convergence problems in the method of configuration interaction. *Physical Review* 97:1474–1489.

Löwdin, P.-O. 1955a. Quantum theory of many-particle systems. II. Study of the ordinary Hartree-Fock approximation. *Physical Review* 97:1490–1508.

Löwdin, P.-O. 1955b. Quantum theory of many-particle systems. III. Extension of the Hartree-Fock scheme to include degenerate systems and correlation effects. *Physical Review* 97:1509–1520.

Löwdin, P.-O. 1957. Present situation of quantum chemistry. *Journal of Physical Chemistry* 61:55–68.

Löwdin, P.-O. 1960. Expansion theorems for the total wave function and extended Hartree-Fock schemes. *Reviews of Modern Physics* 32:328–334.

Löwdin, P.-O. 1963. Introduction. *Reviews of Modern Physics* 35:415.

Löwdin, P.-O. 1965. Discussion on the Hartree-Fock approximation. *Reviews of Modern Physics* 36:196–601.

Löwdin, P.-O. 1965a. Introduction. *Journal of Chemical Physics* 43:S1.

Löwdin, P.-O. 1967. Program. *International Journal of Quantum Chemistry* 1:1–6.

Löwdin, P.-O. 1967a. Nature of quantum chemistry. *International Journal of Quantum Chemistry* 1:7–12.

Löwdin, P.-O. 1973. Obituary – Dr. S.F. Boys. *International Journal of Quantum Chemistry* 7:vii.

Löwdin, P.-O. 1977. Quantum Sciences – some aspects on the American Swedish exchange in quantum sciences, particularly the Uppsala-Florida exchange project. In *Partners in progress. A chapter in the American-Swedish exchange of knowledge*, ed. A. Kastrup and N. W. Olsson, 255–276. Swedish Council of America. St. Paul, MN: North Central Publishing Company.

Löwdin, P.-O. 1980. Opening address at the third international congress of quantum chemistry. In *Horizons of quantum chemistry. Proceedings of the third international congress of quantum chemistry held at Kyoto, Japan, October 29–November 3, 1979*, ed. K. Fukui and B. Pullman, xv–xviii. London: D. Reidel Publishing Company.

Löwdin, P.-O. 1986. Twenty-five years of Sanibel Symposia: a brief historic and scientific survey. *International Journal of Quantum Chemistry* 19:19–37.

Löwdin, P.-O. 1986a. On the state of the art of quantum chemistry. *International Journal of Quantum Chemistry* 29:1651–1683.

Löwdin, P.-O. 1989. Some aspects on the structure and present situation of quantum chemistry as expressed at the Girona Workshop. In *Quantum chemistry: basic aspects, actual trends*, ed. R. Carbó, 605–622. Amsterdam: Elsevier.

Löwdin, P.-O. 1989a. On the long way from the Coulombic hamiltonian to the electronic structure of molecules. *Pure and Applied Chemistry* 61:2065–2074.

Löwdin, P.-O. 1990. Introduction. *International Journal of Quantum Chemistry* 37:317.

Löwdin, P.-O. 1991. On the historical development of the valence bond method and the non-orthogonality problem. *Journal of Molecular Structure THEOCHEM* 229:1–14.

Löwdin, P.-O. 1991a. On the importance of theory in the future development of chemistry. *Journal of Molecular Structure THEOCHEM* 230:1–3.

Löwdin, P.-O., and Y. Öhrn eds. 1977. Proceedings of the international symposium on atomic, molecular, and solid-state theory, collision phenomena, and computational methods. *International Journal of Quantum Chemistry* S11.

Löwdin, P.-O., and B. Pullman, eds. 1964. *Molecular orbitals in chemistry, physics and biology, a tribute to R.S. Mulliken*. New York: Academic Press.

Löwdin, P.-O., J.-L. Calais, and O. Goscinski, eds. 1978. Quantum chemistry—a scientific melting pot. *International Journal of Quantum Chemistry* 12(Suppl. 1).

Lowry, T. M. 1923. Applications in organic chemistry of the electronic theory of valency. *Transactions of the Faraday Society* 19:485–487.

Lundgren, A., and B. Bensaude-Vincent, eds. 2000. *Communicating chemistry. Textbooks and their audiences, 1789–1939.* Canton, MA: Science History Publications.

Mallion, R., R. McWeeny, and B. O'Leary. 1992. Charles Alfred Coulson. *Journal of Molecular Structure THEOCHEM* 259:xv–xx.

Manne, R. 1976. Per Olov Löwdin in the scientific literature 1965–1974. In *Quantum science. Methods and structure. A tribute to Per-Olov Löwdin*, ed. J. L. Calais, O. Goscinski, J. Linderberg, and Y. Öhrn, 25–31. New York: Plenum Press.

Margenau, H. 1944. The exclusion principle and its philosophical tradition. *Philosophy of Science* 11:187–208.

Marinacci, B., ed. 1995. *Linus Pauling in his own words.* New York: Simon & Schuster.

Maruani, J., ed. 1988. *General introduction to molecular sciences.* Vol. 1. Molecules in physics, chemistry and biology. Dordrecht: Kluwer Academic Publishers.

Mayot, M., H. Berthod, G. Berthier, and A. Pullman. 1956. Calcul des intégrales polycentriques relatives à l'étude des structures moléculaires. I. Intégrale tricentrique homonucléaire du type Coulomb-echange. *Journal de Chimie Physique* 53:774–777.

McCrea, W. 1985. How quantum physics came to England. *New Scientist* 17(October):58–60.

McCrea, W. 1986. Cambridge physics 1925–1929. Diamond jubilee of golden years. *Interdisciplinary Science Reviews* 11:269–284.

McWeeny, R. 1954. The valence bond theory of molecular structure. I. Orbital theories and the valence bond method. *Proceedings of the Royal Society of London. Series A* 223:63–79.

McWeeny, R. 1954a. The valence bond theory of molecular structure. II. Reformulation of the theory. *Proceedings of the Royal Society of London. Series A* 223:306–323.

McWeeny, R. 1955. The valence bond theory of molecular structure. III. Cyclobutadiene and benzene. *Proceedings of the Royal Society of London. Series A* 227:288–312.

McWeeny, R. 1955a. On the basis of orbital theories. *Proceedings of the Royal Society of London. Series A* 232:114–135.

McWeeny, R. 1960. Some recent advances in density matrix theory. *Reviews of Modern Physics* 32:335–369.

McWeeny, R. 1961. Quantum organic chemistry. [The Chemical Society] *Annual Reports on the Progress of Chemistry* 58:137–153.

McWeeny, R., ed. 1979. *Coulson's Valence.* Oxford: Oxford University Press.

McWeeny, R. 1986. Back to the valence bond. *Nature* 323:666–667.

McWeeny, R. 1989. Valence bond theory: a tribute to the pioneers of 1927–1935. *Pure and Applied Chemistry* 61:2087–2101.

McWeeny, R. 1990. Valence bond theory: progress and prospects. *International Journal of Quantum Chemistry Symposium* 34:733–752.

McWeeny, R. 1995. Theories and calculations: the search for simplicity. *Proceedings of the Indian Association for the Cultivation of Science* 78B:19–40.

McWeeny, R. 1996. Inside story – some scientific reminiscences. *International Journal of Quantum Chemistry* 60:3–19.

McWeeny, R. 2007. Spiers memorial lecture. Quantum chemistry: the first seventy years. *Faraday Discussions* 135:13–30.

Mead, C. S., ed. 1991. *The Pauling Catalog: Ava Helen and Linus Pauling Papers at Oregon State University*. Corvallis, OR: Kerr Library Special Collections.

Mecke, R. 1925. Zum Wesen der Dublettstruktur einer Klasse von Bandenspektren. *Naturwissenschaften* 13:755–756.

Meckler, A. 1953. Electronic energy levels of molecular oxygen. *Journal of Chemical Physics* 21:1750–1762.

Mehra, J., and H. Rechenberg. 1982. *The historical development of quantum theory*. Vol. 1. New York: Springer-Verlag.

Mehra, J., and H. Rechenberg. 2000. From molecular theory to quantum chemistry. In *The historical development of quantum theory. Vol. 6. The completion of quantum mechanics. 1926–1941. Part 1. The probability interpretation and the statistical transformation theory, the physical interpretation, and the empirical and mathematical foundations of quantum mechanics 1926–1932*. New York: Springer-Verlag.

Mendelsohn, E., M. R. Smith, and P. Weingart. 1988. *Science, technology and the military*. Dordrecht: Kluwer Academic Publishers.

Milne, E. A. 1945–48. Ralph Howard Fowler, 1889–1944. *Obituary Notices of the Fellows of the Royal Society of London* 5:61–78.

Molecular Quantum Mechanics Conference. 1956 Austin, TX: University of Texas.

Mormann, T. 1991. Husserl's philosophy of science and the semantic approach. *Philosophy of Science* 58:61–83.

Morris, P. J. T., ed. 2002. *From classical to modern chemistry. The instrumental revolution*. Cambridge, UK: Royal Society of Chemistry.

Mosini, V. 1999. Wheland, Pauling and the resonance structures. In *Ars mutandi: issues in philosophy and history of chemistry*, ed. N. Psarros and K. Gavroglu, 127–143. Leipzig: Leipziger Universitätsverlag.

Mosini, V. 2000. A brief history of the theory of resonance and of its interpretation. *Studies in History and Philosophy of Science* 31:564–581.

Mott, N. F. 1955. John Edward Lennard-Jones, 1894–1954. *Biographical Memoirs of Fellows of the Royal Society. Royal Society (Great Britain)* 1:175–184.

Mulliken, R. S. 1923. The vibrational isotope effect in the band spectrum of boron nitride. *Science* 58:164–166.

Mulliken, R. S. 1924. Isotope effect in the band spectra of boron monoxide and silicon nitride. *Nature* 113:423–424.

Mulliken, R. S. 1924a. The isotope effect as a means of identifying the emitters of band spectra: application to the bands of the metal hydrides. *Nature* 113:489–490.

Mulliken, R. S. 1924b. The band spectrum of boron monoxide. *Nature* 114:349–350.

Mulliken, R. S. 1925. The isotope effect in band spectra, part I. *Physical Review* 25:119–138.

Mulliken, R. S. 1925a. The isotope effect in band spectra, II. The spectrum of boron monoxide. *Physical Review* 25:259–294.

Mulliken, R. S. 1925b. The isotope effect in band spectra, III. The spectrum of copper iodide as excited by active nitrogen. *Physical Review* 26:1–32.

Mulliken, R. S. 1925c. The isotope effect in band spectra, IV. The spectrum of silicon nitride. *Physical Review* 26:319–338.

Mulliken, R. S. 1925d. On a class of one-valence-electron emitters of band spectra. *Physical Review* 26:561–572.

Mulliken, R. S. 1926. The electronic states of the helium molecule. *Proceedings of the National Academy of Sciences of the United States of America* 12:158–162.

Mulliken, R. S. 1926a. Systematic relations between electronic structure and band-spectrum structure in diatomic molecules. I. *Proceedings of the National Academy of Sciences of the United States of America* 12:144–157.

Mulliken, R. S. 1926b. Systematic relations between electronic structure and band-spectrum structure in diatomic molecules. III. Molecule formation and molecular structure. *Proceedings of the National Academy of Sciences of the United States of America* 12:338–343.

Mulliken, R. S. 1926c. Electronic states and band spectrum structure in diatomic molecules. I. Statement of the postulates. Interpretation of CuO, CH, and CO band-types. *Physical Review* 28:481–506.

Mulliken, R. S. 1926d. Electronic states and band spectrum structure in diatomic molecules. II. Spectra involving terms essentially of the form $B(j^2 - \sigma^2)$. *Physical Review* 28:1202–1222.

Mulliken, R. S. 1927. Electronic states. IV. Hund's theory; second positive nitrogen and swan bands; alternate intensities. *Physical Review* 29:637–649.

Mulliken, R. S. 1928. The assignment of quantum numbers for electrons in, molecules. I. *Physical Review* 32:186–222.

Mulliken, R. S. 1928a. The assignment of quantum numbers for electrons in molecules. II. The correlation of molecular and atomic electron states. *Physical Review* 32:761–772.

Mulliken, R. S. 1929. Band spectra and chemistry. *Chemical Reviews* 6:503–545.

Mulliken, R. S. 1929a. The assignment of quantum numbers for electrons in molecules. III. Diatomic hydrides. *Physical Review* 33:731–747.

Mulliken, R. S. 1930. The interpretation of band spectra. I. *Reviews of Modern Physics* 2:60–115.

Mulliken, R. S. 1930a. Report on notation for spectra of diatomic molecules. *Physical Review* 36:611–629.

Mulliken, R. S. 1931. The interpretation of band spectra. II. *Reviews of Modern Physics* 3:89–155.

Mulliken, R. S. 1931a. Bonding power of electrons and theory of valence. *Chemical Reviews* 9:347–388.

Mulliken, R. S. 1932. The interpretation of band spectra. III. *Reviews of Modern Physics* 4:1–86.

Mulliken, R. S. 1932a. Electronic structure of polyatomic molecules and valence. *Physical Review* 40:55–62.

Mulliken, R. S. 1932b. Electronic structures of polyatomic molecules and valence. II. General considerations. *Physical Review* 41:49–71.

Mulliken, R. S. 1932c. Electronic structures of polyatomic molecules and valence. III. Quantum theory of the double bond. *Physical Review* 41:751–778.

Mulliken, R. S. 1932d. Quantum theory of the double bond. *Journal of the American Chemical Society* 54:4111–4112.

Mulliken, R. S. 1933. Electronic structures of polyatomic molecules and valence. IV. Electronic states, quantum theory of the double bond. *Physical Review* 43:279–302.

Mulliken, R. S. 1933a. Electronic structures of polyatomic molecules and valence. V. Molecules RX_n. *Journal of Chemical Physics* 1:492–503.

Mulliken, R. S. 1934. A new electroaffinity scale; together with data on valence states and on valence ionization potentials and electron affinities. *Journal of Chemical Physics* 2:782–793.

Mulliken, R. S. 1935. Electronic structures of polyatomic molecules and valence. VI. On the method of molecular orbitals. *Journal of Chemical Physics* 3:375–378.

Mulliken, R. S. 1935a. Electronic structures of polyatomic molecules and valence. XIII. Diborane and related molecules. *Journal of Chemical Physics* 3:635–645.

Mulliken, R. S. 1947. Quantum-mechanical methods and the electronic spectra and structure of molecules. *Chemical Reviews* 41:201–206.

Mulliken, R. S. 1949. Quelques aspects de la théorie des orbitales moléculaires. *Journal de Chimie Physique* 46:497–542, 675–713.

Mulliken, R. S. 1954. On approximate computation of molecular energies. In *Symposium on molecular physics. Held at Nikko on the occasion of the International Conference on Theoretical Physics*, 17–21. Tokyo: Maruzen Co., Ltd.

Mulliken, R. S. 1960. Self-consistent field atomic and molecular orbitals and their approximations as linear combinations of Slater-type orbitals. *Reviews of Modern Physics* 32:232–238.

Mulliken, R. S. 1965. Molecular scientists and molecular science: some reminiscences. *Journal of Chemical Physics* 43:S2–S11.

Mulliken, R. S. 1966. Spectroscopy, molecular orbitals, and chemical binding. Nobel Lecture, December 12: 131–160. Available at: <http://nobelprize.org/nobel_prizes/chemistry/laureates/1966/mulliken-lecture.html>. Accessed February 8, 2009.

Mulliken, R. S. 1968. Spectroscopy, quantum chemistry and molecular physics. *Physics Today* 214:52–57.

Mulliken, R. S. 1970. The path to molecular orbital theory. *Pure and Applied Chemistry* 24:203–215.

Mulliken, R. S. 1975. *Selected papers of R.S. Mulliken*. Chicago: University of Chicago Press.

Mulliken, R. S. 1975a. William Draper Harkins, December 28, 1973–March 7, 1951. *Biographical Memoirs. National Academy of Sciences (U. S.)* 47:49–81.

Mulliken, R. S. 1989. *Life of a scientist – an autobiographical account of the development of molecular orbital theory with an introductory memoir by Friedrich Hund*, ed. B.J. Ransil. Berlin: Springer-Verlag.

Mulliken, R. S., and W. D. Harkins. 1922. The separation of isotopes. Theory of resolution of isotopic mixtures by diffusion and similar processes. Experimental separation of mercury by evaporation in a vacuum. *Journal of the American Chemical Society* 44:37–65.

Mulliken, R. S., and C. C. J. Roothaan. 1959. Broken bottlenecks and the future of molecular quantum mechanics. *Proceedings of the National Academy of Sciences of the United States of America* 45:394–398.

National Research Council. 1971. *Computational support for theoretical chemistry*. Washington, DC: National Academy of Sciences.

Nye, M. J. 1989. Chemical explanation and physical dynamics: two research schools at the first Solvay chemistry conferences, 1922–1928. *Annals of Science* 46:461–480.

Nye, M. J. 1992. Physics and chemistry: commensurate and incommensurate sciences? In *The invention of physical science*, ed. M. J. Nye, J. Richards, and R. H. Stuewer, 205–224. Dordrecht: Kluwer Academic Publishers.

Nye, M. J. 1993. *From chemical philosophy to theoretical chemistry. Dynamics of matter and dynamics of disciplines, 1800–1950*. Berkeley: University of California Press.

Nye, M. J. 2000. Physical and biological modes of thought in the chemistry of Linus Pauling. *Studies in History and Philosophy of Science* 31:475–491.

Nye, M. J. 2000a. From student to teacher: Linus Pauling and the reformulation of the principles of chemistry in the 1930s. In *Communicating chemistry. Textbooks and their audiences, 1789–1939*, ed. A. Lundgren and B. Bensaude-Vincent, 397–414. Science History Publications.

Nye, M. J. 2001. Paper tools and molecular architecture in the chemistry of Linus Pauling. In *Tools and modes of representation in the laboratory sciences*, ed. U. Klein, 117–132. Dordrecht: Kluwer Academic Publishers.

Odiot, S., and R. Daudel. 1954. Théorie de la localisabilité des corpuscules. III. Retour sur la signification géométrique de la notion de couche. *Journal de Chimie Physique* 51:361–362.

Odiot, S., and R. Daudel. 1954a. Relation entre le volume occupé en moyenne par an électron dans le cortège d'un atome et le potentiel électrique moyen régnant dans ce domaine. *Comptes Rendus Hebdomadaires des Séances de l'Académie des Sciences* 238:1384–1386.

Ohno, K. 1976. Per-Olov Löwdin – conqueror of scientific, educational and rocky mountains. In *Quantum science. Methods and structure. A tribute to Per-Olov Löwdin*, ed. J.-L. Calais, O. Goscinski, J. Linderberg, and Y. Öhrn, 1–11. New York: Plenum Press.

Ohno, K. 1976a. Publications of Per-Olov Löwdin 1939–1976. In *Quantum science. Methods and structure. A tribute to Per-Olov Löwdin*, ed. J.-L. Calais, O. Goscinski, J. Linderberg, and Y. Öhrn, 13–23. New York: Plenum Press.

Ohno, K. 1978. Quantum chemical computations – personal appraisal of the state of the art. *International Journal of Quantum Chemistry* 12(Suppl. 1):119–125.

Ohno, K. 1995. In memory of Professor Masao Kotany. *International Journal of Quantum Chemistry* 53:451–454.

Orville-Thomas, W. J. 1992. Foreword. *Journal of Molecular Structure THEOCHEM* 259:xiii.

Ostlund, N. S. 1979. Chemistry, computers, and microelectronics: present and future prospects. *International Journal of Quantum Chemistry: Quantum Chemistry Symposium* 13:15–38.

Owens, L. 1986. Vannevar Bush and the differential analyzer: the text and context of an early computer. *Technology and Culture* 27:63–95.

Palmer, W. G. 1959. *Valency, classical and modern*. Cambridge, UK: Cambridge University Press.

Palmer, W. G. 1965. *A history of the concept of valency to 1930*. Cambridge, UK: Cambridge University Press.

Paradowski, R. J. 1972. *The structural chemistry of Linus Pauling*. Ph.D. dissertation, University of Wisconsin.

Paradowski, R. J. 1991. The making of a scientist. Biographical notes and chronology. In *Linus Pauling, a man of intellect and action*, ed. F. Miyazaki, 185–211. Tokyo: Cosmos Japan International.

Pariser, R. 1990. On the origins of the PPP method. *International Journal of Quantum Chemistry* 37:319–325.

Pariser, R., and R. G. Parr. 1953. A semi-empirical theory of the electronic spectra and electronic structure of complex unsaturated molecules. I. *Journal of Chemical Physics* 21:466–471.

Pariser, R., and R. G. Parr. 1953a. A semi-empirical theory of the electronic spectra and electronic structure of complex unsaturated molecules. II. *Journal of Chemical Physics* 21:767–776.

Park, B. S. 1999. Chemical translators: Pauling, Wheland and their strategies for teaching the theory of resonance. *British Journal for the History of Science* 32:21–46.

Park, B. S. 1999a. *Computations and interpretations: the growth of quantum chemistry, 1927–1967*. Ph.D. dissertation, Johns Hopkins University.

Park, B. S. 2000. The contexts of simultaneous discovery: Slater, Pauling, and the origins of hybridization. *Studies in the History and Philosophy of Modern Physics* 31:451–474.

Park, B. S. 2001. A principle written in diagrams: the aufbau principle for molecules and its visual representations. In *Tools and modes of representation in the laboratory sciences*, ed. U. Klein, 179–198. Dordrecht: Kluwer Academic Publishers.

Park, B. S. 2003. The 'hyperbola of quantum chemistry': the changing practice and identity of a scientific discipline in the early years of electronic digital computers, 1945–1965. *Annals of Science* 60:219–247.

Park, B. S. 2005. In the context of "pedagogy": teaching strategy and theory change in chemistry. In *Pedagogy and the practice of science. Historical and contemporary perspectives*, ed. D. Kaiser, 287–319. Cambridge, MA: MIT Press.

Park, B. S. 2009. Between accuracy and manageability: computational imperatives in quantum chemistry. *Historical Studies in the Natural Sciences* 39:32–62.

Parr, R. G. 1960. Introductory note. *Reviews of Modern Physics* 32:169.

Parr, R. G. 1975. The description of molecular structure. *Proceedings of the National Academy of Sciences of the United States of America* 72:763–771.

Parr, R. G. 1977. Erich Hückel and Friedrich Hund – pioneers in quantum chemistry. *International Journal of Quantum Chemistry* S11:29–37.

Parr, R. G. 1990. On the genesis of a theory. *International Journal of Quantum Chemistry* 37:327–347.

Parr, R. G., and B. L. Crawford eds. 1952. National Academy of Sciences Conference on Quantum-Mechanical Methods in Valence Theory. *Proceedings of the National Academy of Sciences of the United States of America* 38:547–553.

Parr, R. G., D. P. Craig, and I. G. Ross. 1950. Molecular orbital calculations of the lower excited electronic levels of benzene, configuration interaction included. *Journal of Chemical Physics* 18:1561–1563.

Parson, A. L. 1915. A magneton theory of the structure of the atom. *Smithsonian Miscellaneous Collections* 65:1–80.

Partington, J. R. 1964. *A history of chemistry*. London: Macmillan Press Ltd.

Pauli, W. 1922. Uber das Modell des Wasserstoffmoleculions. *Zeitschrift fur Physik* 68:177–240.

Pauli, W. 1927. Zur Quantenmechanik des magnetischen Elektrons. *Zeitschrift fur Physik* 43:601–623.

Pauli, W. 1946. Remarks on the history of the exclusion principle. *Science* 103:213–215.

Pauling, L. 1926. The prediction of the relative stabilities of isosteric isomeric ions and molecules. *Journal of the American Chemical Society* 48:641–651.

Pauling, L. 1926a. The dynamic model of the chemical bond and its application to the structure of benzene. *Journal of the American Chemical Society* 48:1132–1143.

Pauling, L. 1926b. The quantum theory of the dielectric constant of hydrogen chloride and similar gases. *Proceedings of the National Academy of Sciences of the United States of America* 12:32–35.

Pauling, L. 1926c. The quantum theory of the dielectric constant of hydrogen chloride and similar gases. *Physical Review* 27:568–577.

Pauling, L. 1927. The influence of a magnetic field on the dielectric constant of a diatomic dipole gas. *Physical Review* 29:145–160.

Pauling, L. 1927a. The electron affinity of hydrogen and the second ionization potential of lithium. *Physical Review* 29:285–291.

Pauling, L. 1927b. The theoretical prediction of the physical properties of many-electron atoms and ions. Mole refraction, diamagnetic susceptibility, and extension in space. *Proceedings of the Royal Society of London. Series A* 114:181–211.

Pauling, L. 1928. The application of quantum mechanics to the structure of the hydrogen molecule and the hydrogen molecule-ion and to related problems. *Chemical Reviews* 5:173–213.

Pauling, L. 1928a. The shared-electron chemical bond. *Proceedings of the National Academy of Sciences of the United States of America* 14:359–362.

Pauling, L. 1928c. The coordination theory of the structure of ionic crystals. In *Festschrift zum 60. Geburtstage Arnold Sommerfelds*. Leipzig: Verlag von S. Herzel.

Pauling, L. 1929. The principles determining the structure of complex ionic crystals. *Journal of the American Chemical Society* 51:1010–1026.

Pauling, L. 1931. The nature of the chemical bond. Application of results obtained from the quantum mechanics and from a theory of paramagnetic susceptibility to the structure of molecules. *Journal of the American Chemical Society* 53:1367–1400.

Pauling, L. 1931a. Quantum mechanics and the chemical bond. *Physical Review* 37:1185–1186.

Pauling, L. 1931b. The nature of the chemical bond. II. The one-electron bond and the three-electron bond. *Journal of the American Chemical Society* 53:3225–3237.

Pauling, L. 1931c. The determination of crystal structure by X-rays. *Annual Survey of American Chemistry* 5:118–125.

Pauling, L. 1932. The nature of the chemical bond. III. The transition from one extreme bond type to another. *Journal of the American Chemical Society* 54:988–1003.

Pauling, L. 1932a. The nature of the chemical bond. IV. The energy of single bonds and the relative electronegativity of atoms. *Journal of the American Chemical Society* 54:3570–3582.

Pauling, L. 1939. *The nature of the chemical bond and the structure of molecules and crystals: an introduction to modern structural chemistry.* 1st ed. New York: Cornell University Press.

Pauling, L. 1940. *The nature of the chemical bond and the structure of molecules and crystals: an introduction to modern structural chemistry.* 2nd ed. New York: Cornell University Press.

Pauling, L. 1947. *General chemistry.* San Francisco: W.H. Freeman and Company.

Pauling, L. 1950. The place of chemistry in the integration of the sciences. *Main Currents in Modern Thought* 7:108–111.

Pauling, L. 1952. Quantum mechanics of valence. *Nature* 170:384–385.

Pauling, L. 1954. Modern structural chemistry. Nobel Lecture, December 12, 1954, 429–37. Available at: <http://nobelprize.org/nobel_prizes/chemistry/laureates/1954/pauling-lecture.html>. Accessed February 8, 2009.

Pauling, L. 1956. The nature of the theory of resonance. In *Perspectives in organic chemistry, dedicated to Sir Robert Robinson,* ed. Sir A. Todd, 1–8. New York: Interscience Publishers.

Pauling, L. 1960. *The nature of the chemical bond and the structure of molecules and crystals: an introduction to modern structural chemistry.* 3rd ed. New York: Cornell University Press.

Pauling, L. 1967. *The chemical bond.* New York: Cornell University Press.

Pauling, L. 1970. Fifty years of progress in structural chemistry and molecular biology. *Daedalus* 99:988–1014.

Pauling, L. 1980. Prospects and retrospects in chemical education. *Journal of Chemical Education* 57:38–40.

Pauling, L., and J. Sherman. 1933. The nature of the chemical bond. VI. The calculation from thermochemical data of the energies of resonance of molecules among several electronic structures. *Journal of Chemical Physics* 1:606–617.

Pauling, L., and J. Sherman. 1933a. The nature of the chemical bond. VII. The calculation of resonance energy in conjugated systems. *Journal of Chemical Physics* 1:679–686.

Pauling, L., and G. Wheland. 1933. The nature of the chemical bond. V. The quantum mechanical calculation of the resonance energy of benzene and naphthalene and the hydrocarbon free radicals. *Journal of Chemical Physics* 1:362–374.

Pauling, L., and E. B. Wilson. 1935. *Introduction to quantum mechanics with applications to chemistry*. New York: McGraw-Hill.

Pauling, L., and D. M. Yost. 1932. The additivity of the energies of normal covalent bonds. *Journal of the American Chemical Society* 18:414–416.

Pauling, L., J. Y. Beach, and L. O. Brockway. 1935. The dependence of interatomic distance on single bond-double bond resonance. *Journal of the American Chemical Society* 57:2705–2709.

Penney, W. G. 1937. The electronic structure of some polyenes and aromatic molecules. III. Bonds of fractional order by the pair method. *Proceedings of the Royal Society of London. Series A* 158:306–324.

Pestre, D. 1992. *Physique et physiciens en France 1918–1940*. Paris: Editions des Archives Contemporaines.

Peyerimhoff, S. D. 2002. The development of computational chemistry in Germany. *Reviews in Computational Chemistry* 18:257–291.

Platt, J. R. 1964. Strong inference. *Science* 146:347–353.

Platt, J. R. 1966. 1966 Nobel Laureate in Chemistry: Robert S. Mulliken. *Science* 154:745–747.

Podolsky, B., and L. Pauling. 1929. The momentum distribution in hydrogen-like atoms. *Physical Review* 34:109–116.

Poitier, R., and R. Daudel. 1943. *La chimie théorique et ses rapports avec la théorie corpusculaire moderne*. Paris: Hermann & Cie.

Pople, J. A. 1950. The molecular orbital theory of chemical valency. V. The structure of water and similar molecules. *Proceedings of the Royal Society of London. Series A* 202:323–336.

Pople, J. A. 1953. Electron interaction in unsaturated hydrocarbons. *Transactions of the Faraday Society* 49:1375–1385.

Pople, J. A. 1965. Two-dimensional chart of quantum chemistry. *Journal of Chemical Physics* 43:S229–S230.

Pople, J. A. 1990. The origin of the PPP theory. *International Journal of Quantum Chemistry* 37:349–371.

Pople, J. A. 1998. Quantum chemical models. Nobel Lecture, December 8, 1998. Available at: <http://nobelprize.org/nobel_prizes/chemistry/laureates/1998/pople-lecture.html>. Accessed February 8, 2009.

Price, W. C., S. S. Chissick, and T. Ravensdale, eds. 1973. *Wave mechanics. The first fifty years.* London: Butterworth Group.

Primas, H. 1980. Foundations of theoretical chemistry. In *Quantum dynamics of molecules, the new experimental challenge to Theorists*, ed. R. G. Woolley, 39–113. New York: Plenum Press.

Primas, H. 1983. *Chemistry, quantum mechanics and reduction.* Berlin: Springer.

Primas, H. 1988. Can we reduce chemistry to physics? In *Centripetal forces in the sciences.* Vol. 2. ed. G. Radnisky, 119–133. New York: Pergamon Press.

Psarros, N., and K. Gavroglu, eds. 1999. *Ars mutandi: issues in philosophy and history of chemistry.* Leipzig: Leipziger Universitätsverlag.

Pullman, A. 1946. *Contribution à l'étude de la structure électronique des molecules organiques. Étude particulière des hydrocarbures cancérigènes.* PhD thesis. Paris: Masson.

Pullman, A. 1947. Influence de l'addition des cycles saturés sur la structure électronique et sur l'activité des hydrocarbures polycycliques. *Comptes Rendus Hebdomadaires des Scéances de l'Académie des Sciences de Paris* 224:120–122.

Pullman, A. 1954. L'interaction de configurations dans un butadiène self-consistant. *Journal de Chimie Physique* 51:188–196.

Pullman, A. 1970. The present image of hetero-aromatics in quantum chemistry – revolution or evolution? In *Quantum aspects of heterocyclic compounds in chemistry and biochemistry. Proceedings of an international symposium held in Jerusalem, 31 March–4 April 1969*, ed. E. D. Bergmann and B. Pullman, 9–31. Jerusalem: Jerusalem Academic Press.

Pullman, A. 1971. Propos d'Introduction. 1970: Bilan et perspectives. In *Colloque International sur les Aspects de la Chimie Quantique Contemporaine*, 8–13 July 1970, Menton, France, org. R. Daudel and A. Pullman, 9–16. Paris: Éditions du Centre National de la Recherche Scientifique.

Pullman, A., and J. Baudet. 1954. Interaction des configurations dans le butadiène calculé par la méthode du champ moléculaire self-consistant. *Comptes Rendus Hebdomadaires des Scéances de l'Académie des Sciences de Paris* 238:241–243.

Pullman, A., and H. Berthod. 1955. L'étude du butadiène par la méthode des 'atomes dans les molécules. *Journal de Chimie Physique* 52:771–774.

Pullman, A., and B. Pullman. 1955. *Cancérisation par les substances chimiques et structure moléculaire.* Paris: Masson & Cie.

Pullman, A., and B. Pullman. 1962. From quantum chemistry to quantum biochemistry. In *Horizons in biochemistry. Albert Szent-Györgyi dedicatory volume*, ed. M. Kasha and B. Pullman, 553–582. New York: Academic Press.

Pullman, A., and B. Pullman. 1973. Quantum biochemistry. In *Wave mechanics. The first fifty years*, ed. W. C. Price, S. S. Chissick, and T. Ravensdale, 272–291. London: Butterworth Group.

Pullman, A., B. Pullman, and P. Rumpf. 1948. Structure électronique du fulvène et du benzofulvène. I- Étude par la méthode de la mésomérie. *Bulletin de la Société Chimique* 15:281–284.

Pullman, A., B. Pullman, and P. Rumpf. 1948a. Structure électronique du fulvène et du benzofulvène. II- Étude par la méthode des orbitales moléculaires. *Bulletin de la Société Chimique* 15:757–761.

Pullman, B. 1948. *Contribution à l'étude de la structure éléctronique des molecules organiques et en particulier des molecules substitutes. Application à l'étude des reactions d' échange isotopique.* Paris.

Pullman, B. 1974. The adventures of a quantum chemist in the kingdom of pharmacophores. In *Molecular quantum pharmacology. Proceedings of the seventh Jerusalem symposium on quantum chemistry and biochemistry held in Jerusalem, March 31th–April 4th 1974*, ed. E. Bergmann and B. Pullman, 9–36. Dordrecht: D. Reidel Publishing Co.

Pullman, B. 1976. Introduction to symposium III – Quantum organic chemistry and beyond. In *The new world of quantum chemistry. Proceedings of the second international congress of quantum chemistry held at New Orleans, USA, April 19–24 1976*, ed. B. Pullman and R. Parr, 133–135. Dordrecht: D. Reidel Publishing Company.

Pullman, B. 1979. Reminiscences. *International Journal of Quantum Chemistry. Quantum Biology Symposium* 6:33–45.

Pullman, B. 1998. *The atom in the history of human thought*. Oxford: Oxford University Press.

Pullman, B., and R. Parr, eds. 1976. *The new world of quantum chemistry. Proceedings of the second international congress of quantum chemistry held at New Orleans, USA, April 19–24 1976*. Dordrecht: D. Reidel Publishing Company.

Pullman, B., and A. Pullman. 1952. *Les théories électroniques de la chimie organique*. Paris: Masson.

Pullman, B., and A. Pullman. 1960. Some electronic aspects of biochemistry. Conference on Molecular Quantum Mechanics, University of Colorado at Boulder, June 21–27, 1960. *Reviews of Modern Physics* 32:428–436.

Pullman, B., and A. Pullman. 1962. Electronic delocalization and biochemical evolution. *Nature* 196:1137–1142.

Pullman, B., and A. Pullman. 1963. *Quantum biochemistry*. New York: Interscience Publishers.

Pullman, B., and A. Pullman. 1963a. Why molecular orbitals. In *Quantum biochemistry*, ed. B. Pullman and A. Pullman, 3–8. New York: Interscience Publishers.

Radom, L. 2008. Pople, John Anthony. *New Dictionary of Scientific Biography* 129–133. New York: Scribner's.

Ramsey, J. L. 1997. Molecular shape, reduction, explanation and approximate concepts. *Synthese* 111:231–251.

Ramsey, J. L. 2000. Of parameters and principles: producing theory in twentieth physics and chemistry. *Studies in History and Philosophy of Science* 31:549–567.

Ramsay, D. A., and J. Hinze, eds. 1975. *Selected papers of R.S. Mulliken*. Chicago: University of Chicago Press.

Ransil, B. J. 1960. Studies in molecular structure. I. Scope and summary of the diatomic molecule program. *Reviews of Modern Physics* 32:239–244.

Ransil, B. J. 1960a. Studies in molecular structure. II. LCAO-MO-SCF wave functions for selected first-row diatomic molecules. *Reviews of Modern Physics* 32:245–254.

Reinhardt, C., ed. 2001. *Chemical sciences in the 20th century. Bridging boundaries*. New York: Wiley-VCH.

Reinhardt, C. 2006. *Shifting and rearranging: physical methods and the transformation of modern chemistry*. Sagamore Beach, MA: Science History Publications.

Rich, A., and N. Davidson, eds. 1968. *Structural chemistry and molecular biology, a volume dedicated to Linus Pauling by his students, colleagues and friends*. San Francisco: Freeman.

Richards, G. 1979. Third age of quantum chemistry. *Nature* 278:507.

Rivail, J. L., and B. Maigret. 1998. Computational chemistry in France: a historical survey. *Reviews in Computational Chemistry* 12:367–380.

Roberts, J. D. 1952. Book review of Coulson's *Valence*. *Chemical and Engineering News* 30:5189.

Robertson, R. 1923. Opening remarks by the chairman. *Transactions of the Faraday Society* 19:483–484.

Roche, J. 1994. The non-medical sciences 1939–1970. In *The history of the university of Oxford. VIII. Twentieth century*, ed. B. H. Harrison, 251–289. Oxford: Oxford University Press.

Rodebush, W. H. 1928. The electron theory of valence. *Chemical Reviews* 5:509–531.

Roothaan, C. C. J. 1951. New developments in molecular orbital theory. *Reviews of Modern Physics* 23:69–89.

Roothaan, C. C. J. 1958. Evaluation of molecular integrals by digital computer. *Journal of Chemical Physics* 28:982–983.

Roothaan, C. C. J. 1960. Self-consistent field theory for open shells of electronic systems. *Reviews of Modern Physics* 32:179–185.

Roothaan, C. C. J. 1991. My life as a physicist: memories and perspectives. *Journal of Molecular Structure THEOCHEM* 234:1–12.

Roux, M., and R. Daudel. 1955. Effet de la liaison chimique sur la densité électronique. Cas de la molécule Li_2. *Comptes Rendus Hebdomadaires des Séances de l'Académie des Sciences* 244:90–92.

Roux, M., S. Besnainou, and R. Daudel. 1956. Recherches sur la répartition de la densité électronique dans les molécules. I. Effet de la liaison chimique. *Journal de Chimie Physique* 53:218–221.

Russell, C. A. 1971. *The history of valency.* New York: Humanities Press.

Russell, C. A., ed. 1985. *Recent developments in the history of chemistry.* Kent: The Royal Society of Chemistry.

Russell, C. A., and G. K. Roberts, eds. 1985. *Chemical history: reviews of the recent literature.* Cambridge, UK: Royal Society of Chemistry Publishers.

Russo, A. 1982. Mulliken e Pauling: Le Due vie della chimica-fisica in America. *Testi & Contesti* 6:37–59.

Salem, L., and H. C. Longuet-Higgins. 1960. The alternation of bond length in long conjugated chain molecules. II. The polyacenes. *Proceedings of the Royal Society of London. Series A* 255:435–443.

Sanchez, M., R. Daudel, P. D. Dacre, R. McWeeny, S. Kwun, and C. Valdemoro. 1977. Using group (or loge) functions to explore the transferability of chemical bonds. *International Journal of Quantum Chemistry* 11:415–426.

Sanchez-Ron, J. M. 1987. The reception of special relativity in Great Britain. In *The comparative reception of relativity*, ed. T. Glick, 27–58. Dordrecht: Springer.

Scerri, E. R. 1997. Bibliography on philosophy of chemistry. *Synthese* 111:305–324.

Scerri, E. R. 2003. Constructivism, relativism and chemical education. *Annals of the New York Academy of Sciences* 988:359–369.

Scerri, E. R. 2007. The ambiguity of reduction. *Hyle* 13:67–81.

Scerri, E. R. 2008. *Collected papers on the philosophy of chemistry.* London: Imperial College Press.

Scerri, E. R., and L. McIntyre. 1997. The case for philosophy of chemistry. *Synthese* 111:213–232.

Schaefer, L. 1991. The mutation of chemistry: the rising importance of ab initio computational techniques in chemical research. *Journal of Molecular Structure THEOCHEM* 230:5–11.

Scherr, C. W. 1955. An SCF LCAO MO study of N_2. *Journal of Chemical Physics* 23:569–578.

Schmidt, O. 1941. Charakterisierung und Mechanismen der Krebs erzeugenden Kohlenwasserstoffe. *Naturwissenschaften* 29:146–150.

Schummer, J. 1996. Philosophy of chemistry in the German democratic republic. *Hyle* 2:3–11.

Schummer, J. 1997. Challenging standard distinctions between science and technology: the case of preparative chemistry. *Hyle* 3:81–94.

Schwarz, W. H. E., D. Andrae, S. R. Arnold, J. Heidberg, H. Hellmann Jr., J. Hinze, A. Karachalios, M. A. Kovner, P. C. Schmidt, and L. Zülicke. 1999. Hans G. A. Hellmann 1903–1938. Part I, a pioneer of quantum chemistry. *Bunsen—Magazin* 1:10–21. [English transl. by Mark Smith and W. H. E. Schwarz, with revisions by J. Hinze and A. Karachalios.]

Schwarz, W. H. E., D. Andrae, S. R. Arnold, J. Heidberg, H. Hellmann Jr., J. Hinze, A. Karachalios, M. A. Kovner, P. C. Schmidt, and L. Zülicke. 1999a. Hans G. A. Hellmann 1903–1938. Part II, a German pioneer of quantum chemistry in Moscow. *Bunsen—Magazin* 2:60–70. [English transl. by Mark Smith and W. H. E. Schwarz, with revisions by J. Hinze and A. Karachalios.]

Schweber, S. S. 1986. The empiricist-temper regnant: theoretical physics in the United States. *Historical Studies in the Physical Sciences* 17:55–98.

Schweber, S. S. 1990. The young John Clarke Slater and the development of quantum chemistry. *Historical Studies in the Physical Sciences* 20:339–406.

Schweber, S. S., and M. Wächster. 2000. Complex systems, modelling and simulation. *Studies in History and Philosophy of Science* 31:583–609.

Servos, J. W. 1985. History of chemistry. *Osiris* 1:132–146.

Servos, J. W. 1990. *Physical chemistry from Ostwald to Pauling, the making of a science in America.* Princeton, NJ: Princeton University Press.

Sidgwick, N. V. 1910. *Organic chemistry of nitrogen.* Oxford: Clarendon Press.

Sidgwick, N. V. 1923. The nature of the non-polar link. *Transactions of the Faraday Society* 19:469–475.

Sidgwick, N. V. 1923a. Co-ordination compounds and the Bohr atom. *Journal of the Chemical Society* 123:725–730.

Sidgwick, N. V. 1927. *The electronic theory of valency.* Oxford: Clarendon Press.

Sidgwick, N. V. 1933. *Some physical properties of the covalent link in chemistry.* New York: Cornell University Press.

Sidgwick, N. V. 1934. Wave mechanics and structural chemistry. *Nature* 7:529–530.

Sidgwick, N. V. 1936. Presidential address delivered at the annual general meeting, April 16th 1936 – structural chemistry. *Journal of the Chemical Society* 149:533–538.

Sidgwick, N. V. 1937. Presidential address delivered at the annual general meeting, March 18th 1937 – hybrid molecules. *Journal of the Chemical Society* 151:694–699.

Simões, A. 1993. *Converging trajectories, diverging traditions: chemical bond, valence, quantum mechanics and chemistry, 1927–1937.* Ph.D. dissertation, University of Maryland.

Simões, A. 1999. Mulliken, Robert Sanderson. In *American national biography.* Vol. 16., 77–79. New York: Oxford University Press.

Simões, A. 2002. Dirac's claim and the chemists. *Physics in Perspective* 4:253–266.

Simões, A. 2003. Chemical physics and quantum chemistry in the twentieth-century. In *Modern physical and mathematical sciences.* Vol. 5. ed. Mary Jo Nye, 394–412. Cambridge, UK: Cambridge University Press.

Simões, A. 2004. Textbooks, popular lectures and sermons: the quantum chemist Charles Alfred Coulson and the crafting of science. *British Journal for the History of Science* 37:299–342.

Simões, A. 2007. In between words: G.N. Lewis, the shared pair bond and its multifarious contexts. *Journal of Computational Quantum Chemistry* 28:62–72.

Simões, A. 2008. Mulliken, Robert Sanderson. *Complete Dictionary of Scientific Biography* 23:209–214.

Simões, A. 2008a. A quantum chemical dialogue mediated by textbooks: Pauling's *The nature of the chemical bond* and Coulson's *Valence*. *Notes and Records of the Royal Society of London* 62:259–269.

Simões, A. 2009. Quantum chemistry. In *Compendium of quantum physics: concepts, experiments, history and philosophy*, ed. D. Greenberger, B. Falkenburg, K. Hentschel, and F. Weinert, 518–523. Berlin: Springer-Verlag.

Simões, A., and K. Gavroglu. 1997. Different legacies and common aims: Robert Mulliken, Linus Pauling and the origins of quantum chemistry. In *Conceptual perspectives in quantum chemistry*, ed. J.-L. Calais and E. S. Kryachko, 383–413. Netherlands: Kluwer Academic Press.

Simões, A., and K. Gavroglu. 1999. Quantum chemistry *qua* applied mathematics. The contributions of Charles Alfred Coulson 1910–1974. *Historical Studies in the Physical Sciences* 29:363–406.

Simões, A., and K. Gavroglu. 2000. Quantum chemistry in Great Britain: developing a mathematical framework for quantum chemistry. *Studies in the History and Philosophy of Modern Physics* 31:511–548.

Simões, A., and K. Gavroglu. 2001. Issues in the history of theoretical and quantum chemistry, 1927–1960. In *Chemical Sciences in the 20th century. Bridging boundaries*, ed. C. Reinhardt, 51–74. New York: Wiley-VCH.

Slater, J. C. 1927. The structure of the helium atom. I. *Proceedings of the National Academy of Sciences of the United States of America* 13:423–430.

Slater, J. C. 1928. The normal state of He. *Physical Review* 32:349–360.

Slater, J. C. 1928a. The self-consistent field and the structure of atoms. *Physical Review* 32:339–348.

Slater, J. C. 1929. The theory of complex spectra. *Physical Review* 34:1293–1322.

Slater, J. C. 1930. Cohesion in monovalent metals. *Physical Review* 35:509–529.

Slater, J. C. 1930a. Note on Hartree's method. *Physical Review* 35:210–211.

Slater, J. C. 1931. Directed valence in polyatomic molecules. *Physical Review* 37:481–489.

Slater, J. C. 1931a. Molecular energy levels and valence bonds. *Physical Review* 38:1109–1144.

Slater, J. C. 1932. Note on molecular structure. *Physical Review* 41:255–257.

Slater, J. C. 1939. *Introduction to chemical physics*. New York: McGraw-Hill.

Slater, J. C. 1954. Work on molecular theory in the solid-state and molecular theory group. In *Symposium on molecular physics. Held at Nikko on the occasion of the International Conference on Theoretical Physics*, 1–4. Tokyo: Maruzen Co., Ltd.

Slater, J. C. 1964. Robert Mulliken of Newburyport. In *Molecular orbitals in chemistry, physics and biology, a tribute to R.S. Mulliken*, ed. P.-O. Löwdin and B. Pullman, 17–20. New York: Academic Press.

Slater, J. C. 1965. Molecular orbital and Heitler-London methods. *Journal of Chemical Physics* 43:S11–S17.

Slater, J. C. 1967. Quantum physics in America between the wars. *International Journal of Quantum Chemistry, Symposium* 1S:1–23.

Slater, J. C. 1967a. The current state of solid-state and molecular theory. *International Journal of Quantum Chemistry* 1:37–102.

Slater, J. C. 1975. *Solid state and molecular theory: a scientific biography*. New York: John Wiley & Sons.

Slater, J. C., and N. H. Frank. 1933. *Introduction to theoretical physics*. New York: McGraw-Hill.

Smiles, S. 1859/2006. *Self-Help: with illustrations of conduct and perseverance*, abridged by G. Bull, with introduction by Sir Keith Joseph Harmondsworth. Middlesex: The Echo Library.

Smith, S. J., and B. T. Sutcliffe. 1997. The development of computational chemistry in the United Kingdom. *Reviews in Computational Chemistry* 10:271–316.

Snelders, H. A. M. 1986. The historical background of Debye and Hückel's theory of strong electrolytes. *Proceeding of the Royal Netherlands Academy of Arts and Sciences* B89:79–94.

Sommerfeld, A. 1923. *Atomic structure and spectral lines* (transl. by H.L. Brose). New York: E.P. Dutton and Co.

Sponer, H. 1925. Anregunspotentiale der Bandenspketren des Stickstoffs. *Zeitschrift fur Physik* 34:622–633.

Sopka, K. R. 1988. *Quantum physics in America. The years through 1935*. American Institute of Physics/Tomash Publishers.

Stanley, M. 2007. *Practical mystic. Religion, science and A.S. Eddington*. Chicago: The University of Chicago Press.

Stranges, A. N. 1982. *Electrons and valence, development of the theory 1900–1925*. College Station, TX: A & M University Press.

Streitwieser, A., and J. I. Brauman. 1965. *Supplemental tables of molecular orbital calculations*. New York: Pergamon Press.

Stuewer, R. H. 1975. *The Compton effect: turning point in physics*. New York: Science History Publishers.

Sugiura, Y. 1927. Uber die Eigenschaften des Wasserstoffmoleküls im Grundzustande. *Zeitschrift fur Physik* 45:489–492.

Sutton, L. E. 1958. Obituary notices, Nevil Vincent Sidgwick 1873–1952. *Proceedings of Chemical Society of London* 310–319.

Sutton, L. E. 1975. *Nevil Vincent Sidgwick. Dictionary of Scientific Biography*, Vol. 12, 418–420. New York: Scribner's.

Symposium on molecular physics. Held at Nikko on the occasion of the International Conference on Theoretical Physics. 1954. Tokyo: Maruzen Co., Ltd.

Syrkin, Y. K., and M. Dyatkina. 1950. *The chemical bond and the structure of molecules*. [translated and revised by A. Partridge and D.O. Jordan]. London: Butterworth & Co.

Tatevskii, V. M., and M. I. Shakhparanov. 1952. About a Machistic theory in chemistry and its propagandists [transl. by I.S. Bengelsdorf]. *Journal of Chemical Education* 78:13–14.

Thomson, J. J. 1907. *The corpuscular theory of matter*. London: A. Constable & Co.

Thomson, J. J. 1913. On the structure of the atom. *Philosophical Magazine* 26:729–799.

Thomson, J. J. 1914. The forces between atoms and chemical affinity. *Philosophical Magazine* 27:757–789.

Tizard, H. T. 1954. Nevil Vincent Sidgwick—1873–1952. *Obituary Notices of the Fellows of the Royal Society of London* 9:236–258.

Uhlenbeck, G. E., and S. Goudsmidt. 1925. Ersetzung der Hypothese von unmechanischen Zwang durch eine Forderung bezüglich des inneren Verhaltens jedes einzelnen Elektrons. *Naturwissenschaften* 13:953–954.

Uhlenbeck, G. E., and S. Goudsmidt. 1926. Spinning electrons and the structure of spectra. *Nature* 117:264–265.

Urey, H. C. 1929. Discussion. *Chemical Reviews* 6:546–547.

Urey, H. C. 1933. Editorial. *Journal of Chemical Physics* 1:1.

van Brakel, J. 2000. *Philosophy of chemistry. Between the manifest and the scientific image*. Leuven: Leuven University Press.

van Brakel, J. 2003. The *Ignis Fatuus* of reduction and unification. Back to the rough ground. *Annals of the New York Academy of Sciences* 988:30–43.

van der Waerden, B. L. 1960. Exclusion principle and spin. In *Theoretical physics in the twentieth century. A memorial volume to W. Pauli*, ed. M. Fierz and V. F. Weisskopf, 199–244. New York: Interscience Publishers.

van Spronsen, J. J. 1969. *The periodic system of chemical elements*. Amsterdam: Elsevier.

Van Vleck, J. H. 1928. The new quantum mechanics. *Chemical Reviews* 5:467–507.

Van Vleck, J. H. 1932. *Theory of electric an magnetic susceptibilities*. London: Oxford University Press.

Van Vleck, J. H. 1933. On the theory of the structure of CH_4 and related molecules. Part I. *Journal of Chemical Physics* 1:177–182.

Van Vleck, J. H. 1933a. On the theory of the structure of CH_4 and related molecules. Part II. *Journal of Chemical Physics* 1:219–238.

Van Vleck, J. H. 1934. On the theory of the structure of CH_4 and related molecules. Part III. *Journal of Chemical Physics* 2:20–30.

Van Vleck, J. H. 1935. The group relation between the Mulliken and Slater-Pauling theories of valence. *Journal of Chemical Physics* 3:803–806.

Van Vleck, J. H. 1964. American physics comes of age. *Physics Today* 17:21–26.

Van Vleck, J. H. 1968. My Swiss visits of 1906, 1926, and 1930. *Helvetica Physica Acta* 41:1234–1237.

Van Vleck, J. H. 1970. Spin, the great indicator of valence behavior. *Pure and Applied Chemistry* 24:235–255.

Van Vleck, J. H. 1971. Reminiscences on the first decade of quantum mechanics. *International Journal of Quantum Chemistry, Symposium* 5:3–20.

Van Vleck, J. H., and A. Sherman. 1935. The quantum theory valence. *Reviews of Modern Physics* 7:167–227.

Vermeeren, H. 1986. Controversies and existence claims in chemistry: the theory of resonance. *Synthese* 69:273–290.

Vermulapalli, G. K. 2003. Property reduction in chemistry. Some lessons. *Annals of the New York Academy of Sciences* 988:90–98.

Warwick, A. 1987. Einstein's theory of relativity and British physics in the early 20th century. Ideas and production. *Journal of the History of Ideas* 7:82–95.

Warwick, A. 1989. *The electrodynamics of moving bodies and the principle of relativity in British physics, 1894–1919*. Ph.D. dissertation, University of Cambridge.

Warwick, A. 2003. *Masters of theory. Cambridge and the rise of mathematical physics*. Chicago: University of Chicago Press.

Weininger, S. J. 1984. The molecular structure conundrum: can classical chemistry be reduced to quantum chemistry? *Journal of Chemical Education* 61:939–944.

Weyl, H. 1931. *Group theory and quantum mechanics*. New York: E.P. Dutton & Co. [First German edition, *Gruppentheorie und Quantenmechanik*. Leipzig: Hirzel, 1928.]

Weyl, H. 1946. Encomium. *Science* 103:216–218.

Wheland, G. W. 1934. The quantum mechanics of unsaturated and aromatic molecules: a comparison of two methods of treatment. *Journal of Chemical Physics* 2:474–481.

Wheland, G. W. 1944. *The theory of resonance and its applications to organic molecules*. New York: John Wiley & Sons.

Wheland, G. W. 1952. Book review of *Valence* by C.A. Coulson. *Journal of the American Chemical Society* 74:5810.

Wheland, G. W. 1955. *Resonance in organic chemistry*. New York: John Wiley & Sons.

Whiffen, D. H. 1994. Leslie Ernest Sutton, 22 June 1906–30 October 1992. *Biographical Memoirs of the Fellows of the Royal Society of London* 40:369–382.

Wilson, E. B. 1976. Fifty years of quantum chemistry. *Pure and Applied Chemistry* 47:41–47.

Wilson, E. B. 1977. Impact of the Heitler-London hydrogen molecule paper on chemistry. *International Journal of Quantum Chemistry* S11:17–28.

Wilson, E. B. 1980. Some personal scientific reminiscences. *International Journal of Quantum Chemistry Symposium* 14:17–29.

Wurmser, R. 1962. Albert Szent-Györgyi and modern biochemistry. In *Horizons in biochemistry. Albert Szent-Györgyi dedicatory volume*, ed. M. Kasha and B. Pullman, 1–7. New York: Academic Press.

Wolfenden, J. H. 1972. The anomaly of strong electrolytes. *Ambix* 19:175–196.

Woody, A. I. 2000. Putting quantum mechanics to work in chemistry: the power of diagrammatic representations. *Philosophy of Science Proceedings* 67:S612–S627.

Woody, A. I. 2003. On explanatory practice and disciplinary identity. *Annals of the New York Academy of Sciences* 988:22–29.

Woolley, R. G. 1978. The quantum interpretation of molecular structure. *International Journal of Quantum Chemistry* 12(Suppl. 1):307–313.

Woolley, R. G. 1985. The molecular structure conundrum. *Journal of Chemical Education* 62:1082–1084.

Woolley, R. G. 1987. Must a molecule have a shape? *Journal of the American Chemical Society* 100:1073–1078.

Woolley, R. G. 1988. Quantum theory and the molecular hypothesis. In *Molecules in physics, chemistry, and biology*, ed. J. Maruani, 45–89. Dordrecht: Kluwer Academic Publishers.

Zahradnik, R. 1971. Some remarks on the present state of the theory of chemical reactivity. In *Colloque international sur les aspects de la chimie quantique contemporaine*, 8–13 July 1970, Menton, France, org. R. Daudel and A. Pullman, 87–119. Paris: Éditions du Centre Nationale de la Recherche Scientifique.

Zerner, M. C. 1999. In memory of B. Pullman 1919–1996. *International Journal of Quantum Chemistry* 75:137–138.

Index

Abegg, R., 50
Academy of Sciences, 32, 64, 120, 202, 207, 237
Academy's Committee on Science and Public Policy, 237, 238
Accuracy, 5, 142, 152, 157, 209, 210, 223, 227–228, 233, 234, 236, 239, 242, 245
Activation, 23–24, 99, 102
Adiabatic transition, 36, 43
Advanced Research Projects Agency (ARPA), 238
Affinity, 6, 29, 32, 54, 77, 121, 123, 125, 181, 255
Air Force Cambridge Research Center, 220
Alvarez, L., 88
American Chemical Society, 18, 68, 69, 202
American Institute of Physics, 218
Americans, 85, 88, 98, 99, 101–102, 156, 160–161, 203
Ammonia, 152, 160
Analogy/ies, 10, 22, 36, 40–41, 43, 58, 70, 88, 119–120, 124, 154, 168, 172, 194, 209–210, 215, 254
Analytic calculations, 24, 224
Anderson, P. W., 208
Annus Mirabilis, 61
Antiparallel, 16–17, 65, 72, 181
Antisymmetry, 28, 89, 152
Appel, K., 212, 260
Appleton, E., 138

Appropriation, 54–55, 184, 246, 249, 250, 255, 257
Approximation(s), 3, 56, 58, 62, 65–68, 77, 79, 81, 84–85, 91–92, 95–96, 101–102, 105, 111, 115, 131–132, 139–141, 160, 166, 170–171, 177, 199, 201, 208–209, 215, 218, 221, 223–227, 230–231, 233, 236, 238, 252
  Hartree-Fock, 236
  Hückel LCAO, 201
  methods, 96, 105, 115, 131–132, 139, 141, 171, 225
  Pariser-Parr-Pople (PPP), 224
  self-consistent field, 139–140
  zero differential overlap, 224
Arbitrariness, 70, 73, 117–118, 180
Argon, 137–138
Armaments Research Department, 227
Armit, J. W., 30
Armstrong, H. E., 78, 109, 250
Army Research Office, 220
Arndt, F. G., 78
Aromatic molecules, 25, 30, 55, 148, 206, 232
Arrhenius, S., 149–249
Arsem, W. C., 51
Atom(s), 10, 14–16, 18–25, 27, 30, 32, 36, 39, 41–44, 46, 48–60, 62–68, 70–78, 80–81, 83–86, 88–94, 99–101, 103, 107, 109–111, 114, 121, 132–133, 135–141, 143–144, 146–151, 153, 157–158, 160, 162–164, 168–169, 172, 179, 182, 194, 202, 206, 210,

Atom(s) (cont.)
  212, 217–219, 234, 238, 247, 249, 251–252, 255–256, 259
  cubic, 49–53
  separated, 36, 43–44, 46, 57, 73, 76, 80, 91, 143, 210
  united, 36, 43–44, 46, 57, 58, 73, 128, 144, 172
Atomic structure, 18, 21, 26, 50, 53–54, 133, 155
Aufbau, 3, 6, 39, 128, 144

Balliol College, Oxford, 167
Barnett, M. P., 203, 204, 260
Barriol, J., 190, 193, 198
Baudet, Jeanne, 190
Bauer, E., 191–192
Bayer, A., 78
Beaven, H. C., 158–159
Benzene, 27, 30, 35, 78–80, 110, 117, 124, 146–147, 152, 161–163, 167, 178–179, 195, 206, 221, 224
Bergmann, E., 195
Berlin, T. H., 203
Berthier, G., 190, 196
Berthod, Hélène, 190
Besnainou, Sylvette, 190
Bethe, H., 95, 154
Binuclear orbit, 10, 55–58, 133
Biology, 2, 126, 138, 174, 193, 201–202, 217–219, 239, 253, 255, 260–261
Birge, R. T., 39, 41, 61, 69, 70, 87
Bishop, R., 256
Bjerrum, N., 11, 13–14
Blackett, P. M. S., 154
Bloch, F., 90–91
Bohr, N., 6, 10–11, 19, 27, 34, 39, 43, 48, 50, 53, 55, 57, 61, 89–90, 93–94, 107, 128, 133, 138–139, 151
Bohr's atomic model, 53–54
Bond(s), 9, 14, 17–18, 22–23, 25, 28–29, 37, 50, 51–53, 58, 61–68, 70, 72–80, 83, 85, 86, 91–93, 96, 99, 101–103, 108–109, 111, 114–115, 117–119, 124–125, 146–147, 149, 151–152, 161–169, 171–173, 177–178, 180–182, 184, 190, 192–196, 201, 206, 233–235, 242, 251
  angles, 67–68, 77, 168
  chemical, 3, 14–15, 25, 31, 47, 50–53, 55, 57–58, 61, 64–65, 67–69, 73, 75–79, 83–84, 86, 90, 102, 104–105, 108, 115–121, 125, 128, 132, 145, 149, 152, 166–167, 178, 181–182, 188–189, 192–194, 199, 205, 230, 251
  covalent, 2, 31, 73, 76, 83, 108, 145–146, 251
  directional character of, 66, 77
  direction of, 66, 168
  double, 25, 27–29, 50, 52, 74, 78, 81, 124, 146, 149, 161–162, 167, 195
  electron-pair, 47, 53, 57, 58, 61–66, 68, 73, 75, 81, 84, 145, 167, 206, 250
  energy, 66, 76–77, 80, 102, 124, 206
  homopolar bond, 14, 17, 37
  length, 163, 165, 166, 231
  nature of chemical, 15, 47, 50, 53, 57, 61, 64, 69, 73, 76, 77, 79, 86, 105, 115–121, 125, 128, 132, 145, 167, 178, 180, 189, 251
  nonpolar, 50, 51, 53, 107
  one electron, 74
  order, 162, 163, 168–169, 182, 189, 203, 206, 231
  partial ionic character of, 76, 78, 251
  π, 29
  polar, 50, 53
  shared electron, 55, 75, 122
  σ, 29
  single, 50, 52, 76, 78, 169
  stability, 32
  three-electron, 74–75, 78, 251
  two-electron, 51
Bonding power, 44, 45, 69, 70
Born, M., 12–13, 16, 22, 26, 34, 55, 87–90, 94, 98, 103, 136, 145, 161, 163
Boron Hydrides, 74, 152
Bose-Einstein statistics, 135

Boulder Conference, 183, 197–199, 201, 223, 226, 228, 230–231, 235
Boundary/ies, 2, 6, 47, 57, 77, 131, 234, 237, 261
Boys, S. F., 207, 209, 227–230, 260
Bragg, Sir L., 55, 134
Bragg, W. H., 55, 62, 132, 145, 147, 154
Brakel, J. Van, 257
Bray, W. C., 48
Bridgman, P. W., 61, 91, 112, 115
Brillouin, L., 194
Brion, Hélène, 190, 194
British Association for the Advancement of Science, 18, 107, 131, 143, 145, 153, 249, 250
Brodetsky, S., 158
Brooks, H., 237
Buckingham, D., 18
Burrau, O., 17, 20, 58, 59
Bush, V., 141, 150
Butadiene, 152, 196, 254
Butlerov, A. M., 121

Calais, J. P., 212, 260
Calculation(s), 3, 5, 9, 10, 12, 15, 20, 24–25, 30, 32–33, 37, 57–59, 64, 71, 75, 80–81, 83–85, 88–90, 93, 95–96, 100–101, 103, 114–115, 120, 126, 132, 136–140, 143, 148–150, 152–153, 155–157, 161–167, 172, 176, 180, 182, 185, 187, 189, 193, 195, 197, 199, 202–211, 220–224, 227–234, 236–241, 245–246, 252–254, 261
  ab initio, 5, 126, 172, 197, 204, 222, 225, 229, 239, 242
  analytical, 5, 116, 255
  semiempirical, 5, 84, 206, 221, 227, 229, 241
Calculators/calculating machines, 204–205, 212, 222, 228, 230, 254
California Institute of Technology, 39, 55, 110
Cancérisation par les Substances Chimiques et Structure Moléculaire, 199, 200

Carbon, 24, 28, 66–67, 79, 112, 147, 149, 162, 168, 180
Carbon atom, 24, 27, 30, 63, 66–67, 78, 162, 168, 179
Carcinogenesis, 193, 199, 200
Carcinogenic activity, 189, 200
Cavendish Professor of Experimental Physics, 134
Censor's Office, 156
Center of Material Sciences and Engineering, 209
Centre de Chimie Théorique, 187, 190, 193
Centre de Mécanique Ondulatoire Appliquée, 193, 198
Centre Européen de Calculs Atomiques et Moléculaires, 196
Centre Inter-regional de Calcul Éléctronique, 196
Centre National de la Recherche Scientifique, 190, 245. *See also* CNRS
Chair of Quantum Chemistry, 216
Chapman, S., 136, 173
Chemical combination, 19, 50, 52, 83
Chemical Elements and their Compounds, 107
Chemical fact, 24, 77, 108, 125, 133, 251
Chemical force, 24, 32, 121
Chemical formula, 23–24
Chemical physics, 27, 105–106, 112–115, 217, 235
Chemical problems, 3, 9, 16, 18, 25–26, 29, 56, 69, 87, 96, 105, 107, 109, 115, 117, 132, 176, 183–184, 199, 218, 239–241, 245–246, 249
Chemical reaction, 9, 21, 23, 43, 106, 146, 149
Chemical reactivity, 49, 124, 169, 200, 206, 254
Chemical Reviews, 30, 64, 88, 101
Chemical Society, 106, 107, 111, 151
Chemical thermodynamics, 247–249, 257–258
Chemistry, 2–9, 12, 17–20, 22–27, 29, 31–33, 36–37, 39–40, 44, 46–49, 53, 55–56, 65, 68,

Chemistry (cont.)
  73, 77–78, 81, 84, 86–87, 91–94, 96, 98, 102, 104–109, 111–119, 121, 123, 125–129, 131–133, 135–136, 138, 143–145, 147–148, 151–153, 158–160, 162, 164, 167–171, 173–177, 180–185, 187–193, 195–205, 209–210, 212–214, 216–218, 222, 224–234, 236–243, 245–261
  computational, 217, 235
  and computers, 217
  inorganic, 18, 33, 53, 126, 148–149, 260
  and mathematics, 183
  organic, 21, 25, 30–31, 67, 78–79, 106–109, 112–113, 118–120, 122–123, 133, 147, 149, 162, 168, 172, 188, 195, 198–200, 225, 227, 231, 235, 237, 239, 241–242, 250, 252, 254
  philosophy of, 5, 247, 261
  physical, 9, 13, 27, 30, 32, 48, 77, 97, 100, 107, 112, 113–114, 116, 134, 167–168, 192, 205, 235, 249–250
  role of theory in, 247, 253, 255, 261
  Second Instrumental Revolution in, 261
  structural, 3, 77, 109, 111, 126, 144, 178, 225, 231
  theoretical, 18, 33, 126, 131, 145, 148–150, 166, 170, 173–174, 177–178, 182–185, 187, 190–193, 196–197, 199–200, 247, 250–251
  theoretical particularity of, 255
Chemists' culture, 19, 38, 47, 65, 107, 120, 122–123, 128
Circulation, 197, 207, 214, 260
Clark, G. L., 19
Classical physics, 16, 28, 120, 145, 205
Claus, A., 78, 80
Clementi, E., 217–218, 243, 260
Clifton College, 158
Closed shell, 20, 30, 43, 83, 101, 238
Clusters of issues, 2–5, 7, 246
CNRS, 190–191, 193, 196, 245, 254
Colby, W. F., 39
Colloque de la Liaison Chimique 191, 193.
  See also Conference, Paris

Complex spectra, 90, 93, 95, 140
Compounds, 19–20, 22, 30, 40, 52, 65, 67, 74–75, 78, 80, 101, 107, 112, 126, 147–149, 188, 195–196, 200, 245
  aromatic, 26–27, 30, 78, 102, 125
  benzenoid, 26–27, 30, 195
  covalent, 75, 148
  nonpolar, 50, 52, 75
  polar, 49, 50, 52, 75
Compressibility, 61, 137
Compromise(s), 122, 213, 233
Compton, A. H., 35, 45, 62, 112–113, 209
Computer(s), 2, 4–5, 7, 126–127, 132, 153, 157–158, 183, 187, 193, 195–196, 199, 204, 207, 209, 213, 215–219, 221–235, 237–243, 245–246, 251–254, 261
  and ab initio computations, 221–222, 226, 233
  and cultures of quantum chemistry, 187, 225
  EDSAC, 157, 228
  hardware, 187, 222, 240–241, 254
  high speed, 2, 157, 224
  program(s), 5, 126, 218, 222–223, 228, 230–232, 235, 237, 239, 241
  and semiempirical approximations, 215, 223, 226
  software, 187, 222, 240–241, 254
  technology, 187, 209, 221–223
Computing Laboratory, 150, 173, 253
Concept(s), 1–3, 18, 21, 23, 27, 37, 41, 44, 47, 51–52, 56, 70, 73, 75, 77, 84, 97, 108, 110–111, 114, 117–119, 121–122, 124–125, 133, 139, 143, 149, 163, 166–169, 173, 175–176, 180–183, 185, 189, 192, 194, 197, 212, 214, 219, 227, 231, 233–234, 246, 248, 251–253, 255–257, 259, 261
  chemical, 86, 109, 117, 168, 184, 190, 193, 206, 215, 234, 240, 242, 259
  resonance, 75, 117, 119, 124, 149, 166, 251
Conceptual framework, 24, 125, 219, 246

Index

Conceptual scheme, 42, 69, 84, 85
Condon, E. U., 58, 59, 62, 90, 217
Conference(s), 1–3, 126, 144, 146–147, 187, 192, 199, 202–208, 217, 221, 230–234, 237, 238–241, 246, 256
  Boulder, 183, 197–199, 201, 223, 226, 228, 230–231, 235
  on Computational Support for Theoretical Chemistry, 1, 187, 237
  in Jerusalem, 195, 197
  Nikko, 210
  Paris, 126, 187, 192, 203
  Sanibel Island, 126, 216–217, 236, 241
  Shelter Island, 126, 202–206, 208–211, 221, 230
  in Sweden, 197
  Texas (Molecular Quantum Mechanics Conference), 221, 233
Conference on Computational Support for Theoretical Chemistry, 1, 187, 237
Configuration Interaction, 92, 171, 196, 206, 208, 210–211, 224, 236
Conjugated systems, 80, 168, 169, 195, 197, 199
Connant, J. B., 121
Consensus, 3, 5, 38, 105, 110, 172, 181, 240, 246, 259
Contexts, 3, 48, 54–55, 219
Contingency, 4, 38, 128
Contingent, 4, 6–7, 246, 256
Coolidge, A. S., 215, 222
Correlation diagram, 46, 83
Coulson, C. A., 1–4, 121, 131–132, 134, 141, 148, 150, 154, 158–185, 189, 191–193, 196, 199, 201, 203–208, 210, 212, 214, 219, 225–227, 229–230, 232–235, 242–243, 249–250, 253, 259, 260
Courant, R., 34
Craig, D. P., 170–171, 175, 206
Crawford, B. L., 203, 206–207, 260
Crick, F., 141
Criterion/a, 4–6, 28, 30, 45, 63, 65–68, 76, 78, 117, 143, 161, 164, 171, 250

Crystal(s), 26, 34, 55, 62, 67–68, 76, 88, 91, 96, 112–114, 136–137, 147–148, 168, 192, 204, 208, 209–210, 212, 219, 221, 251, 260
Culture(s), 4–5, 13, 19, 37–38, 47, 65, 105, 107, 120, 122, 123, 125, 128, 158, 173, 177, 184, 187, 212, 222, 225, 227, 229–230, 235, 246–248, 253–259
Cunningham, E., 134

Dalton, J., 149
Darwin, C. G., 134
Daudel, Pascaline, 191
Daudel, R., 4, 168, 187–190, 192–194, 198–199, 202, 219, 230, 260
de Broglie, L., 188, 193–194, 198–200
Debye, P., 26–27, 34, 89, 108, 113, 145
Del-squared V Club, 135, 138
Department of Theoretical Chemistry, University of Cambridge, 227
Determinant, 62, 87, 89–92, 140, 169, 236
Dewar, J., 78–80, 175
Diamagnetism, 54, 97, 99
Diatomic project, 223, 231
Dickinson, R. G., 55
Dielectric constant, 31, 56, 112
Differential analyzer, 141–143, 150, 154, 157
Dilute solutions, 26, 149
Dirac, P. A. M., 9, 13, 20, 35, 37, 45, 61, 89, 90–91, 94, 96, 131, 135, 138, 145, 150, 154, 157, 159, 176, 185, 207, 208, 223, 256
Disciplinary, 2, 5, 27, 47, 77, 87, 97, 105, 131, 210, 214, 217, 219, 243, 261
Disciplinary emergence, 3, 7
Discourse, 6, 7, 239, 255, 256, 257
Dissertation, 26, 31, 34, 40, 48, 88, 94, 136, 139, 160, 190, 195, 213, 227

Eigenfunction(s), 24, 30, 57–60, 66–68, 75, 81, 110, 140
Eigenvalues, 60, 140, 163
Eistert, B., 78
Electrodynamics, 11, 97, 98, 104, 204

Electron(s), 9, 10–11, 15–18, 20, 21–23, 25, 27–28, 30, 32, 33, 36, 41–45, 47, 49–60, 62, 64–75, 77–78, 80–81, 83–86, 88–92, 95, 97, 100, 101, 103, 110, 117, 122, 125–126, 128, 131–133, 135, 139–141, 143–146, 149–153, 161–164, 166, 168–169, 172, 181–182, 193–194, 200, 203, 205–206, 210, 215, 220–224, 228, 233–234, 236–240, 250–251, 255–256
  bonding, 64, 70–73
  charge density, 20, 189, 231, 233
  delocalized, 193
  jump, 11
  K-, 193
  L-, 22, 193
  localized, 167
  nonbonding, 44, 70, 83
  orbital, 10, 76, 172, 103
  outer, 10, 56
  $p\sigma$, 71
  pairing of, 55, 70, 72, 131, 145, 146
  π, 28, 30, 153, 162, 163, 171, 172, 206, 224
  promoted, 44, 70, 71, 72
  promotion, 41
  ring, 10, 30
  S, 62, 71–72, 90, 101
  sσ, 71
  σ, 193, 238
  spin, 16, 25, 28, 36, 54, 56, 70, 89
  unpaired, 28, 65, 70, 72, 74
  unpromoted, 44
Electronegativity/ies, 76–77, 180, 251
Electronic computers, 2, 5, 157, 215–216, 229–230, 232–235, 242
Electronic structure, 28, 41, 46, 56, 67, 75, 80–81, 95, 110, 125, 133, 138, 144, 146, 149, 160, 162, 189, 190, 194, 200, 203, 206, 213, 215, 225, 227, 228, 251, 254
Electronic Theory of Valency, The, 107
Elementary Wave Mechanics, 104
Energy, 11, 13, 15–17, 19–21, 23, 25, 28, 30, 33, 41, 45–46, 48, 54, 58–59, 61–62, 64, 66, 70–77, 79–80, 92, 93–94, 102, 110, 120, 124, 127, 134, 140, 143, 146, 149, 161–165, 168–169, 171, 176, 180, 194, 204, 206, 209, 232, 238, 241
  cohesive, 204, 210
  of formation, 19, 64, 76, 80
  potential, 28, 31, 44, 75, 140
  of promotion, 44, 71
  resonance, 58, 74, 79, 80, 96, 120, 124, 149, 161
  total, 21, 31, 44, 58, 72, 76, 124, 140, 163
Epistemological, 29, 30, 38, 48, 112, 255–256, 259
Equation, 4, 5, 9, 16–17, 21, 34, 62, 64, 66, 96, 97, 113, 134, 138–142, 150, 154, 158, 166, 169, 177, 206–207, 245, 248
  Schrödinger, 3, 17, 19–20, 22, 33, 37, 19, 37, 52, 111, 125, 128, 140, 166, 183, 215, 228, 236
  secular, 62, 169, 209, 210
Essec Prize of the French League Against Cancer, 200
ETH, 26
Ethylene, 28, 124, 147, 152, 161, 167, 178
Europe, 36, 42, 55–57, 61, 87–89, 94–95, 103, 202, 205, 208, 222, 223
European Office of Air Research and Development Command, 213
Everitt, C. W. F., 16
Ewald, P. P., 13, 14, 27
Exchange, 15–18, 22, 54, 59, 80, 90, 98, 103, 114, 118, 141, 154, 157, 163, 167, 190, 205, 208, 224, 240, 241, 260
Excited state, 24, 43, 147, 171, 194, 222, 228, 254
Exclusion principle, 15–18, 20, 25, 32–34, 37, 42, 44, 54, 58–59, 70, 81, 128, 149, 153. *See also* Pauli principle; Pauli exclusion principle
Explanation, 1–3, 9, 12, 16, 20, 21, 30, 42, 49, 54–55, 58, 62, 64, 66, 72, 75, 85, 94, 97, 101, 108, 111–112, 114, 126, 128, 131, 133,

143–144, 146, 148–149, 161, 180, 182, 192, 195, 200, 201, 205, 252, 253
Eyring, H., 103, 203, 207, 214, 217

Faculty Board of Mathematics, 150
Fairbank, W. M., 16
Fajans, K., 87
Faraday Society, 106, 107, 111, 132, 133, 143, 146, 147, 172–173
Ferromagnetism, 91, 147
First World War, 26, 40, 134, 188
Fischer, Inga, 171, 197, 212
Fock, V., 140, 164, 206, 211, 221, 236
Fowler, R. H., 18, 94, 106, 131–136, 138–139, 145–146, 157, 159, 160
Franck, J., 16, 34
Free radicals, 79, 146, 162
Fröman, A., 212, 260
Fry, H., 21
Fues, E., 31

Gale, H. A., 45
Gallipoli, 134
Garner, W. E., 143
Gaunt, J. A., 135
Geiger, M., 12
General Chemistry, 119
Generalization(s), 23, 41, 65, 70, 75, 101, 150, 203, 250
George Fisher Baker Non-resident Professor/Lecturer, 108, 116
Goudsmidt, S., 57
Graphical method, 33, 142
Great War, 134, 136, 138
Ground state, 15, 22, 28, 66, 92, 94, 99, 101, 110, 144, 164, 171, 215, 222, 228, 233, 234
Group theory, 6, 17, 22–25, 30, 37–38, 89–91, 93, 95–97, 100, 104, 141, 161, 190, 191, 205
Györgyi, A. S., 200

Habilitation, 26, 27, 32
Hacking, I., 6

Hall, G. G., 152, 212, 229, 260
Harkins, W. D., 40, 87
Hartree, D. R., 3, 88–89, 131, 134–135, 138–142, 153–158, 164, 183, 185, 206, 211, 221, 234, 236, 260
Hartree, W., 134, 138, 155, 157
Heilbron, J., 17
Heisenberg, W., 12–13, 15, 18, 27, 34, 45, 88–89, 91–92, 94, 135–136, 138–139, 145–147
Heitler, W., 2–3, 6, 12–18, 20–23, 25, 36–37, 39, 42, 45, 47, 57–60, 64–65, 68–72, 79, 81, 83, 86–87, 89–93, 97–105, 114, 116–117, 127, 131, 141, 143–144, 147, 152, 161, 189, 198, 205, 215, 219, 222, 230, 241, 243, 256, 260
Heitler-London (1927) paper, 18, 37, 64, 101, 127, 152, 205, 230, 256
Helium atom, 18, 57–58, 60, 65, 72, 88–89, 94, 194
Hellmann, H., 31–33, 37, 131, 260
Henri, V., 145, 190
Herzberg, K., 13, 172
Heuristic, 29, 68, 100, 118, 166, 181
Heurlinger, T., 11
High-speed computing, 157, 222
High temperatures, 34, 241
Hilbert, D., 13, 26
Hill, A. V., 134, 138
Hinshelwood, C., 121, 170
Hirschfelder, J. O., 203, 212, 219
Historical, 2–6, 37, 125, 149, 177, 230, 251, 255–257, 259
Historiography, 4, 245, 247
Hoffman, R., 254
Honl, H., 12
Hückel, E., 25–31, 37, 79–81, 87, 96, 102, 119–121, 131, 147–148, 162, 170, 172, 189, 199, 210, 220, 221, 224, 226, 237, 260
Hückel, W., 31
Hückel LCAO approximation, 201
Hulthén, A., 213
Humboldt University, 13

Hund, F., 3, 27–28, 30, 34, 36–37, 39, 41–43, 45, 81, 83, 85, 87, 89, 91–93, 97, 100, 103, 128, 131, 143, 147–148, 172, 260
Husserl, E., 12–13
Hybridization, 60, 62, 64, 77, 146, 167, 180, 182, 206, 251
Hydrogen atom, 14–16, 18, 23, 57, 59, 60, 72, 91–92, 133, 168
Hydrogen molecule, 10, 12, 14–15, 17–21, 57–60, 65, 71, 81, 85, 91–93, 96, 101, 125, 205, 222
  ion, 10, 17, 20–21, 58, 85
Hylleraas, E. A., 88, 217, 218

Imagination, 89, 157, 181, 182, 231
Imperial Chemical Industries, 97
Imperial College, 136, 227
Inaugural lecture, 157, 176, 183, 184, 229, 246
Indistinguishability, 52, 54, 59–60, 74, 141, 193
Ingold, C. K., 78, 121
Inorganic chemistry, 18, 33, 53, 126, 148, 149, 260
Institut de Biologie Physico-Chimique, 193, 200
Institut du Radium, 188, 189, 190, 193
Institute for Advanced Study, 17
Institute for Theoretical Physics, Copenhagen, 13
Institute of Molecular Biophysics, 201, 217
Institute of Technology, Hannover, 31
Institute of Technology, Stuttgart, 31
Instrument, 5, 17, 114, 132, 156–158, 187, 231, 239, 242, 245, 249, 254, 261
Integral(s), 60, 89, 93, 134, 137, 141, 161, 169, 202–203, 206–207, 210, 214, 222, 224, 228, 230, 237–238, 240, 245
  Coulomb, 15, 163
  exchange, 15, 80, 90, 163, 205, 224
  four-center, 203
  molecular, 126, 176, 203–210, 222, 225, 232
  overlap, 13, 68, 180, 203, 220

repulsion, 224
three-center, 162
two-center, 203, 207
International Conference in Physics 1934, 146, 147
International Education Board, 13–14
International Journal of Quantum Chemistry, 218, 241
International Union of Pure and Applied Physics, 147, 205, 208
Internuclear, 16, 21, 32, 36, 44, 46, 57, 83, 92, 171, 209, 218
Interpretation, 11, 13, 17, 19, 24, 34, 37, 40, 42, 46–47, 50, 69, 85, 95, 101, 108, 110, 135, 139, 145, 151–153, 166–169, 206, 215, 219, 225, 232, 259
  chemical facts, 24, 108
Introduction to Chemical Physics, 105, 113, 114
Introduction to Quantum Mechanics with Applications to Chemistry, 119, 251
Introduction to Theoretical Physics, 113
Intuition, 116, 119, 177, 180, 182, 194, 231
Ionization, 31, 54, 140, 194, 223
Isomer(s), 52, 78, 111

Jacobi, C. G. J., 13
James, H. M., 215, 222
Jansen, L., 219
Japan Society for Promotion of Science, 205
Jerusalem Symposia in Quantum Chemistry and Biochemistry, 195, 197
Jewish, 12, 13, 31, 32, 193
John Humphrey Plummer Professorship of Inorganic Chemistry, 148
John Simon Guggenheim Memorial Foundation Fellowship (Guggenheim Fellowship), 55, 58, 89, 95
Joliot-Curie, F., 188, 191, 198, 200
Joliot-Curie, Irène, 188, 200
Jordan, P., 13, 34
Jost, W., 32, 33

Journal, 2, 4, 12, 20, 56, 61, 68–69, 73, 77, 102, 165, 200, 204–205, 208, 218–219, 240–241, 250, 256
Journal of Chemical Physics, 77, 81, 160, 165, 218, 236
Journal of the American Chemical Society, 61, 73, 77, 178
Julg, A., 190, 196, 198

Kant, I., 22
Kapitza Club, 135
Karachalios, A., 25
Karpov Institute, 32, 33
Kasha, M., 201, 217
Kekulé, F. A., 78–80, 117, 119, 124, 149, 181
Kemble, E., 11, 39–40, 45, 61, 87, 94
Kimball, G. E., 203, 214
Kinetic theory, 11, 113, 134, 137, 145, 148
Klein, O., 213
Kopineck, H. J., 207
Kossel, W., 16, 31
Kotani, M., 204, 205, 207–208, 210, 212, 219, 260
Kramers, H. A., 61, 89, 90, 93, 95, 135, 220
Kronig, R., 100–101, 136, 220
Kuhn, T., 86
Kurti, N., 97

Laboratory, 45, 48–49, 51, 55, 98, 107, 112, 127, 134, 150, 157–158, 167–168, 170, 173, 193, 196, 209, 213, 215–216, 219, 224–225, 228, 231, 239, 242, 247, 248, 253–254
  Brookhaven National, 238
  Cavendish, 61, 135
  IBM Research, 217
  Jefferson Physical, 40
  Mallinckrodt Chemical, 215
  Marine Biological, 200–201
  of Molecular Structure and Spectra (LMSS), 126, 202, 216, 219, 220, 222, 231, 253
  Ryerson, 45
Lacassagne, A., 188, 190, 200

La Chimie Théorique et ses Rapports avec la Théorie Corpusculaire Moderne, 188
Ladenburg, A., 78
Lamb, A. B., 61, 73
Language, 4, 10, 13, 19, 24, 30, 43, 69, 81, 85, 101, 105–106, 109, 112, 116, 122, 152, 167, 193, 214–215, 223, 225, 246, 255
Lapworth, A., 78, 132
Larmor, J., 249
LCAO (linear combination of atomic orbitals), 83–85, 201, 222
Le Bel, J. A., 149
Legitimization, 2–3, 5, 105, 192, 239, 255, 259, 260
Lennard-Jones, J. E., 28, 83, 93, 98, 103, 121, 131–132, 134, 136–138, 141, 143–146, 148–153, 159–162, 164–165, 169–170, 172, 183, 185, 191, 204, 225, 227, 260
Les Théories Électroniques de la Chimie Organique, 199
Lewis, G. N., 10, 16, 18–19, 30, 39, 44, 47–55, 57, 64, 65, 70, 74, 75, 77–78, 80, 83, 87, 101, 107–108, 110, 116, 122, 126, 128, 132–133, 143, 152, 181, 243, 247, 248, 249, 250
Lindemann, F., 97
Lindenberg, J., 212, 260
Lindsay, R. B., 154, 155
Loge, 193–195
London, F., 2–4, 6, 12–25, 36–37, 39, 42, 45, 47, 57, 58–60, 64–65, 68–72, 79, 81, 83, 86–87, 90–93, 97–106, 110, 114, 116, 117, 127, 131, 141, 143–144, 147, 152, 161, 189, 191, 198, 205, 213, 215, 222, 230, 241, 243, 256, 260
London, H., 97
Longuet-Higgins, H. C., 168–170, 191, 225, 227, 235, 250, 252
Loomis, F. W., 39, 45
Lord Kelvin (William Thomson), 141
Löwdin, P. O., 1–2, 4, 197, 204–219, 224, 230, 232, 243, 253, 260
Lowry, T. M., 78, 132, 133, 227

Lucas, H., 78, 131
Lucasian Professor in Natural Philosophy, 131

MacInnes, D. A., 202–204, 206
Macroscopic quantum phenomenon, 98, 105
Magnetism, 54, 94, 134, 159
Manifesto, 218, 228
Many-body problem, 22, 23
Margenau, H., 203
Massachusetts Institute of Technology (MIT), 40, 48, 62, 87, 112–113, 141, 156, 202, 204, 209, 211, 217, 222, 225, 228, 253
Mathematical chemistry, 18, 113
Mathematical equivalence, 87, 164
Mathematical Institute, 173, 210, 253
Mathematical Laboratory, 150, 157, 228, 253, 254
Mathematical physics, 27, 34, 131, 133–134, 136, 143, 154, 157, 174, 204, 205
Mathematical problem, 20, 164, 210
Mathematical techniques, 132, 145, 184, 201, 204
Mathematical Tripos, 138
Mathematics, 2, 5, 12–13, 34, 37, 49, 105, 110, 122, 128, 132–134, 136, 149, 150, 154, 158–159, 161, 164, 168, 173, 174–178, 183, 188, 200, 205, 218, 227, 229, 249–250, 253–254, 257, 261
  applied, 1, 3, 31, 33, 131–133, 153, 173–175, 183–185, 210, 226–227, 229, 230, 235, 246
  in chemistry, 49, 132, 258
  role of, 49, 176, 255
Matrix mechanics, 12–13, 17, 20
Matsen, F. A., 221
Mayer, J. E., 117, 203, 208
Mayer, Maria Goeppert, 212
Mayot, M., 190
McCrea, W., 18, 135
McWeeny, R., 168, 219, 227, 229–230, 260
Mecke, R., 41
Meckler, A., 208–209, 228
Meeting(s), 2–3, 18, 27, 42, 56, 62, 68, 69, 84, 88, 107, 108, 121, 126, 131–133, 143–148, 155, 172, 184, 191–192, 197, 202–204, 206, 208–210, 212, 217–218, 230–232, 241–243, 246, 250, 253, 260
1928 Bunsen Gesselschaft, 108
1923 Faraday Society, 107, 132, 133
Meissner, W., 97
Mendelssohn, K., 97
Mesomerism, 78, 117, 120, 189
Metallurgy, 114, 149
Methane, 22–23, 62, 93, 95–96, 152, 160, 162, 168, 205, 228
Method(s), 7, 9, 17, 20–21, 23–24, 26, 30, 32–33, 40, 42–43, 47, 52, 59–60, 62, 68, 70–72, 75, 79–81, 83–86, 88, 90–93, 95–96, 101, 103–106, 108–109, 111–112, 114–115, 117–119, 123–125, 131–132, 134–143, 146–151, 153–154, 158–167, 169–173, 177–181, 184, 189, 192, 195–197, 199–205, 208–212, 211–213, 217, 219–222, 225–226, 229–230, 235–242, 245, 248–250, 252–253, 259–261
  empirical, 58–59, 165
  LCAO-MO, 236
  LCAO-MO-SCF, 222
  molecular orbital, 30, 84–86, 91–93, 96, 101, 118–119, 123–125, 146–149, 151, 161–162, 178, 181, 195, 196, 201, 209–210
  nonempirical, 199, 210
  quantitative, 192, 196
  self-consistent field, 88, 153–154, 158, 164, 196
  semiquantitative, 82, 202
  valence bond, 91–93, 103, 115, 118, 124–125, 146–147, 149, 161, 165, 177, 180–181, 196, 201
Millikan, R. A., 94
Milne, E. A., 134, 173, 174
Minimizing energy, 15, 168
MIT, 40, 48, 62, 87, 112–113, 141, 156, 202, 204, 209, 211, 217, 222, 225, 228, 253
Models, 6, 10, 29, 30, 51, 53, 55, 62, 78, 112, 178, 210, 232
Moffitt, W., 168, 204

Molecular calculations, 136, 193, 206, 237
Molecular fields, 136–138
Molecular orbital theory, 30, 68, 75, 79, 84–86, 91–93, 96, 101, 109, 118–119, 122–125, 146–147, 149–151, 158, 160–164, 166–168, 170–171, 177–178, 181, 184, 189, 190, 192, 195–196, 201, 209–210, 220–221, 235. *See also* Method, molecular orbital
Molecular point of view, 73, 85
Molecule(s), 3–4, 6, 10–11, 16, 18–19, 21, 23, 25–26, 29–30, 32, 34, 36, 39–45, 47–49, 51–56, 58, 62, 64–81, 83–88, 90–95, 99–101, 107–109, 111, 113–115, 117–119, 121, 124–126, 128, 132–133, 135, 137, 143–153, 160–164, 166, 168–169, 171–173, 177, 188–190, 192–196, 199–202, 205–206, 208–210, 212, 215, 217–219, 222–225, 227–228, 230, 232–238, 240, 242, 247, 251–252, 254, 260
  classification of, 95, 146
  diatomic, 10–11, 15, 36, 40, 41–43, 46–47, 57–58, 83, 93, 144, 149, 172, 222, 223
  ethylene, 124, 152
  fulvene, 195, 196
  heteropolar, 223, 251
  H_3, 222
  isosteric, 40–41
  naphthalene, 80, 224
  odd, 54, 74
  polyatomic, 23, 26, 47, 58, 68, 73, 81, 83, 90, 95, 144, 146–147, 161–162, 205, 215, 237
  as united atoms, 43, 44
  water, 10, 66
Monk, G. S., 45
Mormann, T., 13
Mott, N., 98, 135, 154, 159, 204
Moureu, H., 190
Mulliken, R. S., 3, 6, 14, 20, 27–28, 30, 34, 36–37, 39–47, 61, 67–73, 81, 83–88, 91–95, 97–98, 100–101, 103, 125–128, 131, 141, 143–144, 151, 159, 161, 164, 167, 172, 180, 191–192, 196, 202–204, 206–208, 210–214, 216–223, 225–226, 230–231, 235, 242–243, 253, 260
Mulliken, S. P., 40

National Academy of Sciences Committee on Scientific Conferences, 202
National Physical Laboratory, 134
National Research Council, 39–40, 55, 94
National Research Council Report on the Spectra of Diatomic Molecules, 40
National Science Foundation, 217, 220, 230
National Socialist German Workers' Party, 27
Natural sciences, 2, 4, 106, 219
Nature, 41, 99, 109–110, 131, 135, 143, 154, 178
Nazi, 32, 97–98, 156
Néel, L., 208
Neon, 137–138, 160
Nernst, W., 48, 97, 249
Nesbet, R. K., 228
Networking, 3, 197–198, 218, 259, 260
New York Chapter of the National Council of Arts, Sciences and Professions, The, 121
Nitrogen, 22, 74, 106, 137, 147
Nobel prize, 17, 125, 152, 225, 231, 260
Nomenclature, 41, 133, 144, 210, 225
Nonvisualizable, 29, 120
Noyes, A. A., 48, 55–56, 87, 112, 132
Nuclear motion, 36, 40
Nuclear physics, 134, 147, 221
Numerical analysis, 141, 219, 237
Numerical solution, 140, 142, 158, 243, 245
Nikko Symposium, 196, 208, 221, 226

Ochsenfeld, R., 97
Odiot, S., 190
Office of Naval Research (ONR), 202, 220
Ohno, K., 212
Optimists, 93, 97, 201, 210, 242
Orbital(s), 10, 15, 28, 29, 36, 43, 57, 60, 62, 67, 76, 77, 83–86, 89–91, 93, 99, 101, 103, 109, 123, 128, 139, 144, 146, 149–153, 158,

Orbital(s) (cont.)
  161, 163, 170, 172–173, 180–181, 189–190, 194, 209, 211, 221–224, 228, 233, 235, 237
  binuclear, 21, 57
  Gaussian type (GTO), 235
  localized, 83, 91
  molecular, 4, 6, 30, 34, 37, 39, 42–43, 58, 60, 62, 68, 71, 75, 79, 83–86, 91, 92–93, 96, 101, 103, 109, 114, 118, 119, 122–126, 128, 144, 146–147, 149–153, 158, 160–168, 170–172, 177–181, 184, 189, 190, 192, 195–197, 199, 201, 209–210, 220, 221–222, 235–236
  Slater type (STO), 209, 222, 228, 235, 237
Orbital angular momentum, 36, 149
Ordnance Board, 134
Organic Chemistry of Nitrogen, 106
Oseen, C. W., 11
Ostwald, W., 48, 106, 249, 258
Overlapping, 29, 30, 63, 66, 68, 75–76, 94, 114, 161, 167–168, 233
Oxygen, 28, 74–75, 99, 144, 209

Page, L., 39
Paramagnetism, 28, 74, 85, 144
Parameters, 89, 165, 171, 207, 210, 235, 238, 242, 245, 247–249, 254, 258
Pariser, R., 199, 224–226, 230–231, 237, 260
Parr, R., 199, 202–203, 206–208, 224–226, 230–231, 237, 260
Parson, A. L., 51
Pauli, W., 15–18, 20, 25, 29, 32–34, 37, 42, 44, 54, 58–59, 70, 89, 92, 128, 135, 204
Pauli exclusion principle, 16, 20, 34, 37, 42, 70
Pauling, L., 3, 4, 6, 14, 18, 20, 21, 23, 29–30, 32, 36, 39, 47, 55–69, 73–81, 83, 85, 87–89, 92, 94, 96–102, 104–106, 109–113, 115–121, 123–126, 128, 131, 145, 152, 161, 167, 177–178, 180–181, 184, 189, 191–192, 195–196, 204, 214–216, 219, 225, 243, 250–251, 260

Pauli principle, 15, 17–18, 20, 25, 32–33, 44, 58, 59, 128
Pauncz, R., 219
Peierls, R., 92, 154
Periodic table, 20, 42, 46, 51, 77, 223
Perrin, J., 188, 192, 193
Perspectives in Organic Chemistry, 118
Perturbation, 23, 28, 36, 58, 59, 62, 79, 101, 139, 149, 169, 238
Pessimists, 93, 97, 210, 242
Pfander, A., 12
Philosophical problems, 5, 6, 255
Philosophy, 12, 97, 106, 115, 131, 259
Philosophy of chemistry, 5, 247, 261
Philosophy of science, 5, 12–13
Physical chemistry, 9, 13, 27, 30, 32, 48, 77, 87, 106, 107, 112–114, 116, 134, 167, 168, 192, 205, 235, 249, 250
Physical law, 9, 33, 104
Physical Review, 61, 63, 69, 77, 79, 81, 90, 156
Physical Society, 42, 46, 62, 88, 147, 157
Pippard, A. B., 16
Pitzer, K. S., 204
Platt, J. R., 220, 222, 260
Plummer Chair of Mathematical Physics, 134
Polanyi, M., 191
Polyatomic molecules, 23, 26, 47, 58, 68, 73, 81, 83, 90, 95, 144, 146, 147, 161, 162, 205, 237
Polyelectronic atoms, 24, 57, 141
Pople, J., 152, 224–229, 231, 236–237, 260
Practices, 1, 2, 4, 113, 120, 199, 230, 232, 235, 238, 242, 248, 253–254, 257–258, 261
Pragmatism, 21, 65, 115, 116, 131, 252
Prandtl, L., 26
Primas, H., 257
Privatdozent, 34, 56
Proceedings of the National Academy of Sciences, 64, 207
Professor of Mathematical Physics, 131, 157
Pseudopotential, 32, 33
Pullman, Alberte (née Bucher), 4, 168, 188–193, 195–202, 218, 230, 245

Pullman, B., 4, 188, 190–191, 193, 195–202, 212, 229, 230

Quantization, 10, 11, 63–64, 66–68
Quantum biochemistry, 193, 195, 197, 200, 260
Quantum biology, 126, 217, 260
Quantum chemistry, 1–7, 9, 12, 17–18, 25, 29, 31–34, 36–37, 39, 48, 55, 61, 65, 81, 86–87, 92–93, 96, 104–106, 108, 112, 114–116, 118–119, 121, 123, 125–129, 131–134, 143–144, 147–148, 153, 158, 160, 162, 164, 167, 171, 173, 175, 177, 180–182, 184–185, 187–193, 195–202, 204, 209–219, 221–222, 225–227, 229–235, 237–243, 245–247, 250–255, 257–261
  applications of, 252
  characteristics of, 232
  computational, 185, 228, 229
  cultures of, 187, 225
  discourse of, 239, 255
  hyperbola of, 230, 236
  nature of, 219, 233
  as a quasi-laboratory science, 219
  status of, 2, 105, 108, 123, 246
Quantum Chemistry Group, 211–214, 216, 225, 254
Quantum Chemistry Program Exchange (QCPE), 241
Quantum Mechanical Calculations, 3, 9, 143, 165, 231, 241, 252
Quantum Mechanics of Organic Molecules, 118
Quantum number, 25, 36, 41–45, 64, 66, 70, 72–73, 144
Quantum theory, old, 36, 39, 43, 54–56, 94, 133, 139
Quantum Theory of Radiation, 98
Quantum Theory Project, 216–217

Ramsey, J., 257
Randall, M., 48
Ransil, B. J., 222, 223, 231, 235, 260
Rauschning, H., 156

Reductionism, 5, 104, 115, 256, 257, 258
Relativity, 9, 12, 26, 47, 48, 134, 154
Report of the Commission of the Institute of Organic Chemistry of the Academy of Sciences USSR, 120
Resonance, 15, 21, 23–24, 58–60, 66, 67, 73–80, 96, 106, 110–112, 115, 117–121, 123–125, 128, 146, 149, 151, 153, 161–162, 165–167, 169–170, 173, 177, 180–182, 185, 189, 195, 231, 251
  among several valence bond structures, 80, 146
  extension of, 121
  hybrid, 124
  ontological status of, 120–121, 124, 181
  quantum mechanical, 15, 24, 75, 79, 109
Resonance theory, 4, 6, 106, 109, 110–112, 115, 119, 121–126, 167, 180, 189, 215, 251
  arbitrary character of, 126
  extension of, 121
  manmade character of, 124
  popularization of, 106
Reviews of Modern Physics, 69, 218
Rhetoric, 122, 184
Richards, T. W., 48
Richmond, H. W., 134
Robertson, R., 132, 133
Robinson, R., 30, 78
Rockefeller Foundation, 191, 202
Rodebush, W., 19
Roothaan, C. C. J., 196, 203–204, 207, 219–223, 236, 260
Roscoe, H. E., 249
Ross, I. G., 206
Rouse Ball Chair of Applied Mathematics, 173, 210
Roux, M., 190
Royal Society of London, 97, 106, 153, 170
Rüdenberg, K., 204, 207, 222, 224
Rule of eight, 20, 49, 52, 74
Rumer, G., 23, 90, 102
Rumpf, P., 190

Rutherford, E., 50, 107, 109, 134
Rydberg, J., 11, 41

Sanibel Island Conferences, 216
Saunders, F. A., 34, 40, 41
Scerri, E., 257
Scheler, M., 12
Scherr, W. C., 222, 260
Schmidt, O., 188
Schrödinger, E., 12, 14, 16, 18, 20, 22, 24, 42, 55–57, 94, 125, 139
Schrödinger equation, 3, 17, 19, 20, 22–33, 37, 43, 57, 62, 111, 125, 128, 140, 166, 183, 215, 228, 236
Schützenberger, P., 249
Schwarzschild, K., 11
Self-consistent field, 88, 139–141, 153–154, 158, 161, 164, 194, 196, 206, 222, 224, 238
Semiempirical, 3, 5, 7, 24, 31–33, 68, 75–76, 84, 103, 116, 126, 128, 171, 173, 199, 206, 215, 219–227, 229, 235–236, 238, 241–243
 calculations, 5, 84, 206, 227, 229, 241
Shannon, C. E., 194
Shared electrons, 10, 55, 57, 83–84, 133, 144, 181
Shelter Island Conference, 126, 202–206, 208–211, 221, 230
Sherman, A., 37, 46, 93, 96–97, 160, 201, 210, 235, 242
Sherman, J., 80
Shortley, G. H., 90
Shull, H., 204, 207, 212, 219, 260
Sidgwick, N. V., 100, 106–112, 132, 133, 167
Simon, F. E., 97
Simplicity, 162, 181, 184, 201, 215, 233
Simplification(s), 44, 59, 64, 66, 79, 85, 90, 150, 164, 165, 224,
Slater, J. C., 29–31, 36, 45, 61–63, 67, 79–81, 83, 87–93, 95, 97–99, 101–102, 105, 112–115, 117, 140–141, 147, 154–156, 161, 202, 204, 206, 208–213, 216–219, 222, 225–226, 228, 230, 235, 237, 243, 253, 260
Smiles, S., 138

Smithells, A., 250
Sokolov, N. D., 121
Solid-State and Molecular Theory Group, 208–209, 216–217, 253
Some Physical Properties of the Covalent Link in Chemistry, 108, 167
Sommerfeld, A., 11–13, 16, 27, 55–57, 94, 121
Soviet Union, 31–33, 120
Spectra, 10–12, 34, 36, 39–43, 46–47, 66, 68–70, 72, 84–85, 90, 93–95, 107, 126, 128, 135, 139–140, 143–146, 159, 172–173, 195, 202, 216, 219–222, 225, 231, 253
 diatomic, 40, 47
 line, 41, 94
Spectroscopy, 10–13, 34, 36, 39, 40, 54, 112, 143
Spherical symmetry, 36, 164
Spin, 15–18, 20–23, 25, 28, 34, 36, 54, 56, 58, 60, 65, 70, 72, 83, 89, 90–91, 93, 99, 146, 181, 194, 210, 211, 256
Spin theory of valence, 25, 72
Sponer, Hertha, 34, 41, 204
St. John's College, Cambridge, 138
Stationary state, 110, 139, 149
Statistical mechanics, 113–114, 134–135, 138–169
Stein, Gertrude, 86
Stereochemical, 29, 149
Strong electrolytes, 26, 27, 135
Structural formula, 21, 25, 50, 78, 109, 149, 150
Structure(s), 9, 18, 21, 23–24, 26, 28, 30, 34, 36, 40–42, 45–50, 53–57, 67–68, 75, 77–81, 92, 95, 98, 101, 106–114, 117, 119–121, 123–127, 132–133, 138, 140–153, 155, 160, 162, 165–168, 173–175, 180, 182, 188–190, 193–194, 199–203, 206, 209, 213, 215–216, 219–222, 224–225, 227–228, 230–232, 237–238, 249, 251, 253–255
 covalent, 75, 182
 ionic, 75, 182
 Kekulé, 79–80, 117, 119, 124

Lewis, 77–78, 110
resonance, 189
single valence bond, 78, 80, 119
Style of reasoning, 6, 7, 51, 64, 80, 86, 97, 98, 129, 209, 214, 218, 242, 246, 259
Subdiscipline, 1, 2, 4, 6–7, 81, 105, 115, 123, 127, 131, 147, 173, 177, 188, 197, 198, 199, 201, 245, 246, 252, 261
Sugiura, Y., 205, 222
Superconductivity, 23, 97, 104
Superfluidity, 23, 100, 104
Superfluids, 105
Sutton, L., 110, 167, 191, 204
Swedish Natural Science Research Council, 212–213
Swirles, Bertha, 135, 154
Symmetry, 16, 20, 22, 24, 36, 62, 140, 143, 146, 149, 158, 164, 172
Symposium on *Aspects de la Chimie Quantique Contemporaine*, 254
Syrkin, Ya., 32

Tables of Molecular Integrals, 208, 222, 225
Tautomerism, 50, 78, 110, 112, 117
Technische Hochschule Stuttgart, 13
Tetrahedral, 23, 62, 66–67, 81, 149, 160, 162, 168, 180
  carbon, 27, 30, 63, 67
Texas Symposium, 197
Textbook(s), 2, 3, 32, 48, 105, 106, 108, 112–114, 116, 119, 122, 125, 132, 176, 178, 184, 188, 195, 197–201, 210, 214, 243, 246, 260
  Cancérisation par les Substances Chimiques et Structure Moléculaire, 199, 200
  General Chemistry, 119
  Introduction to Chemical Physics, 105, 112–114
  Introduction to Quantum Mechanics with Applications to Chemistry, 119, 251
  Introduction to Theoretical Physics, 113
  La Chimie Théorique et ses Rapports avec la Théorie Corpusculaire Moderne, 188

Les Théories Électroniques de la Chimie Organique, 199
Nature of the Chemical Bond, The, 105, 116–120, 125, 128, 132, 167, 178, 189
Partisan, 122
Quantum Biochemistry, 199, 201
Quantum Chemistry, 33, 198, 214
Quantum Mechanics of Organic Molecules, 118
Some Physical Properties of the Covalent Link in Chemistry, 108, 167
Theoretical Chemistry, 148–149, 198
Theory of Resonance and its Application to Organic Chemistry, The, 106, 122
Thermodynamics and the Free Energy of Chemical Substances, 48
Valence, 48–49, 53, 107, 132, 176–178, 181, 184, 199, 210, 229
Theorems, 25, 31, 38, 116, 145, 174, 250
Theoretical Chemistry, 1, 18, 33, 126, 131, 145, 148–150, 166, 170, 173–174, 177–178, 182–185, 187, 190–193, 196–200, 225, 227, 231, 235, 237, 239, 241–242, 250, 252, 254
Theoretical Physics, 13, 27, 31, 33–34, 39, 55, 94, 98–99, 112–114, 116, 135–136, 138, 153, 156, 169, 170, 173–174, 176, 205, 208
Theoretical Traditions, 9, 251
Theory, 4, 6, 9, 11–13, 17–27, 29–34, 36–39, 41–43, 45, 48–59, 61–62, 64, 66, 68–70, 72, 75, 78–79, 81, 83, 85, 87, 89, 90–91, 93–104, 106–126, 128, 132–137, 139–141, 143, 145–151, 153–154, 158, 160–168, 170–171, 173, 176–178, 180, 182–184, 189–192, 195, 198–199, 201, 205–206, 208–209, 211, 215–221, 224–225, 229, 235, 237–238, 242–243, 247–251, 253–258, 261
  building, 4, 57, 117–118, 258
  chemical, 70, 108, 115, 145, 180, 182, 247, 249, 251
  descriptive, 220
  Lewis's, 53, 70, 83, 101, 107

Theory (cont.)
  magnetochemical, 54
  molecular orbital, 68, 75, 79, 85, 109, 122, 150–151, 158, 160–164, 166–168, 171, 177, 184, 189–190, 192, 201, 220, 221, 235
  perturbation, 59, 62, 79, 139, 149, 238
  phenomenological, 41
  physical, 54, 64, 108, 139, 145, 247, 249
  role of, 5, 247, 253, 255, 261
  qualitative, 119
  quantitative, 220
  semiempirical, 220
  structure/al, 21, 108–109, 111, 119, 122–126, 128, 184, 251, 255,
  valence, 18–19, 25, 42, 45, 48, 50, 55, 68–73, 75, 78, 81, 83–85, 93, 95–96, 101, 103, 107, 109, 118, 128, 132–133, 145–147, 176–177, 202, 206, 251
Theory of Atomic Spectra, The 90
Theory of Electric and Magnetic Susceptibilities, The, 94
Theory of Resonance and its Application to Organic Chemistry, The, 106, 122
Theory of Spectra and Atomic Constitution, 107
Thermodynamics and the Free Energy of Chemical Substances, 48
Thomson, J. J., 49–51, 132, 249
Thought forms, 1, 2, 6, 234, 261
Tiselius, A., 213
Tolman, R., 87
Townes, C. H., 208
Trinity College, Cambridge, 133–134, 136, 159, 160
Triplet State, 28, 209

Ufford, C. W., 204
Ultraviolet, 11, 28, 144, 161
University, 3, 11–13, 16–17, 27, 31, 37, 45, 61, 93, 94, 107, 138, 143, 158, 164–165, 168, 170, 173, 175, 190, 196, 204–205, 207, 212–213, 216–217, 220–222, 221, 228, 233, 240, 253

American, 40
of Bristol, 98, 136, 143, 158
of California at Berkeley, 19, 39, 48, 51
of California at Los Angeles, 122
of Cambridge, 131, 138, 148, 159, 170, 222, 227, 253
of Chicago, 39, 40, 45, 122, 202, 206, 221–223
of Chicago Computation Center, 221
Cornell, 108, 116, 254
of Florida, 213, 216
of Frankfurt, 34
of Gottingen, 12, 26, 34, 136
Harvard, 39, 48, 121, 215
of Jena, 34
Johns Hopkins, 45
of Kiel, 31
of Leipzig, 27
of London, 121, 169, 176
of Manchester, 55, 136, 142, 153
of Marburg, 27
of Michigan, 39
of Munich, 12–13
of Nebraska, 48
New York, 39, 40, 45
of Oxford, 1, 106, 168, 182, 210
of Rostock, 34
of Tokyo, 205
Tokyo Imperial, 204
of Uppsala, 1, 204, 213–214, 253
of Utrecht, 135
Yale, 39, 50
of Zurich, 12
Unshared electrons, 57, 75, 83, 144
U. S. Air Force, 213
U. S. military agencies, 213
Usefulness, 65, 68, 79, 117–118, 126, 167, 182, 199, 249

Valadalen Summer School, 126, 214
Valence, 1, 4, 6, 19, 23, 28–30, 32, 36–37, 40–42, 45, 47–50, 52–55, 59, 61–62, 65, 67–73, 75, 78–81, 83–85, 87–88, 90–93,

95–96, 99–103, 107, 109, 114–115, 117–119, 121, 123–125, 128, 131–133, 143–147, 149–151, 160–162, 164–165, 168, 171–172, 176–178, 180–181, 184, 189–190, 192–193, 195–196, 199, 201–202, 206, 210, 229, 235, 251
  contra, 50
  directed, 63, 67, 81, 149, 151
  normal, 50
  saturation, 18
Valence and the Structure of Atoms and Molecules, 48, 107, 132
Valence bond (VB) method, 91–93, 103, 115, 118, 124–125, 146–147, 149, 161, 165, 177, 180–181, 196, 201
Van der Waals forces, 14–16, 25, 145
Van't Hoff, J., 29–30, 149, 247–249
Van Vleck, J. H., 18, 20, 36–37, 46, 62, 81, 84–85, 87–90, 93–98, 100, 102, 128, 160, 162–163, 201, 204, 208, 210, 217, 235, 242–243, 260
Variational method, 140, 149, 163, 171
Veterinary College, Hannover, 31
Visualizability, 13, 19, 25, 29, 120, 129, 177, 232–233, 258

Waller, I., 208, 213
Watson, J., 200
Wave function, 15–17, 20–21, 59–60, 62–63, 66, 68, 76, 79, 85, 89–92, 97, 139–141, 149, 152, 157–158, 161, 163–164, 171, 177, 180, 189, 202, 211, 217, 222–225, 227, 233–234, 236, 252
Wave mechanics, 24, 37, 56, 92, 104, 109, 111–112, 114, 135, 139, 145, 149, 150, 176
Weizmann, C., 195, 197
Wentzel, G., 56, 94, 222
Werner, A., 67, 149
Weyl, H., 17, 23, 88, 141
Wheeler, B. I., 48
Wheland, G. W., 79, 80, 101, 106, 118, 120–124, 178, 195, 204
Whittaker, J. M., 135

Wiener, N., 88
Wigner, E., 22, 23, 86, 89, 90, 92, 141
Wilkes, M. V., 157, 228
Wilson, A. H., 135
Wilson, C. T. R., 154
Wilson, E. B., 18, 116, 207, 214, 251, 252
Wood, R. W., 45

Zeitschrift fur Physik, 56, 136

We thank Antonia Pavli for help with the index.